HERBAL TECHNOLOGY:
Recent Trends and Progress

M. Daniel
Arun Arya
Vinay M. Raole
Department of Botany,
The Maharaja University of Baroda,
Vadodara, 390 002, India

SCIENTIFIC PUBLISHERS (INDIA)
P.O. BOX 91 JODHPUR

Published by:
Pawan Kumar
Scientific Publishers (India)
5-A, New Pali Road, P.O. Box 91
Jodhpur - 342 001
E-mail: info@scientificpub.com
www.scientificpub.com

Print, 2019

ISBN: 81-7233-479-6

Lasertype set: Rajesh Ojha
Printed in India

PREFACE

This book is a sequel to the recently published book "**Herbal Technology: Concepts and Approaches**"(M. Daniel, 2006, Satish Serial Publishers, New Delhi). As defined there and as also conveyed by the various conferences and workshops held here in the Botany Department of the Maharaja Sayajirao University of Baroda on this new concept, **Herbal Technology** encompasses all the myriads of ways of utilizing the multifarious potentialities of plants (other than the genetic manipulations) for human welfare. Though some scientists tend to refer the avenues related to Medicinal Plants only in the circumscription of Herbal Technology (and Dr. Banerji calls the same as Phytotechnology- see the first chapter of this book), we find it proper to include Natural Dyes, Biopesticides, Biofertilizers and Biofuel into this novel and highly utilitarian aspects of Plant Sciences.

The Medicinal plants are an area plagued by a multitude of problems such as the correct botanical identity, name changes, biomarkers (both chemical and pharmacognostic), active principles, cultivation, extraction, formulations, pharmacology, pharmacodynamics and pharmacokinetics. The scope of these plants in Aromatherapy, Herbal Cosmetics, Herbal Veterinary Medicine is not properly assessed. The emerging fields of antioxidant therapy and Metabolomics as well as the need of documentation of the knowledge accrued from the traditional knowledge library to prevent biopiracy and their utility needs careful attention. Therefore the section on **Medicinal plants** forms the first and largest division. In this part, the concepts of Bioprospecting and Chemoprospecting are brilliantly explained with examples by Dr. Banerji. Alongwith an account of the plants mentioned in Vedas, there are a good number of attempts on ethnobotanical studies from various parts of the country and care is taken to include only the novel properties/uses of plants. A number of papers dealing with biomarkers, both pharmacognostic and phytochemical, on a good number of medicinal plants form a major share in this section. The plants studied thus are *Ficus benghalensis, F. religiosa, Woodfordia floribunda, Lagerstroemia flos-reginae, Sida veronicaefolia, Euphorbia thymifolia, E. hirta, Ammania baccifera, Lawsonia inermis, Artocarpus heterophyllus, Jatropha curcas* and *Jatropha gossypifolia*. The utility of micromorphological characters, which otherwise were ignored as biomarkers till today, in at least eleven medicinal plants and as well as the presence of characteristic silica bodies in three medicinal grasses are explained in two papers. And we are happy to state that almost all these works are from our laboratories. Tissue culture practices as well as evaluation of the clinical properties on animals are the other aspects dealt here. Papers on phytochemicals in modifying the properties of microbial pathogens and medicinally useful mushrooms and their cultivation are the other topics of interest.

A lot more is desired in the field of **Natural dyes.** The botany, chemistry and uses of many plants which yield these compounds are not clearly known. The scope of utilizing natural dyes as colors for food and pharmaceutical preparations, in cosmetics like face powder and lipsticks, in paints and their application in painting, is yet to be pursued. The various dyeing practices prevalent in different parts of country are to be standardized with precision and the various components of different common dyes and their effects on the health of the consumer are to be made known and this will add the value of clothes dyed by natural dyes. This is an area where at present only the textile chemists are active and absolutely no Phytochemist or Botanist is seen anywhere in this arena. Here in this book, Natural Dyes form the second section and it covers the application of dyes from six plants such as Rohira, Katha, Ravenchi wood, Annatto, Babool, banana on various textiles. A novel method of utilizing microencapsulated fragrances of Lavender, Rose, Musk, Lemon, Sandalwood, Tropical Paradise and Bouquet on various textiles is presented in another paper.

Biopesticides is another virgin area which will pay great dividends to the common man and the scientific community. India can boast of more than 300 Angiosperms having pesticidal properties (Daniel, 2006). Since these pesticides are plant extracts, they contain many other non-pesticidal compounds such as minerals, co-factors, vitamins, hormones, anti-oxidants and even nutrients and all these preparations will directly benefit the recipient plant. In this third section, papers on Biocontrol agents, control of toxicogenic fungi by phytochemicals and antifungal activity of plant extracts are included.

The fourth section is on **Biofertilisers.** They are biologically active organisms/products including bacteria, algae or fungi, with the ability to provide nutrients to plants. They form an important alternative source of plant nutrition and help in improving plant health. They belong to two types; (1) nitrogen fixing and (2) phosphate solubilising. Nitrogen fixing biofertilizers fix atmospheric nitrogen into forms which are readily usable by plants. These include *Rhizobium, Azotobactor, Azospirillum*, blue green algae and *Azolla*. While *Rhizobium* requires symbiotic association with the root nodules of legumes to fix nitrogen, others can fix nitrogen independently. Phosphate solubilising micro-organisms (PSMs) secrete organic acids which dissolve rock phosphate and tricalcium phosphates which are readily absorbed by plants. PSMs are particularly valuable as they are not crop specific and can benefit all the crops. The papers presented here include the utility of marine algae, blue green algae and AM fungi.

Biofuel, also a highly rewarding area and which is the need of the day, includes the various alternate sources of fuel such as petrocrops, biocrude, liquid resins, vegetable oils and plant oil methyl esters, biogas, alcohols and biological hydrogen. The utility of biogas as well as a number of new sources of fatty oils have been presented in the papers presented in this section.

Herbal technology as a subject is in its stages of infancy. The various vistas identified in this subject offer great potentials if and when carefully researched and nurtured. The principal players here are the Botanists and Phytochemists. This book, along with the first book mentioned in the beginning, are aimed to awaken the scientists and intelligentsia on the enormous possibilities existing in these areas and on the need to fight diseases, pollution and fuel crisis, which are the main problems facing man today, in an eco- friendly way. Each of the topics included here is designed in such a way to rouse the curiosity of the student/reader and a conscientious worker can add many different dimensions in every topic. It is hoped that the subject of Herbal Technology will find a place in the curricula of graduate and post-graduate classes, will initiate a great deal of interest in the scientific community and will lead to many research programmes in each of the topics, be that Biopesticides or Biofuel. The full potential of the plants, when realized and manipulated in an ecofriendly manner will make the life in this planet (we have only one planet to live!) much more soothing, comfortable and in tune with the nature.

27, March, 2007

M. Daniel
Arun Arya
Vinay M. Raole

CONTENTS

B. NATURAL DYES

E. BIOFUEL

1

HERBAL TECHNOLOGY : CONCEPTS AND PROSPECTS

M. Daniel

Phytochemistry Laboratory, Dept. of Botany, M.S. University of Baroda, Vadodara – 390 002

ABSTRACT

Herbal technology encompasses all the advancing technological frontiers (except the genes) to tap the myriads of modes of manipulating the plants around us. It includes the fields of herbal medicines, natural dyes, biopesticides, biofertilizers and biofuel. These avenues, when fully developed, will make the world a much better living place.

INTRODUCTION

Herbal technology includes all the advancing technological frontiers (except the genes) to tap the myriads of modes of manipulating the plants around us. Through the plants were known to provide us food, fodder, fiber and shelter, a large number of technologies developed of late have opened up new areas with a very high potential of development and include the fields of herbal medicines, natural dyes, biopesticides, biofertilizers and biofuel. These avenues when fully developed will make the world a much better living place.

Herbal Medicines

Herbal medicine is one of the fields very much in disarray and least standardized. There is confusion and ignorance in every area like identity of correct medicinal plant, time of harvesting, method of extraction, active components, storage of drugs, pharmacodynamics, shelf life and cultivation of herbs. New uses are discovered for a number of known medicines. For e.g. the hypoglycemic and estrogenic activates of *Achyranthes*, hypoglycemic properties of *Aegle*, antimutagenic activity of *Emblica*, immunomodulatory, antiulcerogenic and antitumour activities of *Ocimum*, hypoglycemic and contraceptive potential of *Phyllanthus*, hepatoprotective activities of *Piper* and antiplaque activities of *Sida* are some of the new properties located recently. The major / active components of a number of drugs also are revised. For example ephedrine in *Sida*, madecassoside in *Centella*, aloesone and alocutins of *Aloe*, liriodondrin in *Boerhavia*, wedelolactone in *Eclipta* and phyllanthin from *Phyllanthus* are comparatively new reports.

The efficiency of herbal medicines is attributed now to the synergistic action of a wide variety of chemically and physiologically active components, rather than a single component. The activities of a large number of apparently simple compounds which are otherwise common in plants also are being discovered. Thus chlorogenic and caffeic acids are known to be "integrase inhibitors"; glutathione (a tripeptide) is known to detoxify the kava lactones in Kava and ferulic and gentisic acid are now pain relievers.

Many improvisations are to be sought on the cultivation, harvesting time, mode of extraction and shelf life of compounds. Since a large number of compounds found in plants are survival mechanisms formed due to the response of the plant to a specific environment, every medicinal plant is to be grown in its natural surroundings or in a climate/soil simulating the original ecosystem. Use of synthetic fertilizer is to be avoided since they favour rapid vegetative growth, but poor formation of adaptational chemicals. The concept of "Standardised plant" is a method to ensure that a drug plant contains a specific concentration of phytochemical in question. The harvesting times (both in terms of the age and nature of the plant and the time of the day of harvesting) are to be ascertained. The harvested material is to be cleaned thoroughly, diseased material removed and to be dried in shade. The material thoroughly dried, should be stored in air tight containers preferably in nitrogen. Such way aflatoxins can be avoided.

Since every medicinal herb contains a wide variety of compounds differing in solubility, stability etc., the solvents used for extraction are to be standardised. Standardised extracts are measures ensuring a definite concentration of certain compounds in an extract. Solvents used should be chemically pure and also the water used should be deionised (using hard water for extraction will lead to high mineral content in extracts and thus in formulations). Shelf life of good number of compounds is yet to be quantified.

Correct and strict quality control measures are not available for many of the Indian drugs. The pharmacognostic characters to assess the genuineness of a drug in powered sample or in extract from are not available for many drugs as also the biomarkers needed for a chemical analysis of the drug. In plant drugs where the biomarkers are known, these compounds are not available commercially.

Substitute drugs also are to be legalized if they contain same compounds. The data on the pharmacognosy and phytochemistry on the adulterants will help to locate the adulteration.

Natural Dyes

Natural dyes from another great avenue to be developed and nurtured which will pay great dividends in terms of improving the quality of life. Till today only a few textile chemistry/dyeing departments were using natural dyes on an experimental scale. The dyes used, usually, are some of costly medicinal drugs like madder, turmeric saffron etc. and only very few cheap sources have been attempted as sources of dyes. Large scale surveys on the plant world for dyestuffs were not carried out in right earnestness. The availability of cheap and easily available dyes will make a great difference in the economy of natural dyeing industry. The proanthocyanins and iridoids are two groups of pigments for the future as they are available in a large number of plants.

Natural dyes in food and pharmaceuticals are another area of great promise. Since many of the dyes are antioxidants they may protect the components of food and medicines and thus prolong their shelf life, besides giving the antioxidant benefits. Natural dyes used in wall paints and as a medium in paintings hold great potential. The area of cosmetics also is not exploited properly. Natural antioxidant colors in nail polishes, lipsticks and rouges will revolutionize the cosmetic industry. These compounds will provide additional health benefits in addition to the beautifying action.

Biopesticides

Biopesticides include all the naturally occurring compounds which control pests (biochemical pesticides), micro-organisms that control pests (microbial pesticides) and pesticidal substances produced by plants containing added genetic material (Plant incorporated protectants or PIPs). Theses were approximately 195 registered biopesticides compounds and 780 products, by the end of 2001 and the number is increasing day by day.

A large number of phytochemicals are found to have pesticidal properties. Some of the important compounds are pyrethrins, rotenones, nicotine, acetogenins, limonoids, quassinoids etc. Pyrethrins, produced by *Chrysanthemum* sp, are a group of wide – spectrum insecticides. They are mixed terpenoids found in flower of *Chrysanthemum cinerariaefolium*. Rotenones are a group of isoflavonoid derivatives found in species of *Derris, Pongamia, Lonchocarpus* etc. Nicotine is a volatile alkaloid produced by species of Tobacco while acetogenins are a group of polyketides possessing tetrahydrofuran moieties, found in *Annona* and related plants. More than 250 different acetogenins have been isolated from 26 plants of Annonaceae. Limonoids and quassinoids are tetranortriterpenoids widely prevalent in members of Meliaceae, Rutaceae, Simaroubaceae and Cneoraceae. Azadirachtin from neem is a celebrated member here. Allyl sulphides and related sulphur containing compounds from garlic, cocculidine form *Cocculus,* piperine from *Piper*, aconitine from *Aconitum* and asarone from *Acorus* are some of the widely used pesticides. In addition, a large number of plants such as *Calotropis, Lantana, Vitex, Ipomoea, Spilanthes, Aristolochia* etc are among the many plants used effectively in many villages of India, through the active compounds are never pinpointed.

Microbial pesticides belong to bacteria, viruses, fungi and protozoa. Apart from the well-known *Bacillus thuringiensis* (Bt) many other species of *Bacillus* such as *B. sphaericus, B. popilliae* as well as of *Agrobacterium, Serratia* and *Pseudomonas* are used to combat many pests.

Viruses such as nuclear polyhedrosis viruses, granulosis viruses and non – occluded baculoviruses (eg. *Anagrapha, Autographa, Helicoverpa* etc.) are effective against many insect pests.

The fungi thus used belong to the genera *Lagenidium, Phlebiopsis, Pythium, Beauveria, Metarhizium, Entomophaga, Zoophthora, Poecilomyces, Normurarea, Ampelomyces, Candida, Geocladium* etc. *Nosema* and *Thelohania* are the protozoan pesticides.

A number of nematodes also are found effective against pests. They are *Steinernema, Heterorhabditis* and *Romanomermis*. In addition, a number of other agents such as pheromones, parasitoids, predators and microbial by-products also are used against certain pests.

Biofertilizers

They are biological active products including bacteria, algae or fungi, having the ability to provide nutrients to plants. They form an important alternative source of plant nutrition. They belong to two types; (1) nitrogen fixing and (2) phosphate / sulphur solubilising. Nitrogen fixing biofertilizers fix atmospheric nitrogen in to forms which are readily absorbed and used by plants. These include *Rhizobium, Azotobactor, Beijerinkia, Frankia, Azospirillum*, blue green algae and *Azolla*. While *Rhizobium* requires symbiotic association with the rood nodules of legumes to fix nitrogen, others can fix nitrogen independently. In addition a number of aerobic nitrogen fixing bacteria such as *Clostridium, Desulphovibrio*, purple sulphur bacteria, purple non-sulphur bacteria and green sulphur bacteria also are employed as biofertilizers. Phosphate – solubilising micro-organisms (PSMs) secrete organic acids which dissolve rock-phosphates and tricalcium phosphate which are readily absorbed by plants. PSMs are particularly valuable as they are not crop specific and can benefit the entire crop.

Arbuscular Mycorrhizal Fungi (AMF) are another group of fungi which exhibit soil enhancing properties. More than 50% of plant families form relationships with AMF. They take sugars from the plants and the plants gain nutrients, generally phosphate from fungi.

Biofertilizers have an added advantage that they can decompose organic material and oxidize Sulphur in the soil. On application, they enhance the growth of the plants, increase the yield and also improve the soil fertility and reduce pollution. They also provide certain growth promoting substances like hormones, vitamins, aminoacids etc. In contrast to chemicals which are to be supplied repeatedly, biofertilizers once applied, supply nitrogen, phosphorous or sulphur continuously throughout the entire period of crop growth in the field under favorable conditions. These organisms multiply in course of time and there is no toxicity at higher doses.

Rhizobium strains are used on crops like pulses while *Azotobactor* and *Azospirillum* are used for non-legume crops (including dry land crops) and cereals respectively. Blue – green algae and *Azolla* are best – suited for rice and crop in wet lends. Phosphate solubilisers benefit all crops. AMF / VAM are found useful for trees and ornamentals.

The potential demand for biofertilizers is of the order 3.4 lakh tones and the country's annual production had reached only 13,000 tonnes. Though the awareness among Indian farmers is very high, many constraints such as quality of the products, poor storage facility and inconsistent field performance of inoculants are the factors impeding their production which are to be solved in the coming years.

Petroleum substitutes / Biofuel

The presence alkanes, aromatics, pyrrole rings, pristane etc in petroleum indicates its origin from prehistoric plants. The high molecular weight polyisoprenoids such as rubber and gutta in many laticiferous plants enabled Calvin to propose all the laticifers as Petrocrops. The plants which are extensively worked out thus are *Euphorbia resinifera, E. lathyris, E. tirucalli, Grindelia camphorum* and *Taraxacum. Copaifera longdorfii* and *C. multijuga* are two trees which contain liquid resin in the xylem canals. Trees of 1 mm growth can yield up to 20 liters of resin in 2-3 hours which can be directly used as diesel. *Hardwickia pinnata* of a growth 2.5 m. can yield 12 –40 gallons of resinous biofuel at a time. Another tree, *Dipterocarpus turbinatus* will yield approx. 40 gallons of liquid fuel. *Botyrococcus braunii* is a chlorococcan alga which is tipped as a source of petroleum in that the wall of this plant contain 70% hydrocarbons (C_{25} – C_{31}) similar to crude oil. *Chlorella pyrenoidosa* can be catalytically hydrogenated in presence of cobaltomoly-bdate and 50% of biomass is converted to clear golden oil

Biodiesel, a chemically modified form of vegetable oil, is the craze in India today. In advanced countries where there is an excess of edible oil like soyabean oil, these oils are methylated or ethylated to manufacture biodiesel by any enterprising person. In India non-edible oils such as those obtained from *Jatropha curcas, Pongamia pinnata, Calophyllum* or *Hevea* are used as sources of such fuel. The toxicity of *Jatropha* goes against the wide- scale cultivation of this plant. But the plant world is rich in seed oils in that almost every plant contains a good amount of oil in its seed and a concerted effort has to be done to harvest these seed oils, which otherwise are wasted in the ground / soil.

Alcohols like methyl and ethyl alcohols also are used as fuel in combination with petrol or diesel. Methanol can be manufactured by pyrrolysis of wood / lignin and in W. Germany a mixture of Methanol and petrol (15:85) is used as a fuel. Ethanol, which can be manufactured from sugar / starch, is another choice. In Brazil, ethanol : Petrol (20 : 80) is used as gasohol. Biogas (natural or from biological waste) is another fuel.

Hydrogen is the ideal fuel. And photolysis of water in to gaseous hydrogen and -OH radicals takes place in all the plants in daytime during photosynthesis. The only hitch is that this hydrogen is not released, but readily accepted by NAD. The enzyme hydrogenase will prevent this combination and cause the gaseous hydrogen to escape. Hydrogenase is present in *Scenedesmus, Chlorella, Anabaena, Escherichia coli, Citrobactor, Closterium* and *Spirulina* etc. *Azolla* containing *Anabaena azollae* releases H_2 but this H_2 combines with N_2 to form ammonia. Some legumes also are found to release H_2 during day time. Experiments simulating photosynthesis using chlorophyll are being conducted to harvest *Biological Hydrogen* from water.

Herbal Technology is in fact Applied Botany. Every branch of HT is a potential treasure and the career opportunities are legion. Both herbal medicines and natural dyes make man healthier while biofertilizers and Biopesticides make plants healthier. The world will be a better living place when all the potentialities of wide variety of plants around us are tapped and channelised.

REFERENCES

Daniel, M. 2006. *Medicinal plants: Chemistry and Properties*, Science Publishers, New Hampshire, USA.

Daniel, M. 2006. *Herbal technology: Concepts and Approaches*, Satish Serial Publishers, New Delhi.

A

MEDICINAL PLANTS

2

PHYTOTECHNOLOGY - VALUE ADDITION TO BIODIVERSITY THROUGH CHEMOPROSPECTION

A. Banerji

Centre for Biotechnology, Amritapuri Campus, AVV, Kollam - 690 525 (Kerala)

ABSTRACT

The economic potential of biodiversity can be enhanced by developing appropriate technologies for phytotherapeutics, advanced agrochemicals, high-value fine biochemicals etc. In spite of being the biggest supplier of raw materials, India's share in the global herbal market is a mere 2%. Therefore concurrent to bioprospection, efforts should be made to isolate and characterize the active principles from the natural products for value addition. Phytochemical investigations during last several decades have resulted in the isolation of an array of well-characterized compounds with exotic molecular architectures. Thus, repository of potential biomolecules and traditional knowledge base will be useful for the modern discovery programs through the library of natural products.

INTRODUCTION

Plants occupy the most prominent position in the biosphere. They are the genetically controlled bioreactors of nature. During the long course of history of the earth, elegant chemical processes have evolved which furnish an array of molecular architectures using very simple molecules as starting materials and under extremely mild reaction conditions. Through the remarkable process of photosynthesis, the green plants (autotrophs) convert carbon dioxide (inorganic carbon) to the reduced form of carbon (*i.e.*

carbohydrates), using water and abundantly available light energy. In the process, oxygen is set free which is indispensable for the sustenance of life in our eco-system. The energy for almost all the biological activities are directly or indirectly derived from the reduced carbon compounds, thus synthesized. The scale of the photosynthetic activity is stunning. Each year, tens of billion tons of carbon are taken from the atmospheric CO_2 and are converted into incredibly complex primary and secondary metabolites such as nucleic acids, carbohydrates, proteins, amino acids, heterocycles, isoprenoids, phenolics, alkaloids, lipids *etc.* through remarkable biochemical reactions. An insight into the technologies which Nature uses and may be called as *"Natural technology"*, will offer entirely new concepts for the development of eco-friendly technologies; it may be termed as *phytotechnology* when based on plants (Banerji, 1992)

Plants, as renewable source of chemicals, have been recognized by man even before the dawn of civilization. Mainly through intuition and keen sense of observation, the ancient man had discovered many useful properties of plant products. Timber has been in use for construction purposes, cellulose as fibre, cereals for food, natural fats and vegetable oils as industrial products and phytochemicals as drugs, nutraceuticals, dyes, cosmetics etc (Fellows, 1992). Even some of the animals have learnt to use certain natural products to their advantage (Rodriguez *et al.,* 1985). The oriental systems of medicine have evolved out of the wisdom and experiences accumulated through generations. In spite of the fact that about 80% of the world population still use phytotherapy for their primary health care, this branch of science has not received the attention it deserves (O'Neill and Lewis, 1993).

Indian Scenario

Varied climatic conditions and diverse geo-ecology have contributed greatly towards the biodiversity of India. With mere 2.4% of the global land, India possesses about 8% of total biodiversity. The Western Ghats have been designated as one of the hot spots of global biodiversity. A rich repository of flora with high endemism is found in Himalayas (Purohit and Dhar, 1997). In addition to the rich biodiversity, India has very well documented traditional knowledge systems. This body of knowledge can be very useful for the selection of prospective plants for drug discovery by providing lead molecules without going through expensive mass screening programmes. Food crops and fibre are the most important products from plants. However, little attention has been paid to the unexplored potential of non-crop plants, which can provide an array of substances of economic value. Recent researches in phytochemistry have revealed that in addition to conventional products, plants can offer many wealth creating products such as high-value chemicals (e.g. drugs, agrochemicals, dyes), a variety of biopolymers, feedstock chemicals and bio-fuels (Murphy, 1995).

Advances at the interfaces of chemistry, material sciences, biology and medicines have triggered intense activity towards development of novel materials having biotic or abiotic properties. Access to molecular diversity is a prime requirement for any discovery program. Materials can be obtained either artificially by synthesis or from natural sources. Progress in the organic chemistry has made most of the natural products accessible by synthesis though not necessarily commercially viable. However, the success rate for drug discovery for synthetic compounds has been rather dismal (1 in 5000 to 10,000 compounds) and takes very long (7-15 years). In addition to adverse effects on the environment, the synthetic approach is becoming more and more expensive. Though there are wide variations in the cost estimations for the drug development, it lies between $200 to 800 million for the discovery of one drug. About half of the cost goes for the discovery of lead molecules. The concern over undesirable side effects of increasing number of drugs demands development of safer medicines. To improve the success rate, it is imperative that more compounds with diverse chemical structures should be available for screening. Compounds created by the living organisms show enormous chemo-diversities and bioactivities. Therefore, there is great awareness on the potentialities of biodiversity, not only for environmental/ ecological reasons but also as *resource for wealth creation*. The chemical and biological information contained in the natural products have to be utilized judiciously. Systematic inventorisation of the bioassets is the first necessary step towards value addition to biodiversity. Isolation, characterization and pharmacological evaluation of the active principles from the raw materials or extracts contribute to several hundred-fold increase in the value addition. Need to discover the active principles of the plants was referred by Paracelsus as early as 1524. For example, the global sales of vincristine and vinblastine, the active principles of *Catharanthus roseus* is about $100 million while that of taxol from *Taxus* sp. is $1 billion per year. Therefore, considering the untapped economic potential of natural products, *serious efforts should be made towards isolation, characterization and pharmaco-logical/clinical evaluation of the active principles*, thus enhancing the value of biodiversity by the *product development* such as phytotherapeutics, nutraceuticals, advanced agrochemicals, fine biochemicals, and industrial chemicals. The innate taxonomic diversity of the flora throws a *challenge to the Indian scientists to transform the bioresources into economic wealth through appropriate technology and intellectual property.* If adequate attention is not paid, the free accessibility of information through bioinformatics could be exploited by the technologically advanced countries for economic gains by value addition and protection through intellectual property rights.

Of the world's 25 best selling pharmaceutical agents, about half are derived directly or indirectly from natural sources (Kinghorn and Balandrin, 1993). Cardiac glycosides from *Digitalis purpurea,* tranquilizer reserpine from *Rauwolfia serpentina*, analgesics morphine and codeine from *Papaver somniferum* are some of the examples of drugs which were discovered several decades ago but are still in use. The global market for plant – based medicines is about US$ 16 Billion with a growth rate of 30%. The Indian contribution in the global market is estimated as less than 2% (Subramanian, 2002). This is because only raw materials are exported from India while

China exports processed products and has a larger share (40%). In addition to pharmaceuticals, plant-derived products contribute greatly towards wide range of industries such as fine chemicals, agrochemicals, cosmetics, industrial chemicals etc. The natural products also provide advanced raw materials for the high-value drugs such as steroids. In addition, the natural product research has stimulated the organic chemists to synthesise active principles and their analogs with a view to provide compounds with improved bioactivities and other desirable properties. Taking cues from nature, organic chemists have been able to mimic the synthesis of several compounds following biosynthetic approach (Todd, 1961). Biomimetic synthesis have great relevance towards the development *Green chemistry*.

Ethnomedicine in Drug Discovery

Some of the ancient remedies are documented in the oriental literature such as Ayurveda, Unani, Amchi, Kampo, Jamu and Chinese systems of medicine. In addition, considerable knowledge on medicinal plants and remedies is available with the tribes and natives (Cox, 2000). By word of caution, it may be stated that tribal or folk medicine is often not recorded with authenticity and accuracy (Daniel, 2004). Unfortunately, many of the claims are associated with exaggerations or myths and could not be substantiated. Misinformation on plants makes correct identification and validation of the claims difficult. Even people having genuine knowledge of traditional remedies keep them closely guarded secrets and are reluctant to reveal information.

Leads from ethnomedicine have contributed greatly in the drug discovery programs. It will continue to play important role in the future programs also by providing pharmacologically active molecules or important lead compounds. Evidence-based phytomedicine (placebo controlled double blind randomized clinical trials) using isolated compounds or standardized extracts, is gaining acceptance in several countries. Arogyapacha *(Trichopus zeylanica),* Asmol (a mixture of biorationally selected herbs for asthma), Vijaysar *(Pterocarpus marsupium,* anti-diabetic), Salai guggul *(Boswellia serrata,* anti-inflammatory), *Janakia arayalpathra* (Anti-tumour), guggul *(Commiphora mukul,* cholesterol lowering), *Phyllanthus urinaria* (anti-tumour) are some examples of evidence-based phytomedicines developed in recent times in India (Roy Chaudhury, 2004). It may be mentioned that besides Asia, practice of folk or traditional medicine is also prevalent in other parts of the world such as Africa, Russia or South America. Useful knowledge can also be imbibed from the folk medicine of other parts of the world. A comparative study shows commonality between them and the Indian folk-lore/traditional medicine. Contrary to general belief, the safety of plant based medicines cannot be taken for granted. In fact fatal cases have been reported due to toxicity of plant products or formulations (Chaffin, 1998). Therefore, safety aspects should also be considered for the popularization of phytomedicine. Bioprospection of the indigenous flora reveals that much of the sources or related plants mentioned in other countries/cultures also occur in India. Inclusion of these through cross-cultural interactions will increase our bioresource base. Some of the plants (or closely related species) used therein could provide useful leads for prospecting for drug development in India. For example, use of *Onichium auratum*, which occurs in Himalayas, is not recorded in the traditional Indian system but it features in the Chinese traditional medicine (known as Gouji) (Zhong *et al.* 1999). Its anti-neoplastic, antispasmolytic and myocardial nutritional properties have been established. The active principles (rearranged sesquiterpenes - onitin and onitisin and their glycosides) have been identified recently; incidentally, these constituents were characterized in India many years ago (Banerji *et al.*, 1974). Many examples may be cited where cues taken from the traditional medicine of other countries resulted in the discovery of outstanding drugs or other bioactive molecules.

Bioassay – Directed Approach

There is increasing use of traditional medicine in the health-care systems of countries such as Germany, United Kingdom, France, U.S and China. Considering the recorded global flora (estimated to be 250,000 species), only a miniscule fraction is in use in traditional medicine. Therefore, it will not be wise to neglect the vast majority (95%) of the unexplored flora that does not feature in the traditional systems of medicine or folklore. It has been mentioned in the ancient Indian literature such as Ashtangahrydaya that no plants are devoid of some or the other therapeutic properties. This gets credence by the modern theories. The phytochemicals have evolved due to functional necessity, perhaps on response to evolutional or ecological pressures. It is not surprising that, they exhibit varied bioactivities. The pharmacophoric diversity offered by the natural products (both medicinal and non-medicinal) is unmatched to the limited structural diversity obtainable through the conventional or combinatorial synthesis.

Revival of interest and awareness of natural products as repository of novel pharmacophores have resulted in the discovery of entirely new range of bioactive molecules (phytotherapeuticals). Many of the recent drugs were discovered by the bioassay aided isolation from the plants collected at random with no known previous history of bioactivity. Mass screening of plants for bioactivity by conventional (whole organism-based) bioassays is often very cumbersome, expensive and time consuming and has been the major bottleneck in the discovery of new drugs. Advancements in molecular medicine, development of remarkably sensitive techniques for isolation and characterization, availability of cell-lines, receptors and enzymes (through biotechnology), have contributed to the development of High Throughput Screening (HTS) technology (Broach and Thorner, 1996) which allows quick high volume primary screening for leads using very small quantities of test samples. HTS has led to concomitant increase in the demand for test compounds for screening. Availability of disease-based targets (*e.g.* enzyme inhibitors, anti-metabolites, antioxidants etc.) in conjunction with automated HTS, allows primary screening of a large number of extracts or isolated compounds expeditiously. This has accelerated the pace of drug discovery by expanding the resource base. Some of the outstanding bioactive compounds, which were discovered by biorational approach in recent years are: epothilones from the myxobacterium *Sorangium cellulosum* (anti-cancer), otteliones from the Indian aquatic weed, *Ottelia alismoides* (taxol-like properties), cholestane glycosides from *Ornithogalum saundersiae* (taxol-like activity), betulinic acid from *Ziziphus microphylla* (inhibits oncogene activation) etc. It may be mentioned that most of these plants do not feature in folk/traditional medicine. These discoveries would have been missed without the bioassay directed approach. One of the obvious limitations of target-oriented assays is that these cover very specific range of activities, 'inactive compounds' from such assays may be useful candidates for some other bioactivities.

Chemoprospection-Discovery through Chemo informatics

The modern drug research has become very expensive in terms of time and money. Therefore serious thoughts are given in improving the cost effectiveness for the development of prescription drugs. Advancements in modern biology have created a new awareness in the biotechnical potential of natural products. Modern natural product research is purpose oriented and is guided by biological activities (Kubinyi and Muller, 2004). The bioassay-guided approach has the advantage that the active compounds, when present even as a minor component, can be isolated and characterized. As a result of innovations in technology for isolation and characterization, vast libraries of well-characterized natural products with exotic molecular architectures have been created. Therefore, an alternative approach to drug development would be what may be termed as *chemoprospection* in which the selection of molecules from the library is based on structure-guided approach. This is well illustrated by the discovery of pro-apoptotic activities of betulinic acid, from the extracts of the plant, *Ziziphus mauritiana*. Though known for several decades, its pro-apoptotic properties were discovered only recently by bio-assay directed fractionation. This discovery prompted studies on other related triterpenoids. Since then many triterpenes were subjected to bioevaluation and found to possess such activities (Taraphdar *et al.*, 2001). Through chemoinformatics, it is now known that many plants, including medicinal plants (mentioned in Ayurveda and other indigenous systems e.g. *Ocimum* sp) contain pro-apoptotic triterpenes. Thanks to bioinformatics, the biological activities of several molecules are accessible from the data-bases or literature. Therefore it should be possible to make an intelligent guess for the possible bioactivity of the extracts through the knowledge of molecular profile of the plants. **Dereplication** (direct identification using the key data such as MS, IR, UV, NMR *etc.*) of the known chemical compounds(s) from the available data base or literature goes a long way in avoiding duplication of efforts for re-isolation and characterization of compounds, which may not be of interest. With the advent of metabolomics (Weissman, 2003) recently, the molecular profiles of the organisms will be useful in relating their syntheses to specific gene(s).

Occurrence of plants is dependent on several factors such as climate, ecology, geographical location *etc.* Therefore, the accessibility of plants is often restricted. This limitation can be somewhat obviated by chemoprospection. Once the active molecule(s) (single or mixtures of compounds) of the plants are characterized, it may be possible to find out alternative indigenous plant sources containing the same or related compounds through bio- and chemoprospection. For example, the anti-cancer drug Camptothecine was initially isolated from the Chinese plant *Camptotheca acuminata*. Phytochemical investigation showed that the same compound could be obtained from an abundant Indian plant, *Mappia foetida (Nothapodytes foetida)* (Govindachari and Viswanathan, 1972). The discovery of the new source of camptothecine had a great impact in the development of clinically important camptothecine derivatives with improved bioactivity. Another example of chemoprospection-based discovery is that of phytoestrogens. It was by chance that estogenic activity of isoflavones was discovered when it was observed that the sheep, which grazed on sweet clover (*Trifolium subterraneam*), developed sterility. The cause of sterility was traced to the presence of isoflavonoids, which are known to function as estrogen antagonists. Recently, it has been found that some of the isoflavonoids stimulate a signal transduction pathway leading to apoptosis and now feature in the cancer chemoprevention and hormone replacement therapy. Soya (*Glycine max*) provides good dietary source of phytoestrogens. Similarly, highly functionalized coumarins, calonolides, isolated from the rare Malaysian plant, *Calophyllum lanigerum* provides examples of non-nucleosidic lead compounds (reverse transcriptase inhibitors) with anti-HIV activities. Chemoprospection of the flora reveal the presence calanolide analogs in many Indian species. Chemotaxonomy also plays role towards drug discovery and value addition. The well-known example is that of taxol. Taxol is present in the bark (non-renewable source) of 30-40 year old Western pacific yew, *Taxus brevifolia* that does not occur in India. The Indian species, *Taxus baccata* contains very low levels of taxol but the leaves (renewable source) contain compounds such as baccatin or deacetyl baccatin. In spite of having the basic carbon framework, these compounds do not show appreciable cytotoxicity but can be transformed in to taxol or its derivative by semi-synthetic procedures. This route is followed for the commercial production of taxol and its derivatives. Some other examples of value addition by chemoprospection are: *Gloriosa superba* as an alternate source of colchicines, *Coptis teeta* and *Coscinium fenestratum* as sources for berberine and *Dalbergia* sp as source of isoflavonoids (phytoestrogens). In spite of their ubiquitous presence, tannins did not attract much attention till recently because of difficulty in isolation and characterization. Remarkable topoisomerase inhibitory activity of an ellagitannin, acutisimin A (Quideau *et al.*, 2003) isolated from red wine, has been reported recently. Acutisimin A and related compounds were isolated earlier from many plants such as *Quercus* sp. Therefore this discovery opens new frontiers towards development of topoisomerase inhibitors. Tannins occur widely in plants. It is interesting to find that many of the ayurvedic formulations are prepared from water extracts and are rich in tannins. Thus cues from ethnomedicine, chemotaxonomy and phytochemistry can help in the acceleration of the drug development programs. Some examples of alternate sources for bioactive compounds are presented in the Table 1.

Table 1. Examples of Recent Drug Discoveries and Alternate Plant Sources

Sl. No	Name of the plant and origin	Reported Activity	Active Principle	Evidence-based Bioactivity	Possible Indigenous Alternative sources
1	*Alternanthera philoxeroides* S. America	Folk	Sulphated polysaccharides, High β-carotene and iron content	Anti viral. Inhibition of viral attachment	*A. sessilis* *A. repens* *A. philoxeroides* *Tinospora cordifolia*
2	*Ancistrocladus korupensis* Cameroon	Tribal Medicine	Michellamines Dimeric naphthyliso-quinolines	Anti-HIV. Inhibition of reverse transcriptase	*A. heyneanus* *A. tectorius*
3	*Artemisia annua* China	Folk, Anti-Fever	Artimisinin	Anti-malarial	*A. absinthium* *A. nilgirica* and many sp.
4	*Boswellia serrata* India	Anti-inflammatory	Boswellic acid	Anti-inflammatory	Other sources for boswellic acid and analogs.
5	*Calophyllum lanigerum* Malaysia	NIH-Bioassay based fractionation	Calanolides	Non-nucleosidic Anti HIV agent	*C. apetalum* *C. inophyllum* *C. soulattri* *C. tomentosum*
6	*Camptotheca accuminata* China	NIH-Bioassay based fractionation	Campothecin and derivatives	Anti-tumor Topoisomerase I inhibitor	*Mappia foetida* (*Nothapodytes foetida*)
7	*Commiphora mukul* India	Ayurvedic medicine, Lipid disorders	E- and Z-guggulsterones	Antagonist ligand for FxR. Chlesterol lowering	*C. berryi*, *C. caudata*, *C. roxburghii*
8	*Cephalotaxus harringtonia*		Harringtonine Homoharring-tonine	Antitumour	*Cephalotaxus graffiti*, *C. mannii*.
9	*Combretum caffrum*		Combrestatin A- 4	Anti-tumour, Inhibitor of microtubular action	*C. acuminatum*, *C. decandrum*, *C. pilosum*
10	Cotton seed China	Epidemiological al info. Male anti-fertility	Gossypol	Male anti-fertility	*Thespesia populnea*, *Gossypium* sp

Sl. No	Name of the plant and origin	Reported Activity	Active Principle	Evidence-based Bioactivity	Possible Indigenous Alternative sources
11	Cruciferous edible plants		Isothiocyanates Sulfophrane	Anti-oxidants, Induction of quinone reductase	Numerous potential plants
12	*Eleutherobia* sp. Marine soft coral	Ecological observation	Eleutherobin	Anticancer, microtubule stabilization	Soft corals from Indian ocean
13	*Ervatamia microphylla*	NIH Bioassay directed fractionation	Complex Alkaloid	Inhibits oncogenes	*E. dichotoma* *E. divaricata* *E. heyneana*
14	*Glycyrrhiza glabra* India	Diuretic, tonic and emetic	Glycyrrhizin	Anti-HIV	*Abrus precatorious*
15	*Hippophae rhamnoides* China, Russia	Folk information, *Amchi* medicine (Tibetan)	Vit. C, Carotenoids, flavanoids, Ellagitannins Ursolic acid	Nutraceuticals. Chemoprevention	*H. rhamnoides, H. tibetania, H. salicifolia*
16	*Homolanthus mutans,* Samoa Africa	Samoa Tribes	Prostratin	Anti HIV	*Euphorbia hirta* and other sp.
17	*Maesa lanceolata* Africa, Ayurveda	Folk, Cholera, Asthma	Maesinin	KDO transferase inhibition, COX inhibitor	*M. perrottetiana* *M. indica* *M. argentena etc.*
19	*Maytenus ovatus*		Maytansine	Anti tumour	*Celastrus emarginatus* *C. senegalensis* *C. spinosus*
20	*Morinda citrifolia* Polynesia	Folk	Antioxidants, anthraquinones	Chemoprevention	*M. tinctoria* *M. angustifolia* *M. umbrellata* etc.
21	*Musa paradisiaca*	Banana variety, Kerala	Diarylheptanoid	Inducers of quinone reeducates, Cancer chemoprevention	Many varieties are available
22	*Ochrosia*		Ellipticine	Anti tumor	*O. elliptica* *O. oppositifolia*
23	*Ornithogalum saundersiae* USA	Ornamental, No earlier history. Bioassay directed fractionation	Steroid glycosides	Human lung cancer similar to taxol	Many sp. are available
24	*Ottelia alismoides* Aquatic Indian weed, USA	No history NIH Bioassay directed fractionation	Otteliones	Taxol like properties	*O. alismoides* is available in India

Sl. No	Name of the plant and origin	Reported Activity	Active Principle	Evidence-based Bioactivity	Possible Indigenous Alternative sources
25	*Pancratium littorale*		Pancratistatin Alkaloid	Anti-tumor	*P. triflorum* *P. zeylanicum*
26	*Pfaffia paniculata* Brazil		Beta-ecdysone, nortriterpenoid, pfaffic acid, sitosterol, sigmasterol, pantothenic acid adaptogenic	Anabolic, anti-inflamatory, antimutagenic. Sports medicine	Many Indian plants contain ecdysoids
27	*Phyllanthus acuminatus*		Terpene glycoside Phyllanthoside	Antitumor Inhibitor of protein synthesis	*P. amarus*, *P. uraria* *P. acidus*, *P. aderaspatensis* *P. reedi*, *P. simplex*.
28	*Phytolacca americana*, USA	Bioassay directed fractionation	Single chain RIPs	Inhibition of HIV-1 integrase	*P. acinosa* *P. dioica*
29	*Pyrus malus*	NCI bioassay directed fractionation	Dihydrochalcone, quercetin, chlorogenic acid Antioxidant	Colon cancer	*P. communis* *P. pyrifolia* *P. pashia*
30	Red wine (grapes)	Epidemiological	Resveratrol, anthocyanins, acutissimiin A and related tannins	Anti-oxidants, Induction of quinone reductase. Topoisomerase inhibitor	Numerous potential tannin-rich plants
31	*Rhodiola rosea* Himalayas, India, Russia	Folk, Ginseng like activity	Tannins flavonoids, phenylpropanoid glycosides	Anti-stress activity, Radio protection	*R. imbricata*, *R. heterodontata* . Cold deserts of Himalayas
32	*Sedum rosea* Himalaya, India	Folk		Anti-stress activity	*Sedum ewersii* sp available in W. Himalayas
33	Soy (*Glycine max*) and other dietary sources	Epidemiological	Genistein and related isoflavonoids	Induction of quinone reductase, oestrogen receptor inhibitor	*Iris* sp., *Dalbergia*

Advances in the techniques of isolation and characterization of phytochemicals have resulted in the creation of vast libraries of well-characterized compounds, which have not been adequately evaluated for different bioactivities. Therefore in addition to bioprospection, attention should be given for the elaboration of the pharmacological profiles of the vast untapped source of already known biomolecules. Further, the library of natural products will provide novel molecular scaffolds for the production of natural product-like molecules through combinatorial approach (Rivkin *et al.*, 2005). Combinatorial biosynthesis (predestined modification of biosynthesis by genetic engineering) holds considerable promise in providing library of entirely novel molecules (natural product-like molecules, hybrid molecules) (Hopwood *et al.*, 1885). Awareness of traditional knowledge base, advancements in the chemical and biological sciences and instrumental technology have brought about a quiet but very significant revolution in the modern approach to natural product research. With a systematic and sustainable tapping of biosources and *efforts towards value addition*, biodiversity rich country such as India can provide lead to future plant based industries and *phytotechnology*.

CONCLUSION

The global herbal medicine is a US$ 16 bn industry and is rising at the rate of 30%. Though India is the biggest supplier of raw materials, its share in the global market is only 2%. Therefore there is urgent need for value addition to the herbal industry through the development of appropriate technology. Documentation of the entire Indian biodiversity is a Herculean task and will take long time. But if concurrent efforts for its utilisation are not made, the available information may come handy to the technologically advanced countries, for exploitation after value addition through isolation and characterization of active molecules and protection through intellectual property rights. Thus chemical profiles of natural products can be very useful in enhancing the bio-resource base. It will also help in relating the synthesis of metabolites with specific gene(s) through metabolomics. Awareness of traditional knowledge base, advancements in the chemical and biological sciences and instrumental technology have brought about a quiet but very significant revolution in the natural product research. It is no more confined to mere isolation and cataloging the compounds but it is directed by biology.

It is a highly interactive multi-disciplinary area of research. Plants as source of drugs and other useful chemicals are known since antiquity. Due to limitations of then available techniques, the earlier work on natural products was mainly confined to the studies on major components. With the aid of modern techniques, it is possible to isolate and characterize even the minor constituents. This has led to an exponential increase in the number of well-characterized natural products. Thus, a vast library of natural products and their analogs is awaiting utilization. Concurrent to bioassay-directed isolation of natural products from medicinal plants, screening the judiciously selected known compounds through chemoprospection will complement the current drug discovery programs (old compounds/new activity).

REFERENCES

Banerji, A. (1992) *Bioelectrochemistry and Bioenergetics* 27: 105-113.

Banerji, A., Ramakrishnan, G. and M.S. Chadha (1974) *Tetrahedron. Lett.* 1369

Broach, J. R. and J. Thorner (1996) *Nature* 384, (Supplement 7 November) 14.

Chaffin, Z.T (1998) *American Family Physician,* 58 : 1-89.

Cox, PA, (2000) *Science,* 287 (5450) : 44.

Daniel, M. (2004) *Curr. Sci,* 87: 274.

Fellows, L. (1992) *Lancet* 339 : 1330.

Govindachari, T. R and N. Viswanathan (1972) *Phytochemistry,* 11: 3529.

Hopwood, D.A. *et al.,* (1985) *Nature* 314: 642.

Kinghorn, A.D. and M.F. Balandrin (1993) *Human medicinal agents from plants* (Washington: American Chemical Society).

Kubinyi, H. and G. Muller (2004) *Methods and Principles in Medicinal Chemistry*, 22, Wiley-VCH, Weinheim.

Murphy, D.S. (1995) *Chem. Br.,* 31: 300.

O'Neill, M.J. and J.A. Lewis (1993) In *Human Medicinal Agents from Plants;* (A. Kinghorn and D. Balandrin Eds.) American Chemical Society, Washington. 48-55

Purohit, A.N. and U. Dhar (1997) *Proc Indian Nat. Sci. Acad.* B63: 187-209.

Quideau, S. *et al.* (2003) *Angew Chem Inl Ed Engl* 42: 6012.

Rivkin, A., Chou, T.C. and S.J. Danishefsky (2005) *Angew. Chem. Int. Ed.* 44: 2838.

Rodriguez, E. *et al.* (1985) *Experientia,* 41: 419-420.

Roy Chaudhury, R. (2004) *Herbs for Healing,* New Dawn Press, New Delhi 2004.

Subramanian, T.C. (2002) In *International Seminar on Medicinal Plants & Spices-* Patents & Exports, Chennai.

Taraphdar, A.K. Roy M.R.K. Bhattacharya (2001) Natural products as inducer of apoptosis, *Curr Sci.,* 80 : 1110.

Todd, A. (1961) *Chemistry of Natural Products*; 1960 IUPAC Symposium Australia (London : Butterworths) 359-366.

Weissman, K. (2003) *Chem. World.* 24.

Zhong, X. *et al.* (1999) *Chinese Herbal Material,* 22: 662-663.

3

MEDICINAL PLANTS MENTIONED IN ATHARVA VEDA

Arun Arya

Department of Botany, Faculty of Science,
The Maharaja Sayajirao University of Baroda, Vadodara 390002

ABSTRACT

The famous French philosopher Voltaire had said "The Vedas are the most precious gift for which the West has ever been indebted to the East." Rig Veda is the oldest book of Aryans. Max Muller estimated its period 1500-1200 BC; Rajaram places it to be 3700 BC. There are four Vedas. Sanskrit the language used in *rucha* (verses) of the hymns of Veda developed after refinement of languages. The Atharva Veda is full of Ayurveda and herbal products. There are 5987 mantras (*rucha*). Thirty seven *sukta* of chapter (*Kand*) 4 is devoted to medicine and medicinal plants. 3rd mantra of this chapter describes the insecticidal properties of guggul (*Commiphora wightii* Arnott.). Second mantra of 38th sukta emphasizes the importance of guggul and tells us that with the use of it all diseases disappear like deer. Guggul is used in treating rheumatism, obesity, neurological and urinary disorders.

"Yam bheshjasya guglo surbhirgandho asnute," is a part of 1st verse in 38th sukta of 19th chapter, which means the sweet smell of guggul makes us free from diseases. Oleo-gum resin obtained form guggul is known for its medicinal use. Importance of gum and resin obtained form chir pine (*Pinus longifolia* Roxb.) pakar (*Ficus infectoria*), flame of the forest (*Butea monosperma* (Lam.) Kuntze) are mentioned in the 5th mantra of the 5th part of 5th chapter. Vedas tell that Rishis/ Saints pray to god so that these come out of the tree and deposit there. These have wound healing property. 15th sukta of 5th chapter is devoted to uses of madhula or mulhatti (Jesthimadhu). It is botanically called

Glycyrrhiza glabra L. Vedas say that it makes our voice sweet. Drug consists of dried roots. It has been confirmed that chemical Glycyrrhizin is present which is an expectorant, laxative, and anti-inflammatory agent. Plants like Mahua (*Madhuca latifolia*), white sarson (*Brassica* sp.), til (*Sesamum indicum*), dub (*Cynodon dactylon*), pipli (*Piper longum*), satavari (*Asparagus racemosa*) and king of herbs called *soma* are also mentioned in Atharva Veda. Paper describes some of the important plants mentioned in Atharva Veda and their properties.

INTRODUCTION

The early home of Aryan was in south Russia in the **Steppes** between Danube, the Volga and the Urals. During 1600 BC Aryans left this home land and dispersed east and west in large groups. Saraswati also called Sadaneera (ever flowing) is mentioned in Rig Veda while there is no mention of sindhu or Indus and Ganga. It is later in the vedic period the land inhabited began to be called Saptasindhavah, the land of seven rivers (Agnihotri, 2000). These were Sutudri (Satlaj), Vipas (Beas), Parushri (Ravi), Asikni (Chenab), Vitasta (Jhelum), Sindhu and Saraswati. Aryans settled along the rivers. Early vedic Aryans were primarily pastoral, when they settled in Punjab, they cut jungles, built their villages, grazed their cattle in jungles and planted barley in the land close to their habitation. Eighteenth mantra of 2nd part of 8th chapter of Atharva Veda says -

"Shivam te stam virhiyavavabala savdomadho"

Which means use of paddy rice and barley in our meals may keep our body free from all diseases. In the next mantra also it is adviced to consume these cereals obtained through agriculture practice. The other important cereal wheat (*Triticum aestivum*) was not an important part of daily diet at that time.

It was customary to take food after reciting 83rd mantra of 11th chapter of Ayajur Veda.

"Om annpate anyashya no dehi Anmivashyah sushimina"

O God! This food may be health giving and keep ourselves free form diseases. Drug development based on natural products is a fast growing area combining the strength of knowledge base of traditional systems such as Ayajurveda with the dramatic power of combinatorial sciences and high throughout screening will help in generation of structure activities libraries. Ayurvedic knowledge and experimental database can provide new functional leads to reduce time, money, and toxicity- the three main hurdles in drug development.

Man has an intimate relationship with nature. Most of the world civilization originated and flourished in the vicinity of plants. Biotic form useful to man earned it awe and reverence. Nutritional and medicinal properties of number of plants are described in Vedic text.

"Amanyuta no virudho bhavanti" Atharva Veda 12:3:31

Medicines may be useful to us. Another mantra says -

"Atho kranom bheshajam yathasat sathayanah" Atharva Veda 8:7:22.

With the help of medicine may you live upto hundred years.

"Prastranati stambi nirek shunga...Purush jivani" Atharva Veda 8: 7: 4

God has produced numerous medicinal herbs. With the better use of these we may lead a healthy long life.

"Ayavtishy manushya bhesajam bhishjo vidh", Atharva 8: 7: 26

All the plants (herbs) necessary for the medicine may be available to us. Third mantra of 12th part of first chapter says we should be in harmony with trees and hills to be free form diseases. Ayajur Veda also tells *"Satam jivem sharadah satam"* we may live upto 100 years; we may live even more than 100 years. Use of rice and barley is recommended in 18th mantra of 2nd sukta of 8th chapter. 15th mantra of same sukta advocates three thing, taking of rice in food, living on hills (in pure and clean air) and be in an open air, living with sun and moon is good for health.

PLANTS MENTIONED IN ATHARVA VEDA

1. Guggul (*Commiphora wightii* Arnott.)

Guggul is the oleo-gum resin obtained form *C. wightii* of family Burseraceae, the plant is a small tree with 1-3 foliate leaves with shining rhomboid ovate leaflets. The red coloured drupe contains two woody pyrenes. 37th sukta of chapter 4th is devoted to medicine and medicinal plants. 3rd mantra of this chapter describes the insecticidal properties of guggul. 2nd mantra of 38th sukta emphasizes the importance of guggul and tells that with the use of it all the diseases disappear like deer.

"Yam bheshjasya guglo surbhirgandho asnute"

is the first verse of 38th sukta of 19th chapter which means that the sweet smell of guggul makes us free from diseases.

The resin contains about 0.37% essential oil consisting of myrcene, dimyrcene and polymyrcene, a diterpene hydrocarbon and sterols like guggulosterons I, II, and III (Patil *et al.*, 1972), Sesamin and cholesterol.

Guggul is used in treating rheumatism, obesity, neurological and urinary disorders. Guggulsterol are found to be anti-inflammatory, and hypocholesteremic (Sabnis and Daniel, 1990). Guggul is also used as major component of the *Dhup* and *Havan samagri* (A mixture of plant produce used to put on holy fire). Because of its antimicrobial property it was an important component to be applied to Egyptian mummies which helped them from any kind of deterioration by microbes.

2. Peepal (*Ficus religiosa* L.)

Ficus religiosa is the member of family Moraceae. The tree reaches much large proportion and largest among Indian fig trees. The leaves are leathery, broadly ovate, suddenly narrowed at the apex and formed a long tail. The base is heart shaped. Hindus consider the tree as the most sacred and worship every morning after completion of bath. Lord Buddha attained and enlightenment under this tree in Bodhgaya, Bihar and the tree is also called as bodhi tree or Bo-tree.

"Aswatho devsadnah" Atharva Veda 6: 95 :1

Peepal is residing place of Gods. Another mantra says -

"Tanwath ni shridih satrun" Atharva Veda 3:6:2

It means O Ficus! You kill the demons (Evil spirits).

"Aswatho darbho.....havih" Atharva Veda 8:7:20.

Peepal, Kush (a grass) soma and cereals are pious medicines. 4th verse of 37th part of 4th chapter says -

"Yatrashwatha nyagrodha mahavriksha shikhandinah,
tat pare tap sparsah pratibudha abhuta"

which means plantation of peepal, ficus (Vat) and shikandi (A kind of Jasmine) is insecticidal.

Peepal is one of the drug most commonly used as an antiseptic and is beneficial in treatment of ulcers. It relives burning sensation and reduces accumulated fat in obesity. Peepal bark is astringent, cooling and aphrodisiac. Fruits are laxative and digestive. Leaves and tender shoots are purgative and seeds are refrigerant and laxative(Babu and Madhavi, 2001).

3. Mahua (*Madhuca indica* Gmel.)

It is a large tree of family Sapotaceae. The kernel yield 55-65% of whitish yellow oil consisting of glycerides in the composition, palmitic acid 23-27%, stearic acid 19-25%, oleic acid 41-45% and linoleic acid 9-13%. Atharva Veda says

"Anjanasya madhudhasya kusthasya naldashya cha"
Atharva veda 6:1.2:3

Which means arjun (*Terminalia arjuna*), Mahua (*Madhuca indica*) and Kuth and Khas Khas (*Vetiveria zizaniodes*) act as refrigerant. Rabindranath Tagore (1928) tries to be intimate with *Madhuca* and addresses her by her Indian name mahua.

"O! Mahua, the day I shall meet my love
I shall also call her by your name"

4. Shatavari (*Asparagus racemosa* Willd.)

Asparagus is a stem climber of family Liliaceae. The plant has cladodes and small scaly leaves. The root tubers are medicinally important. The roots yields four saponins (0.1-0.2%) based on sarsapogenin designated as shatavarin I, II, III and IV, plenty of starch, asparagines and s-methyl methionine. Root is used as diuretic, demulcent, aphrodisiac, and general tonic. (Sabnis and Daniel, 1990).

"Madhyen yaksham badhate" Atharva Veda 19:36:2

Middle part of shatavari is used for cure of diseases verse 4 and 6 also describes uses of asparagus.

"Shatam viranjanyat satam yakshman pavpat" Atharva Veda 19:36:4

Means *Asparagus* produces 100 brave sons and cures 100 diseases.

5. Vad (*Ficus benghalensis* L.)

Banyan is native of India. It belongs to family Moraceae. It is an enormous tree, and attains to a height of 30 to 45 meter and sending down roots from the branches. These prop roots enter the ground and form trunks. Fourth verse (37th sukta of 4th chapter) describes a tree Naygrodha (*Ficus*) as major insecticidal tree.

6. Babul (*Acacia nilotica* Delile)

Acacia is a native of Africa and Asian tropics. It belongs to family Mimosaceae. Bark is a source of tannin in India. The plant has bi-pinnate leaves. Small yellow flowers in 20-60 flowered heads and long pods, contracted between seeds. Mantra 4-6 of 5th sukta of 5th chapter of Atharva Veda describes gum obtain from plants like *A. nilotica* and *Butea monosperma.*

It is red and shining and has wound healing properties. According to Laddha and Jain (2005) gums are considered to be pathological products formed upon injury of the plant or owing to unfavourable or stressed conditions such as drought. Some researchers, however, presume that gums are normal metabolic products of certain plants. Gum is cooling, emollient, expectorant, liver toxic, aphrodisiac, haemostatic, antipyretic, and toxic.

7. Shami (*Acacia suma* Buch. Ham.)

Shami is member of family Mimosaceae. Second mantra of 30th sukta of 6th chapter tells the elegance of shami tree. It should grow with its 100 of branches. Next mantra says O! Shami tree! You increase the length of hair as mother nurtures her son.

8. Palash (*Butea monosperma* (Lam) Kuntze) (Syn. *B. frondosa* Koenig ex Roxb.)

It is a native of India and belongs to family Fabaceae. It is a medium sized deciduous tree with crooked trunk and branches looking ugly except flower period. The leaves are trifoliate, used for making plates. Due to vermilon bloom the plant is called as flame of the forest. Rabindranath Tagore (1986) wrote

"The santhal woman juries up and down the gravelled path under the simool tree, a coarse grey sari closely twines her sleder limbs, dark and compact; its red border sweeping across the air with the flaming red magic of palash flower."

Gum has numerous uses 3-9th mantras of 5th sukta of 5th chapter are devoted to gum. 5th mantra says —

"Bhadrat plakchhani stisthasya swatha tkhadiradwat,
Bhadranya grodhat parnatsa na yehrudh anti" Atharva Veda 5:5:5

Importance of gum and resin obtain form Pine (*Pinus roxburghaii* Roxb.) pakar (*Ficus infectoria*), flame of the forest (*B. monosperma*) are mentioned in this mantra. Veda tells that *rushis* pray to God so that these come out of the trees (exuded) and deposited there. These have wound healing property. The wood of this plant is soft but not durable. The gum obtained form the bark, known as Bengal kino or Butea kino is used as an astringent. The leaf juice is prescribed for curing sterility in women. Powdered seeds are used as anthelmentic remedy and to cure skin diseases. This plant also serves as the most important host for the lac insect.

9. Brahmi (*Bacopa monnieri* Wettst.)

It belongs to family Scrophulariaceae. It is a small herb growing in marshy land near perennial streams. Mandukparni and matsyakshi are two species these along with vacha (*Acorus calamus*) rhizome are good for enhancing memory. It has bacoside which are useful as nervine tonic and cooling effect. 11th verse of 37th sukta of 4th chapter describes the use of brahmi for cure of veneral diseases.

10. Arjuna (Anjan) *Terminalia arjuna* W. & A.

This is a large tree belonging to family Combretaceae. Plantation of arjuna tree is suggested in 5th verse of 37th sukta of 4th chapter. To remove disease causing agents. Atharva Veda says - "Vabhro

"Arujn kandshya avashya to palalya tilashya tilpinjya,
virutchetri yahash anyap shetriya muchuta" Atharva Veda 2:8:3

Use of stem of arjuna tree, glumes of barley and flowers of til (*Sesamum indicum*) helps in cure of hereditary diseases.

Another mantra says *Terminalia* enhances memory and gives longevity.

"Atho harit bheshjam" Atharva Veda 4:9:3.

It is a cure for jaundice.

11. Pippali (*Piper longum* L.)

Piper longum belonging to Piperaceae is a slender creeping under shrub with erect and subscadent branches, rooting at nodes. The fruits are small and fused into cylindrical spike like cones. They contain alkaloids chavicine, free piperine, piperethine, piperidine and piperlongumine. The fruits are more aromatic and sweeter than black piper.

Mantras 1-3 of 101 sukta of 6th chapter describe the uses of long piper. It is used to cure joints pains. Atherva Veda says -

"Yam jivmashn vamhai na sa rishyati purushah" Atharva Veda 6:109:2

The person who uses long pepper does not fall sick. it is useful in fever, influenza, bronchitis and bronchial asthma (Das, 1993). It is used for indigestion. Along with bark of *Terminalia arjuna* it cures heart diseases.

12. Licorice (*Glycyrrhiza glabra* L.)

Commonly known as mulhathi. The plant belongs to family Fabaceae. The glandular, hairy, herbaceous, perennial, one meter in height bears pinnate leaves having 9-17 ovate leaflets, pale blue flowers in spike and a 3-4 seeded red brown glabrous pod. Glycyrrhizin is a saponin obtain from root. The roots are fibrous, chocolate brown and yellow inside. It is 50-150 times sweeter than sugar. The two phenolic glucosides, isoliquritin and liquritin accompanying this saponin also are sweet (Sabnis and Daniel, 1990). Obtaining root by diging the soil is said in 1st mantra, 34th sukta of 1st chapter.

"Madhume madhula karah" Atharva Veda 5:15:1.

Licorice has given sweetness to my voice. The roots are used in soar throat. Glycyrrhizin is an expectorant, laxative and anti-inflammatory. It has been also recommended for stomach ulcers, anemia and blood vomiting.

13. Varuna (*Crataeva nurvala* Buch.-Ham.)

It is a small sized tree of family Capparaceae. Leaves are compound, 3 leaflets on a common stalk. Flowers are white; fruit is berry, globose with woody rind. Leaves and bark are used medicinally. Bark extract which contains tannins and also saponin is used as a laxative and appetizer. It is also reported to be useful in infections of the urinary organs (Randhawa, 1983). To remove urinary stone, bark of this tree is boiled in water and decoction is given to the patient in a dose of 30 ml. 1 tds (Dash, 1993).

"Varno varyata yam devo vanaspati aykshmo yo,
asminnavasht stamu deva avivaran" Atharva Veda 6:85:1

First mantra of 68 sukta says varuna is good for mucus, urinary problem and its anthelmentic. Second and third mantras also describe uses of this plant. It is good for loveliness and is a blood purifier.

14. Mustard (*Brassica campestris* L.)

It belongs to family Brassicaceae. It is a globose, glabrous with the few bristles at the base and leaves are broad and coarsely dentate at edges. Flowers are yellow. The oil is obtained for seeds. They contain 30-45% fixed oils. It is yellow in colour and possesses little pungent taste. It contain glycerides of erueic (50-55%), oleic (20-23%), linolic (15-20%) linolenic (1-3%)

palmitic and lignoceric/ behenic (1% each) acids. The oil is thermogenic, anodyne and anti-inflammatory. Veda says —

"Tamoshadhe twam nasyashyah kamal manjivam" Atharva Veda 8:6:9.

Brassica should make vagina of the lady disease free.

"Aryan shwakiskido bajah pingo aninashat" Atharva Veda 8:6:6.

Yellow coloured mustard kills the painful disease causing agent. Another mantra 24 of the same chapter emphasized that white and yellow *Brassica* kills the pathogens complitly. Efficacy of 5 different fixed oils was tested against two fungal pathogens and fixed oil of mustard was found effective (Arya, 1988).

15. Durva (*Cynodon dactylon* Pers.)

Dub grass or Durva belonging to family Poaceae is a prostrate, extensively creeping perennial grass. It is highly branched and rooting at every node. The leaves are narrow and linear, whole plant is used as medicine it is haemostatic and cooling Veda says -

"Teb sahastrakanden pari nah pahi vishwate" Atharva Veda 2:7:3

Third and 4th mantra is devoted to dub grass. It says grass may protect us from all sides. Use of hundreds of stems of grass may inspire us to be away from any type of anger.

16. Vacha (*Acorus calamus* L.)

It is commonly known as sweet flag, vacha belong to family Rananculaceae. It is a semi aquatic rhizomatous herb. Rhizome is cylindrical, light brown externally and spongy white within. Rhizomes are nervine tonic, digestive, wormicidal, thermogenic, aromatic and intellect promoting. It is mentioned in mantra.

The aromatic principle from rhizome is insecticidea and gungicidal. The effiecacy was proved against *Cochliobolusverruculosus, Penicillum citrirum, Aspergillus ustus* and *Chaetomium cybiforme* by Shah and Arya (2000)

Vedas refer use of many plants according to Majumdar (2004) the number of medicinal planst is 150 and in charka samhita 526 plants are mentioned. Ayurveda is sceicne of life. It is one of the ancient medical sciences in the world, adopt a holistic approach. It treats a human being as a complex whole and stricves for harmony between a human being and the entire environment that surrounds him. In ayurveda, presently about 1,000 single drugs and 8,000 compound formulations are of recognized merit. Ayurvedic compoumds formulations are divided into two groups:

1. Kasthausadhi (Predominanly plant drug)
2. Rasausadhi (Predominanly metals and minerals)

Tenth mantra of 3rd chapter (31st sukta) says with the use of herbal medicines we may lead a happy healthy life. First mantra of 96th sukta in 6th chapter describes importance of Soma medicine.

"Ya Oshadhyah Somaragyi barhi sativichakshdah" Atharva veda 6:96:1

Soma is the king of herbs. Medicines are cure for hundreds of diseases.

"Mumuchana oshadhyo agne vaishwanara dadhi,
Bhumi santnavtirati yasam rajah vanaspatih" Atharva veda 8:7:16

The king of herbs, soma vanaspati you keep us free from diseases and be readily available to us.

According to (Wassen, 1968) the soma herb is a mushroom, *Amanita muscaria*, family Amanitaceae.

Researches are needed to find out other medicinal plants maentioned in Atharva Veda and other vedic texts. According to Gangadharan (2005) in the new era of drug development based on natural products, there can be two paths: One the traditional (western) methods of drug discovery based on active principles and hit and trial method and the another is intentional based on sound understating of the paradigm of natural products action in the body based on Ayurvedic principles. The later one has got greater chances of success and also has lesser time bar for development as it is based on the clinical activity.

Table 1. Plants mentioned in Atharva Veda

Botanical Name	Common Name	Family
Acacia nilotica	Khair	Mimosaceae
Aegle marmelos	Bael	Rutaceae
Asparagus racemosus	Shatavar	Liliaceae
Bacopa monnieri	Brahmi	Scrophulariaceae
Brassica campestris	Sarson	Brassicaceae
Butea monosperma	Palash	Fabaceae
Commiphora wightii	Guggul	Burseraceae
Crataeva nurvala	Varun	Capparaceae
Cynodon dactylon	Durva	Poaceae
Dalbergia sissoo	Shaisham	Fabaceae
Desmostachya bipinnata	Kusha	Poaceae
Ficus benghalensis	Vat	Moraceae
Ficus glomerata	Udambar	Fabaceae
Ficus religiosa	Pipal	Moraceae
Glycyrrhiza glabra	Mulhathi	Fabaceae
Madhuca indica	Mahuda	Sapotaceae
Pinus roxburghii	Chir	Pinaceae
Piper betel	Paan	Piperaceae
Piper longum	Pippali	Piperaceae
Terminalia arjuna	Arjun	Combretaceae
Vetiveria zizanoides	Khaskhas	Poaceae

REFERENCES

Agnihotri, J.P. (2000). In: *Ancient and medieval history of Indian agriculture* (Eds. Chaoudhary, S.L., G.S. Sharma and Y.L. Nene) Rajasthan College of Agriculture, Udaipur, 28-31

Arya, A. (1998). *Indian Phytopathology,* 41(2): 214-219

Babu, S.S. and M. Madhavi (2001). *Green remedies, Healing Power of herbs,* Pustak Mahal, Delhi, 243 pp.

Dash, B. (1993). *Ayurvedic cures for common diseases,* Hind Pocket Books, Delhi.

Dwivedi, K. (1988). *Atharva Veda subhashitavali,* Vishwabharti Anushandhan Parishad.

Gangadharan, G.G. (2005). *Chronicle Pharmabiz,* March 24th, 2005 p. 17

Laddha, K.S. and M.B. Jain (2005). *Indian Drugs,* 42(3): 178.

Majumdar, A. (2004). *Ayurveda, The ancient science of healing.* Macmillan India Ltd., Delhi, 426 pp.

Patil, V.D., U.R. Nayak and S. Dev (1972). *Tetrahedron,* 28: 2341.

Randhawa, M.S. (1983). *Flowering Trees,* National Book Trust, India, New Delhi.

Shah, N.R. and A. Arya (2000). *Studies on some fungal biodeteriogens,* Bhartiya Kala Prakashan, Delhi, 208 pp.

Siddhantalankar, H. (1990). *Atharva Veda Bhashyam,* Bhagwati Prakashan, Delhi, Vol. I & Vol. II.

Sabnis, S.D. and Daniel, M. (1990). *A Phytochemical approach to economic botany,* Kalyani Publishers, Ludhiana.

Tagore, R. (1986). *Poems* (Visva-Bharati Publications, Calcutta): 157

Wasson, R.G. (1968). *Soma-divine mushroom of Immortality.* Harcourt Brace Jovanovich, New York.

4

KORKU USES OF SOME LESSER KNOWN MONOCOTYLEDONOUS ETHNOMEDICINAL PLANTS OF MELGHAT (DIST. AMRAVATI)

***Prabha Y. Bhogaonkar and Vinod D. Devarkar*[*]**

Dept. of Botany, Government Vidarbha Institute of Science and Humanities, Amravati.
[*]Shri Chattrapati Shivaji College, Omerga

ABSTRACT

A survey of Korku traditional health practices in Melghat was carried out in collaboration with Amravati Zillah Parishad under Integrated Tribal Upliftment Programme named 'Madhughat Prakalpa'. The survey was carried out during 1997-1999. About 47 villages were visited. Eighty Korku persons engaged in traditional health practices were interviewed. Survey resulted in recording 169 plant species used by Korkus to treat about 59 ailments and some used as poisons, antidotes, cosmetics and tonics. Of the 169-ethno-medicinal species recorded, 138 belong to 50 families of dicotyledons and 31 species belong to 11 families of monocotyledons. When compared with the available literature, out of 31 monocotyledonous ethnomedicinal herbs, 11 were found to be used in a novel way.

INTRODUCTION

Melghat is one of the nine Tiger habitats initially selected in the county during 1973 for launching 'Project Tiger' scheme. Melghat is a southern offshoot of the Satpura range. The tract is very hilly and consists of a succession of hills and valleys in a confused pattern with constant and abrupt variations in altitude, aspects and gradient; even within a single compartment. The entire region consists of Deccan traps with occasional thin top soil. The average rainfall is usually higher on the main ridge of Gavilgarh amounting to 1784 mm. Some of the valleys receive up to 2500 mm rainfall in a year. Maximum temperature reaches 35°C and minimum, especially in certain valleys going below freezing point. Melghat forests are dry deciduous.

Over 53 million tribals belonging to over 550 tribal communities that come under 227 linguistic groups (Anon., 1994) inhabit the Indian subcontinent. Melghat Sub-Division of Amravati District comes under category I, *i.e.* where tribal population is over 50%. Tribal population of Melghat is about 80% of its total. Korkus, Gonds and Nihals are the scheduled tribes residing in Melghat forests. Majority of them are Korkus.

MATERIAL AND METHODS

About 80 Korku medicine-men belonging to 47 villages were interviewed for traditional knowledge about herbal medicine. To confirm the diagnosis of diseases, the surveying team was assisted by doctors working at Amravati Zillah Parishad Primary Health Centers located at different villages in Melghat. Field trips were made with medicine men. The plants prescribed by these persons were collected from the field. Medicinal and other uses of plants were noted only after observing the plants in nature. Wherever possible, vernacular and Korku names of the plants were noted (given in parenthesis). Plants were brought to the laboratory, identified with the help of standard floras and processed for herbarium specimens. Voucher specimens are deposited in Herbarium of Botany Department, Government Vidarbha Institute of Science and Humanities, Amravati. For comparison, recorded uses noted from earlier literature are mentioned along with local uses.

Habenaria gibsonii Hook. F. (Orchidaceae.) VDD 78.

Leafy herbs, stem with closely appressed sheaths in lower part. Leaves ovate-oblong. Flowers greenish-white, in lax racemes. Petals 2-partite. Lip 3-partite. Capsules oblong, beaked. Rare. Makhala valley and core area of Tiger Reserve.

Local use : Tuber powder with milk taken as tonic.

Recorded Uses: Tender shoots and leaves used as vegetable (Manadhar, 1993).

Curcuma inodora Blat. (Zingiberaceae) V. *Jangli halad* K. *Bedi halad*, PYB 208.

Rhizomatous herbs. Perennating by root stock; fibrous roots ending in small tubers. Leaves radical, broadly elliptic-lanceolate. Flowers red-purple with dark yellow spot on labellum; in cymose spike; bracts rosy-pink with purple tips. Fruit 3-valved capsule; seeds arillate.

Most common in Melghat.

Local use : Fresh rhizome paste applied over cuts as strong antiseptic. The smoke of dried rhizome used to hypnotize a person, some use it in *Tantrik, Vashikarana* and *Mayajal Kriyas*.

Recorded uses : Root paste in diarrhoea and snake-bites (Mudaliyar *et al* 1987, Shah and Gopal, 1982).

Curcuma pseudomontana Grah. (Zingiberaceae) *V. Jangli halad* K. *Kukus mucus* PYB 209.

Perennial herbs with root stocks. Fibrous roots ending in tubers. Leaves elliptic-lanceolate to oblong. Flowers pale yellow with bright-yellow lip, in lateral as well as terminal dense spikes; flowering bracts green, bracts of coma purple-pink. Capsule ellipsoid-ovoid, glabrous.

Frequent, on moist hill slopes along stream banks.

Local use: Used in *Tantrik* activities by *Mantrik* for hypnotic effects.

Recorded uses: Rhizomes and tubers for leucoderma (Kothari and Murthy, 1996).

Canna indica L. (Cannaceae) V. *Sabbajaya* VDD 66.

Rhizomatous perennials. Leaves elliptic-lanceolate. Flowers scarlet or yellow, in terminal panicles. Upper staminodes and lip variously brightly coloured. Capsule, globose, trigonous, echinate.

Commonly cultivated.

Local use: Root powder is an appetizer.

Recorded uses: Decoction of root in fever and dropsy, also used as ear-drops (Nadkarni, 1995).

Ensete superbum (Roxb.) Cheesm. (Musaceae) *V. jangli kela, chowani*, K. *Jangli kela*

Tree like herbs. Leaves large, oblong, narrowed to the base. Inflorescence spadix; flowers in 2-dense rows, arranged in spikes; bracts deep red or purple. Fruit oblong, nearly dry when rip. Seeds many, brown or black.

On all higher precipitous slopes, on rocks.

Local use: Water from the stems or petiole/sheathing base is used to nullify the effect of alcoholic drinks. Seeds are given to develop permanent sterility in males.

Recorded uses: leaves used on boils to remove pus (Rai, 1985)

Dioscorea hispida Dennst (Dioscoreaceae) V. *Baichabdi, Dukarkand, Tikar* K. Kulu. VDD 220

Very large, climbing herbs. Tubers subglobose; stem prickly. Leaves compound; petiolate, prickly; leaflets 3, ovate-oblong, petiolate. Flowers greenish white. Fruit triquertous, winged, wings oblong.

Common throughout, especially at higher elevations.

Local use : Root smoke for *'Bhut-pret Jagruti'* in *Tantrik kriya* (To awaken ghosts). Leaf juice or decoction mixed with liquor to enhance its effect.

Recorded uses: Tuber is laxative, toxic, and useful to heal the wounds. Decoction of tuber is an alterative, diuretic and used in chronic rheumatism, also as arrow-head poison.

Dioscorea pentaphylla L. (Dioscoreaceae) V. *Shendelvel* K. *Babra* VDD 219

A climbing perennial. Tubers oblong. Stems prickly near the base only. Leaves 3-5 foliolate, elliptic-lanceolate, ovate or obovate. Flowers greenish white. Males in solitary or fascicled paniculate racemes and females in axillary spikes solitary or 2-3 together. Capsule quadrately oblong, apiculate.

Frequent especially in moist localities at higher elevations.

Local use: Young tuber used as blood purifying tonic.

Recorded uses: Tubers in rheumatism. Flower and buds taken in constipation (Bhatt and Sabnis, 1987).

Dioscorea pentaphylla L. var. ***pentaphylla*** (Dioscoreaceae) V. *Babra* K. *Munguswel* VDD 263

Tuberous twining herbs. Stem usually with 2 prickles at each node. Leaves 5 foliolate, elliptic-lanceolate, ovate or obovate. Flowers white or greenish-white; male flowers in recemes towards branch endings; female flowers in pendulous spikes. Fruit a capsule, retuse at both ends. Frequent at higher elevations on Chikhaldara plateau.

Local use: Tubers used against intestinal worms. Leaf juice given internally to cure stomach ache.

Recorded uses: Tubers are used to treat swelling and as a tonic. (Anon., 1948).

Asparagus racemosus Willd. var. *javanicus* Baker (Liliaceae) *V. satavari, Satmuli*, K. *Satawar* VDD 124.

Candent, spinous much branched under shrub. Root nodulose, fasciculated. Leaves reduced to recurved spines. Cladodes green, needle shaped, in groups of 4 or 5. flowers white; in long racemes. Fruit globose berry; red when ripe.

Fairly common throughout in forest undergrowth among grasses.

Local use: General tonic for males. Root extract is used in treatment of behavioral disorders.

Note : Asparagus racemosus var. *racemosus* is an established medicinal plant. However, this use of *A. racemosus* var. *javanicus* appears to be new.

Drimia indica (Roxb.) Jessop. (Liliaceae) *V. jangali Kanda* VDD 123.

Bulbous perennial herbs; bulbs yellowish white. Leaves linear. Flowers purplish-green, on long scapes, 4-30 flowered racemes. Capsule ellipsoid, brownish. Rare.

Local use : Bulb in honey kept in air tight bottle for a year. This medicated honey is recommended for maintaining the normal blood pressure. One spoonful honey is taken every morning. Fresh bulb used against snake poison.

Recorded uses : Tuberous roots are expectorant and diuretic, emetic and cathartic, abortifacient, stimulant, useful in liver disorders, cough, asthma, cardiotonic and bronchitis. (Chopra *et al.*, 1969)

Alocasia indica (Lour.) Koch. Schott. Araceae (*V.& K. Kali pothi*). VDD 159.

Stout herbs; stem short, under ground. Leaves ovate triangular with deep sinus at base, acuminate, glaucous green; petioles purple or dark red. Spathes long, bright yellow, fragrant. In marshy places. Also cultivated around villages.

Local use : Rhizome powder given orally with honey to dissolve intestinal tumors.

Recorded uses: Leaf juice astringent. Rhizome mild laxative, diuretic and useful in anasarca, ear ache, piles, fistula, ascites, rheumatism, constipation, aphthae, oedema and colic (Nadkarni, 1995).

Dendrocalamus stictus Nees. Poaceae (V. *Halpa, Veddar, Math.* K. *Indo*)

Arborescent, tufted bamboo. Culms much tall. Leaves linear-lanceolate; petiole small. Flowers in large, branching panicles of dense globular heads. Spikelets spinescent. Fruit ovoid, beaked, hairy; grain brown.

Gregarious in Melghat.

Local use : The exudates collected from cuts made in young shoots mixed with powder of *Andrographis paniculata* is given in diabetes.

Recorded uses : Seeds cooked as rice during scarcity.

REFERENCES

Anonymous (1948-1962). *The Wealth of India* . I – V. CSIR, Delhi.

Anonymous (1994). *Ethnobiology in India- A Status Report*

Bhatt, R.P. and S.D. Sabnis (1987). *J. Econ. Taxon. Bot.* 9(1) : 139-145.

Chopra, R.N., D. Chopra and B.S. Verma (1969). *Supplement to Glossary of Indian Medicinal Plants*. CSIR, New Delhi

Kothari, M.J. and S. Moorthy (1996). In *Ethnobiology in Human Welfare* ed. S.K. Jain, 403-407

Mandhar, N.P. (1993). *J. Econ. Taxon. Bot.* 17(2): 373-383.

Mudaliyar, S.K., B.G. Kulkarni and B.D. Sharma (1987). *J. Econ. Taxon. Bot.* 15(3): 557- 563

Nadkarni, A.K. (1995). *Indian Materia Medica (Vol. I & II).* Popular Prakashan Pvt. Limited.

Rai, M.K. (1985). *J. Econ. Taxon. Bot* 7(2): 385-387

Shah, G.L. and G.V. Gopal (1982). *J. Econ. Taxon. Bot* 3: 353-364.

5

MEDICO-BOTANICAL SURVEY OF VANASPATIBET OF WADAWAL-NAGNATH

Sharad D. Biradar, Sunil Modak and Prakash K. Kanthale*

Dept. Botany, Dnyanopasak College, P.B. 54 Near Zilla Parishad, Parbhani 431 401
**Department of Botany, Nutan mahavidyalaya Selu, Dist. Parbhani (MS) 431 501*

INTRODUCTION

In Marathawada there are about 1650 species of flowering plants out of which 500 species are used as a source of medicines (Naik, 1998). Jungles in Marathwada region have already been perished and so is the position of medicinal plants in this area. During the last decade it has been found that about 25 species of medicinal plants have disappeared and a number of other species are on the verge of doom. The main reason behind this fall is the lack of knowledge among the medicinal practitioners of 'Ayurveda'. These people are not accustomed to recognise the medicinal plants. Even graduates in 'Ayurveda' are unable to recognise them. There are number of books written to help and guide medicinal practitioners. However, the books lack systematic information to identify medicinal plants.

There are few geographical area in Marathwada which are rich in vegetation of medicinal plants. One of the place where medicinal plants are available on large scale is *'Wadwal-Nagnath'*, located in Chakur Tahsil of Latur district of Marathwada region. The tribal folk and rustics of the area and near by area have been using various plants and their parts as medicines during *"Uttara Nakshtra"* (2nd week of September). The knowledge of the use of the plants has been followed traditionally by trial and error method.

The unscientific methodology and the knowledge of medicinal plants, their uses is being passed on from generation to generation. The use of the plants for medicine is not well documented. Its use is only verbally passed on by rustics and tribals to their next generation. Hence, it is essential to make a documentation of such important information on medicinal plants.

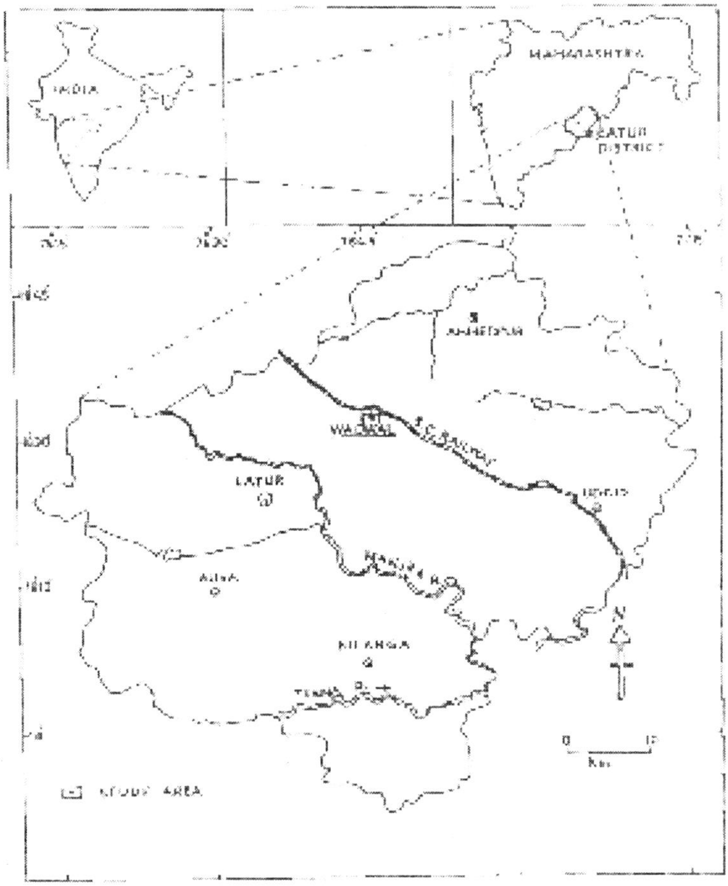

LOCATION MAP OF "WADWAL NAGNATH AREA"
TEHSIL - CHAKUR OF LATUR DISTRICT IN MARATHWADA

Therefore, present study is planed to document the medicinal plants and evolve methods of systematic identification of the plants of medicinal value. Study will also be focused on the medicinal properties (uses) of different plants.

MATERIALS AND METHODS

Flowering plants were collected in different seasons and preserved by usual herbarium techniques. Identification of plants was done by using renowned floras, Cooke (1908), Gamble (1918) and Naik (1979, 1998).

Medicinal uses of important plants (except crop plants) were collected by approaching to local people, tribals and from medicinal books and Journals such as. Chopra *et al.* (1956), Jain (1985), Sinha (1996) and Vartak and Mandavgane (1981).

Abrus precatorius L.

Local Name : Gunj

Family : Fabaceae

Medicinal uses : The extract of leaves is used in the treatment of obesity.

Abutilon indicum (L.) Sweet

Local Name : Petari, Shikka

Family : Malvaceae

Medicinal uses : The leaves and roots are used to treat fever.

Achyranthus aspera L.

Local Name : Aghada

Family : Amaranthaceae

Medicinal uses : It is used against dog bite.

The leaf extract is used for toothache, arthritis (rheumatism)

Aegle marmelos (L.) Corr.

Local Name : Bel

Family : Rutaceae

Medicinal uses : A decoction of the bark of 'Bel' and 'Jambhul' is given as remedy for menstrual irregularities.

Argemone mexicana L.

Local Name : Piwaladhotra, Bilayati

Family : Papaveraceae

Medicinal uses : Paste of root is used against snake bite by applying to the injured area.

To cure partial headache the juice of leaves or roots may be put in the opposite nostrils.

Azadirachta indica A. Juss.

Local Name : Kadulimb

Family : Meliaceae

Medicinal uses : It is used as antiseptic. A paste of bark is applied over the wounds.

Balanites aegyptiaca (L.) Del.

Local Name : Hinganbet

Family : Balanitaceae

Medicinal uses : It is used to enhance the milk production in cattles

Fruits are used as substitute for soap for washing clothes.

Blepharis repens (Vahl.) Roth.

Local Name : Hadson

Family : Acanthaceae

Medicinal uses : The extract of stem is given to treat patients who are suffering from chronic fever taken for three days.

Bombax ceiba L.

Local Name : Katesawar

Family : Bombacaceae

Medicinal uses : The extract of leaves and garlic bulb is given to person for 21 days to improve sperm count or impotency.

The extract of the plant is used as tonic.

Cadaba fruticosa (L.) Druce

Local Name : Kalitakli

Family : Capparaceae

Medicinal uses : It is used against rheumatism or arthritis. The leaves should be taken for three days

Caesalpinia bonduc L. Emend Dandy et Excell.

Local Name : Gajga, Sagargota

Family : Caesalpiniaceae

Medicinal uses : Leaf juice is given to cure haemorrhage and leprosy.

Capparis zeylanica L.

Local Name : Waghati

Family : Capparaceae

Medicinal uses : Leaf juice should be taken in the morning before taking food to treat jaundice. Leaves should be used to treat diarrhoea in animals

Cassia fistula L.

Local Name : Bahawa

Family : Caesalpiniaceae

Medicinal uses : Pods are used for the treatment of constipations

It has great antiviral activity inhibiting several kinds of viruses (Babar *et al.*, 1979)

Crotalaria retusa L.

Local Name : Khulkhula

Family : Fabaceae

Medicinal uses : Mixture of extract of leaves and *Withania* roots is given by adding cow milk against white discharge or leucorrhoea.

Dioscorea bulbifera L.

Local Name : Jathashankar

Family : Dioscoreaceae

Medicinal uses : It is used in deshi ghee (Made from cow milk) against gastro intestinal disorders.

Diplocyclos palmatus (L.) Jeffrey

Local Name : Shivlingi

Family : Cucurbitaceae

Medicinal uses : Fruits are used in the treatment of impotency.

Echinops echinatus Roxb.

Local Name : Katechubuk or (Bramhadandi)

Family : Asteraceae

Medicinal uses : The seeds are used to improve or enhance sperm count.

Eclipta albab (L.) Hassk

Local Name : Maka

Family : Asteraceae

Medicinal uses : It is used to cure chronic fever.

It is used for controlling high fever, rheumatic pains, jaundice etc. (Vartak *et al.,* 1981).

The seeds of 'Bawacha' with 'Maka' are used in combination to treat leucoderma and leprosy.

Enicostema axillare (Lam.) Raynal

Local Name : Nai

Family : Genetianaceae

Medicinal uses : It is used to treat chronic fever.

Erythrina suberosa Roxb.

Local Name : Pangram

Family : Fabaceae

Medicinal uses : The powder of bark is given for 21 days to women suffering from white discharge (leucorrhoea).

Hybanthus enneaspermus (L.) F. Muell.

Local Name : Ratanpurush

Family : Violaceae

Medicinal uses : A person does not have male child in such cases a mixture of powder of 'Ratanpurush' and 'Dhorgunj' is given along with milk and sugar candy for 21 days.

To treat premature ejaculation, the mixed powder of roots of 'Ratanpurush', Talinkhana' and Bramhadandi' will be given along with sugar candy for 21 days.

Jatropha curcas L.

Local Name : Mogalierand, Ranerand

Family : Euphorbiaceae

Medicinal uses : The latex is applied to cure various teeth diseases.

Lawsonia inermis L.

Local Name : Mehndi

Family : Lythraceae

Medicinal uses : The paste of leaves is applied to sole of legs to treat the cracks in the sole.

Pergularia daemia (Forsk.) Choiv.

Local Name : Utran

Family : Asclepiadaceae

Medicinal uses : The juice of leaves is applied externally against skin diseases

Solanum virginianum L.

Local Name : Bhuiringni

Family : Solanaceae

Medicinal uses : It is used to treat scabies.

The fruits are used in animal diseases.

Trianthema portulacastrum L.

Local Name : Wasu

Family : Aizoaceae

Medicinal uses : Root extract is given in the treatment of jaundice, one teaspoon does is given for 3 days. The root extract is poured in the eyes in adults to treat jaundice.

Trichosanthes cucumerina L.

Local Name : Ranmirchi, Ranpadwal

Family : Cucurbitaceae

Medicinal uses : The extract of leaf is used against snake bite.

DISCUSSION AND CONCLUSION

During **"Uttra Nakshtra"** people of **Wadwal-Nagnath** area and nearby area come to **"Vanaspatibet"**. They keep fast (With taking traditional food) for three days and during this period of fasting the food they eat is only plant parts. Thereafter, they eat cooked rice which is prepared in water from nearby lake.

Majority of plant found on **Wadwal-Nagnath"** area are also available elsewhere in Marathwada region. Inspite if this fact, people from Marathwada and adjoining areas of the state visit **"Wadwal-Nagnath"** during **Uttara-Nakshtra**. The people having some kind of disease or ailment eat plant parts on their own. The plant part which they eat includes roots, leaves, stem barks, fruits etc. Such users claim that when they repeat the practice of consuming plant parts for two to three successive years during the same period (Uttara Nakshtra) their disorders get corrected and they become disease free. This fact is confirmed by asking them question with regard to plant parts they eat and the type of disease they have.

It is sure that plants of this area are definitely having medicinal properties unlike similar plants available in other areas of Marathwada. This difference in the properties of plants of this area and those of other areas may be the type of soil on which they grow and other condition. During **Uttara-Nakshtra** plants of **Wadwal-Nagnath,** which are said to have medicinal properties probably acquire these properties by virtue of peculiar soil and prevailing climatic conditions in the area. However, for confirming these facts further studies are required on determining the chemical composition of soil, other climatic conditions and chemical composition of plant parts in general and during **Uttara-Nakshtra** in particular. Such studies would pinpoint the reasons behind the medicinal value of plants of this area during of **Uttara-Nakshtra.**

REFERENCES

Babbar, O.P. *et al.* 1979. *Indian J. Expt. Biol.* 17:451-454.

Chopra, R.N., S.L. Nayar and I.C. Chopra 1956. *"Glossary of Indian Medicinal Plants"*, C.S.I.R. New Delhi.

Cooke, T. 1901-1908. "*The Flora of the Presidency of Bombay*", Vols. I & II. London B.S.I. Reprint 1958 Calcutta.

Gamble, J.S. 1918. "*Flora of the Presidency of Madras*", Pt. I-XI. London. B.S.I. Reprint 1958 Calcutta.

Jain, S.K. 1985. "*Medicinal Plants*", 5th Ed. National Book Trust, New Delhi.

Naik, V.N. 1979. "*The Flora of Osmanabad*", Venus Publishers, Aurangabad.

Naik, V.N. and Associates 1998. "*Flora of Marathwada*" Amrut Prakashan, Aurangabad.

Naik, V.N. 1998. "*Marathwadyateel Samanya Vanaushadhi*" Amrut Prakashan, Aurangabad.

Sinham, R.K. 1996. "*Ethnobotany*", Ina Shree Publishers, Jaipur.

Vartak, V.D. and R. Mandavgane 1981. *J. of Univ. of Ponna Science Technology;* 54: 91-99.

6

OBSERVATIONS ON ETHNOMEDICINAL PLANTS OF DANTA FOREST IN NORTH GUJARAT

K.C. Patel and A.S. Reddy*

Smt. S.M. Panchal Science College, Talod (N.G.).
**Department of Biosciences, Sardar Patel University, Vallabh Vidyanagar (Gujarat)*

ABSTRACT

The present paper deals with the study of wild plants used for ethnomedicinal purposes. The ethnomedicinal data of 31 wild plant species are collected from rustic people as well as tribes residing and practicing in various areas of Danta range forest of Banaskantha district in North Gujarat. The collected plant species are frequently used for curing various human diseases. The useful plants are arranged alphabetically and their vernacular name(s), family, herbarium specimen number(s) with locality of each species are followed by its scientific names. The description of ethnomedicinal uses of such plants, their useful parts, doses, duration and formulation are detailed along with the name of resource person.

INTRODUCTION

During our plant collection trips on ethnobotanical and floristic point of view, we collected data on folk medicines of some plants used as local medicine. Some selected wild plants are used by tribal and rural people for curing diseases. These observations are based on interviews with knowledgeable persons, who are practicing herbal therapy and are much familiar with forest areas. This paper present data on 31 wild medicinal plants collected from the forest areas of the Danta taluka in North Gujarat during 1996 to 2002. 'Local Vaidyas' are knowledge rich tribal or rural people living in rural or forest areas who know certain traditional herbal remedies for particular illness. In cases of illness, village people and tribals go to the local vaidyas and take the herbal drugs which are available locally and grown in the forest areas and seasonally collected by the experts. Some times the drugs given in the form of dried powder of plant parts are stored earlier. It has also been observed that the wild ethnomedicinal plants are naturally propagated but local tribe do not interfere.

The dry deciduous scrub type *Boswellia* forest 5/E2 (Champion and Seth, 1968) of Danta having the range of Aravalli hills in North Gujarat. Banaskantha district, in northern part of the Gujarat State, lies between 71° to 73° East longitude and 23° 35' to 24°34' North latitude. Danta taluka has got two forest ranges viz. Danta range forest and Ambaji range forest. The former one is situated on eastern part of the district.

MATERIALS AND METHODS

During the course of botanical studies, field surveys were conducted covering various areas of Danta range forest. Various medicine-men among tribal and rural areas were contacted and interviewed to record different wild plants used for various diseases. Such informations were verified through available ethnobotanical literature, repeated queries with another 'local vaidya' and experimental work. The plants are identified with the help of Gujarat Flora (Shah, 1978). Voucher specimens are deposited in the Herbarium department of Sardar Patel University, Vallabh Vidyanagar.

Enumeration

Ethnomedicinal plants are arranged alphabetically. Their vernacular name(s), family, herbarium specimen number(s) with locality of each species are followed by its scientific name. The description of ethnomedicinal uses of plant parts, doses, duration, and formulation are detailed along with the name of resource person (s).

Achyranthes aspera L. (Anghedi) Amaranthaceae (Kundel), 10512 (Pataliya).

About 2g of plant ash is taken with honey thrice a day to cure asthma and cough (Lebabhai). Same use is described by Katewa *et al.* (2001) & Shukla *et al.* (2001a) who suggested roots and seeds for asthma. Fresh mature stem is used as toothbrush for dental diseases (Shamalbhai and Malabhai). Leaf poultice is applied externally to the scorpion-sting. Leaf paste is applied on wounds and abscess. About 20g fresh flowering twigs with leaves are crushed and filtrate is taken with honey twice a day to cure swelling of spleen and for removal of kidney stones. Fresh leaves are crushed into paste and bind over bleeding wounds (Kantibhai). Flowering spike is crushed with sugar; pills of red gram of size are made out of it; and two pills are given internally thrice a day in case of rabies. About a teaspoonful of seed powder is given internally twice a day in case of snakebites (Karimbhai). Fruit powder is sniffed to cure any type of headache (Malabhai). A teaspoonful of seed powder with buttermilk is taken once in a day in case of bleeding piles (Devadattbhai).

Agave americana L. (Ketki, Ramban) Agavaceae 4084 (Shiawada), 4107 (Harivav), 4450 (Mal).

About a cup of fresh rhizome juice is taken thrice a day for abortion. Leaf juice is applied on muscular swellings. Inhaling the fragrant flower relieves headache (Shamalbhai).

Alangium salviifolium (Linn.f.) Wang (Ankol) Alangiaceae 4064 (Vashi), 4093 (Harivav), 4125 (Savaipura), 4285 (Zer).

Dry inner bark powder is sprinkled on wounds for healing. Fresh young twigs are used as toothbrush to get relief from toothache (Vimlaben).

Argemone mexicana L. (Darudi) Papaveraceae 2388 (Piplavalivav), 4117 (Savaipura), 4257 (Kundel).

Seed oil is applied externally in case of headache (Lebabhai). About 20 drops of seed oil on a lump of sugar given internally once a day in case of stomach complaints (Shamalbhai).

Barleria prionitis L. (Pilo Kantashelio) Acanthaceae 4210 (Kanbiyavas), 4279 (Zer), 10601 (Vadvera).

Plant ash mixed with cooked rice water (conjee) is administered in case of dropsy and cough. Fresh root juice is applied on scorpion sting (Shamalbhai). A teaspoonful of fresh leaf juice is mixed with sugar and given thrice a day to the children in case of any type of fever (Karimbhai).

Butea monosperma (Lam.) Taub. (Khakhro, Kesudo) Papilionaceae 2385 (Piplavali vav), 4137 (Divadi), 4272 (Zer), 4298 (Pith).

Fresh root is rubbed on a stone with few drops of water and the resultant smear is poured into eye for relief in eye diseases. About 100g of dried flowers are pounded in a bucketful of water. Taking bath with this water cures venereal and skin diseases (Shamalbhai). Young leaves are roasted and bound on abdomen to relieve pain. Fresh/dried flowers and pods are pound in water that used for bathing to cure jaundice (Lebabhai). Seed paste is applied on ringworm. Flowers of this species and leaves of *Limonia acidissima* (KOTHI) are pounded in water and bath is taken to cure sunstroke in children (Dhanabhai).

Caesalpinia crista L. (Karkas) Caesalpiniaceae 4312 (Vashi), 10631 (Vadvera).

About 50 leaflets are crushed with water so as to make a cup of juice and is given with milk to remove kidney stones (Dolatsinh). Juice of about 10 leaflets is given daily for a week to cure hookworm and fever (Shamalbhai).

Mixture of leaf poultice made from this species, *Calotropis procera* (AAKADO) and *Ricinus communis* (ERANDI), is applied on muscular swellings (Durgaben). Roasted seeds are taken with jaggery as carminative (Kantibhai). A teaspoonful of fresh pod decoction is taken for a month to cure bleeding piles (Devdattbhai).

Calotropis procera (Ait.) R. Br. (Akado) Asclepiadaceae 2356 (Mal), 4223 (Vejalwada).

Yellow coloured old leaves are streamed; juice of these leaves is mixed with half amount of rape oil and boiled. Cool down, mixed yellow sulphur half amount of oil and kept this mixture under sun light for an hour. This paste is applied on psoriasis (Devdattbhai). Apical buds of this plant, leaves of *Ricinus communis* (ERANDI) and *Datura metel* (DHATURO), in equal are crushed. The mixture is made into pills each of a pea seed size. One pill a day is taken regularly with tea or coffee to cure rheumatism. Dried stem smoke is inhaled through nostril to cure migraine (Shamalbhai). Quarter teaspoonful of young dried fruit powder is taken regularly once a day with honey for asthma (Devdattbhai).

Cardiospermum halicacabum L. (Kag-Doliyu) Sapindaceae 2199 (Khaiwad), 4188 (Kanbiyavas), 4399 (Rangpur), 10603 (Vadvera).

Leaves crushed with castor oil are taken internally in rheumatism and lumbago. If jaggery is added to this mixture; it is used for sore eyes (Chandubhai). Young branches are boiled in vegetable oil and used for massage on body in bilious affections (Shamalbhai).

Cassia occidentalis L. (Sundro, Kasundri) Caesalpiniaceae 4132 (Divadi), 4384 (Rangpur), 4455 (Mal), 10710 (Trishulghat).

Roots are used as diuretic. Leaves are used as purgative (Karimbhai). Few drops of leaf juice are dropped into eye for relief in eye diseases (Shamalbhai). Fresh leaf juice is taken internally and applied externally in case of itch and other coetaneous diseases both in human beings and animals (Chandubhai).

Cassia tora L. (Puvadio) Caesalpiniaceae 4482 (Mal).

A teaspoonful of leaf decoction is given twice a day to children, suffering from fever during teething period (Karimbhai). Mature leaves are fried with castor oil and applied externally on ulcers. Seeds are crushed with urine of cow and applied on eczema (Shamalbhai). Fresh leaves are crushed and applied externally on gout, sciatica and pain at joints (Dhanabhai).

Celastrus paniculatus Willd. (Malkangani) Celastraceae 4133 (Divadi), 10728 (Zer).

Boiled leaves are applied externally on swellings and fracture (Dhanabhai). About 5g of seed oil along with hot milk is taken regularly for rheumatism and paralysis (Shamalbhai).

Chlorophytum borivilianum Sant. & Fernanad (Dholi-Musli) Liliaceae 2354 (Mal), 10589 (Khaiwad).

Three pieces of fresh tubers are eaten with buttermilk as a remedial medicine to cure diarrhea and kidney stones (Chandubhai). About 2-3 fresh tubers are eaten with milk and sugar everyday to cure leucorrhoea. 'Sukhai'- A local sweet cake is prepared out of the dried root tubers of this species, jaggery and ghee. It is taken regularly to cure physical weakness and increase male fertility (Shamalbhai).

Cissus repanda Vahl. (Pani-Velo, Jal-Velo) Vitaceae 4358 (Pamodara).

About 2.5 cm piece of stem is crushed with few drops of water and the gelatinous mixture is taken with empty stomach every morning to cure calculus, acidity, goiter and burning of urinary tract (Dhanabhai and Lebabhai). Fresh stem bark is crushed with the rhizome powder of *Curcuma longa* (HALDAR) and applied on swelling of joints (Lebabhai).

Cocculus hirsutus L. Diels (Vevdi, Jal-Jamni) Menispermaceae 2377 (Khaiwad), 4095 (Harivav).

A teaspoonful of root decoction is taken thrice a day to cure stomach pain and fever (Dineshbhai). About 10g of fresh roots and 3g of black pepper crushed into goat milk and given once in a day to cure rheumatism. About 10 fresh leaves are crushed with 2-3 drops of water and little amount of sugar to form a gelatinous paste and is taken regularly to cure acidity and gonorrhoea (Chandubhai). A cup of fresh root juice is taken early morning to cure gout. During administration leaf 'kari' (name of a soup made of gram-flour, buttermilk, salt, spices etc.) is given to the patient once a day (Shamalbhai). Fresh leaves are crushed with few drops of water to make the paste and applied on scalp as cooling agent. This practice should be continued up to a month (Karimbhai).

Cynodon dactylon (L.) Pers. (Dharo) Poaceae 4383 (Rangpur), 4439 (Gorad), 4496 (Pataliya).

A cup of fresh leaf juice of white coloured whole plant is taken regularly once a day to cure migraine and increase sexual fertility in women (Shamalbhai).

Echinops echinatus Roxb. (Shulio, Utkanto) Asteraceae 4076 (Shiawada), 4098 (Harivav), 4187 (Kanbiyavas).

About 10 pieces of fresh roots are crushed and mixed in a bucketful of warm water and used for bathing for a week to cure leprosy. Fresh roots bought generally on Saturday and tide over neck to the child suffering from whooping cough. It is tide at waist of lady to prevent abortion and assuring fertilization. After 9 months this root is tide with 'choti' of woman for easy delivery and it is removed after delivery. A teaspoonful of root powder is taken with milk and sugar regularly once a day to cure dysurea.

Euphorbia hirta L. (Rati-Dudheli) Euphorbiaceae 4032 (Ambaghat), 4320 (Vashi), 4442 (Gorad), 4485 (Pataliya), 10694 (Trishulghat).

Approximately 200g of fresh whole plant is crushed with about 20g of black pepper and mixed in a glass of water. This mixture is taken to remove kidney stones. Immediately after few minutes of administration of this mixture about 300 ml of buttermilk is given to the patient. This practice is repeated for 2-3 times if necessary with a gap of two days. Roots are crushed with cow milk; a teaspoonful of this mixture is given to the child to ease bowel movements (Devdattbhai). Fresh root pieces are kept in ear cavities of a patient, who is suffering from malaria (Malabhai). Young flowering shoots are crushed with sesame oil, warmed and applied on of rheumatic swellings.

Flacourtia indica (Burm.f.) Merr. (Kanti) Flacourtiaceae 4015 (Vagol), 4266 (Kundel), 4299 (Pith).

Fresh roots and leaves are crushed with water; a cup of this mixture is taken thrice a day in case of snakebite (Shamalbhai). Fresh stem bark paste mixed in warm sesame oil and applied externally as a treatment for rheumatism and gout (Lebabhai & Devdattbhai).

Grewia flavescens Juss. (Dhamadi) Tiliaceae 4351 (Pamodara), 10629 (Vadvera), 10693 (Trishulghat).

A cup of root juice is taken regularly for easy delivery. A teaspoonful of stem bark juice is taken twice a day to cure gonorrhoea (Lebabhai).

Hibiscus ovalifolius (Forsk.) Vahl. (Chikani Bhindi) Malvaceae 4411 (Rangpur), 4432 (Gorad).

About 15 cm long fresh and mature two twigs are soaked in water over night in an earthen pot. This water is consumed next day morning as a treatment for leucorrhoea and menorrhagia. This procedure is followed for two weeks (Durgaben)

Indigofera tinctoria L. (Gali, Bahufali) Papilionaceae 4276 (Zer), 4341 (Vashi).

Whole plant with flowers, fruits and seeds are powered. A teaspoonful of this powder is taken with milk twice a day regularly to increase sperm count in men (Devdattbhai). A teaspoonful of root powder is taken for relief in stomach complaints (Shamalbhai).

Limonia acidissima L. (Kothi) Rutaceae 4003 (Ambaghat), 4289 (Zer).

A teaspoonful of stem gum is taken with honey thrice a day to cure diarrhoea and dysentery. Fresh leaf paste is applied on abscess (Devdattbhai). About 20 leaves are crushed and pounded in bucketful of water. It is kept in sun light for an hour. Bathing with this water cures sunstroke and fever in children. A cup of leaf juice is taken with cow milk twice a day to cure jaundice. Half a teaspoonful of seed oil is taken twice a day in case of mouse bite (Dhanabhai). A cup of leaf juice is taken regularly in case of venereal diseases. Ripe fruit pulp of one fruit is taken with honey regularly to cure asthma, acidity and hiccup (Shamalbhai). Leaf juice is aromatic and act as carminative (Chandubhai). About 10 fresh leaves of this species and 20 fresh leaves of *Kirganellia reticulata* (KAMBOI) are crushed with 20ml of water. Filtrate is given once in a day till three months to increase fertility in ladies (Vimlaben).

Luffa acutangula (L.) Roxb. var. *amara* (Lam.) Cl. (Kukad-Vel, Jungli-Turiya) Cucurbitaceae 4150 (Divadi), 10592 (Vadvera).

Dried fibrous fruits are de-seeded and filled with water. After an hour this water is collected and inhaled through the same nostril of the paining side in case of migraine (Devdattbhai). A teaspoonful of fruit juice is taken twice a day for jaundice (Dalpatsinh). Roasted fresh fruit pulp is applied on head in case of headache. Fresh fruit pulp is applied externally on piles (Kantibhai). A young fruit is crushed in 100 ml of water and filtrate is taken in the morning for five days to cure rabies and diarrhea (Shamalbhai). Whole dried fruit is made into powder and sniffed once a day, to cure jaundice. Pericarp of young fruit is chewed to remove germs in teeth (Lebabhai).

Maytenus emarginata (Willd.) D. Hou (Vikalo, Kankayo) Celastraceae 4022 (Vagol), 4157 (Shiawada), 4409 (Rangpur).

About 5 fresh leaves and 5 fresh fruits are taken everyday morning to remove physical weakness and for blood purification (Shamalbhai). About 10 fresh leaves are chewed thrice a day to get rid of mouth ulcers (Devattbhai). Ash of leaves is mixed with ghee and applied externally on pustule (Vimlaben).

Moringa concanensis Nimmo (Jangli-Saragavo) Moringaceae 4232 (Vejalwada).

A cup of root and stem bark juice is taken thrice a day for abortion (Dhanabhai). Stem gum and dried roots are powdered and make 'sukhadi' out of this powder, ghee and jaggery. About 2 sq. inch piece of this preparation is taken regularly to cure gout (Dalpatsinh).

Phyllanthus fraternus Webster (Bhoyn-Amli) Euphorbiaceae 4228 (Vejalwada), 4429 (Gorad), 4500 (Pataliya).

A cup of root juice is taken with cooked rice water (conjee) regularly once a day to cure leucorrhoea. About a teaspoonful of flowering shoot powder is taken with honey regularly once a day for long and healthy life (Shamalbhai). A teaspoonful of young shoot infusion is taken thrice a day to treat dysentery. Fresh leaves are crushed with buttermilk and a teaspoonful of this mixture is taken twice a day to cure itching (Devdattbhai).

Solanum indicum L. (Ubhi-Ringni, Dorli) Solanaceae 4104 (Harivav), 4124 (Savaipura), 4193 (Kanbiyavas), 4327 (Vashi), 4426 (Gorad).

Fresh roots are kept between paining teeth to cure toothache. A cup of juice of flowering twigs with leaves is taken twice a day in case of 'Datura' poison (Leababhai). A teaspoonful of stem juice is taken thrice a day in difficult parturition and toothache. A ripe fruit is roasted and eaten with honey at night for insomnia (Dhanabhai). Ripe fruits are roasted and applied externally on abscess and elephantiasis (Suramabhai). Young fruits are used as vegetable regularly once a day cures kidney stones and lung diseases. Fruit smoke is sucked for removing worms in teeth (Malabhai & Shamalbhai).

Tecomella undulata (Sw.) Seem. (Ragat-Rohido, Roydo) Bignoniaceae 2392 (Ambaghat), 4024 (Vagol).

Fresh root bark crushed with water of cooked rice (conjee). A cup of filtrate is given orally once a day to cure leucorrhoea (Vimlaben). About 2g dried root powder is take with ghee twice a day to cure chest pain and removal of muscular pain. A teaspoonful of stem bark powder is taken with honey once a day to cure muscular swellings, muscular pain and for curing blood cancer (Shamalbhai).

Terminalia crenulata Roth. (Sadad) Combretaceae 4337 (Vashi).

About 10g of fresh stem bark is crushed and boiled with water, filtrate is taken regularly early in the morning to cure swelling of joints (Becharbhai). A cup of stem bark decoction is given internally thrice a day in case of diarrhoea and intestinal ulcers (Chandubhai).

Urginea indica (Roxb.) Kunth. (Jangli-Dungli, Jangli-Pyaz) Liliaceae 10537 (Khaiwad).

Bulbs are crushed into paste and applied externally on abscess and eczema. Boiled bulbs are applied externally to the soles of cracked feet (Lebabhai). Fresh bulbs are crushed and applied externally on scorpion sting (Shamalbhai).

CONCLUSION

The data shows that the informers have sound information about the wild plants. The tribals are very inhibited and not interested to share traditional knowledge with common men. Only few intimate people can get such knowledge. Further investigations on pharmaceutical and therapeutic aspects are very much desired for the discovery of new herbal drugs.

ACKNOWLEDGEMENTS

Authors are very much thankful to all the informers for kindly sharing their knowledge with us and also to the Head, Department of Biosciences, Sardar Patel University for providing working amenities.

7

CONSERVATION THROUGH PROPAGATION OF THE STATE FLOWER *SAUSSUREA OBVALLATA*: AN UNDER-EXPLOITED MEDICINAL HERB OF UTTARANCHAL

Mitali Joshi and Uppeandra Dhar

G.B. Pant Institute of Himalayan Environment and Development
Kosi-Katarmal, Almora 263 643 (UA), INDIA

ABSTRACT

Utilization of medicinal plants in Indian System of Medicine (ISM), needs scientific inputs for conservation and sustainable utilization of this heritage and can contribute significantly to India's own health care needs. This would require both *in situ* and *ex situ* conservation strategies. While issues of *in situ* conservation are addressed through efforts of protected area network, *ex situ* conservation needs cost effective protocols. The efforts are being made to find out protocol for *in vitro* cultivation of *Saussurea obvallata,* an important herbal plant of Uttaranchal region. Micropropagation trial showed good results when Kin (1.0 μm) in combination of NAA (0.25 μm) was tried, good number of shoots were produced per explant.

INTRODUCTION

The alpine zone of Himalaya support diverse and unique vegetation. It also harbours a rich repository of medicinal and aromatic herbs. While some of the herbs are of known pharmacological importance, a large number of these are used in indigenous medicines and drugs by local inhabitants of the area to meet their primary health care needs. Keeping in view the history of utilization of medicinal plants in Indian System of Medicine (ISM), scientific inputs for conservation and sustainable utilization of this heritage can contribute significantly to India's own health care needs. This would require both *in situ* and *ex situ* conservation strategies. While issues of *in situ* conservation are addressed through efforts of protected area network, in order to address *ex situ* conservation, it is imperative to develop cost effective and easy to use propagation protocols of prioritized plants species. The present study is a step in this direction, wherein attempts have been made to develop propagation techniques for *Saussurea obvallata* (DC.) Edgew. (Family Asteraceae), an endangered and near endemic medicinal herb of Indian Himalayan Region (IHR).

Saussurea obvallata constitutes one of the important herbal drugs of Traditional Himalayan Medicinal System (THMS). The species is widely used by the inhabitants for preparation of traditional medicines, for example, flower heads are boiled in butter oil (ghee) and used in the treatment of bone-ache, intestinal ailments, urinary tract problems, cough and cold (Negi *et al.*, 1999). The rootstock is well known for its antiseptic and healing properties and used in cuts and bruises (Kirtikar and Basu, 1984). In addition, the flower has got a high religious value in the region. On account of the multiple uses of the species, it is being indiscriminately harvested from the wild populations. According to the CAMP (Sastry and Chatterjee, 2000) it is one of the top ranking priority species for conservation.

This plant is propagated through seeds and vegetative perennial rootstock in nature. However, propagation protocols for the species are not reported. Considering all these facts the present study envisages to: (1) Develop various seed storage and germination protocols using simple techniques, (2) Develop various *in vitro* protocols that include both direct micropropagation as well as indirect shoot regeneration through callus.

MATERIALS AND METHODS

Detailed tests were conducted on mature achenes (seeds) of *S. obvallata* collected from three distantly located populations (Hemkund-4210 m asl; Latakharak – 4000 m asl; Neelkanth – 4200 m asl). Whole inflorescence was harvested manually in last week of September (2001-2002) from 25 randomly selected apparently healthy individuals. Seeds were manually removed from the flower heads and immediately after collection, seeds were air dried at room temperature (22 ± 4°C) for three days and then pre-treated uniformly for all the experiments.

Seeds were counted in six inflorescences of marked individuals from each population and average number of seed production per individual was estimated. To estimate the variation in seed morphology, pooled seeds of these inflorescences were used. Ten seeds each in three replicates of each population were weighed and mean for each replicate was calculated to determine the average seed weight. Similarly seeds were further analyzed for length and diameter using electronic vernier calipers.

Seed storage and viability test

Dried seeds were stored in domestic fridge, temperature maintained at 4°C and room conditions with daily and seasonal temperature fluctuations, in brown paper bags. Seed samples were drawn in three replicates of 10 seeds each from every condition at storage interval of 0, 6 and 12 months for Tetrazolium (TZ) viability test (Agrawal *et al.*, 1973). Further the seed germination test was done in plastic petriplates containing single layer of moistened filter paper (Whatman No. 1). As and when necessary the filter papers were moistened with distilled water using a spray bottle. Petriplates were exposed to 14 h light and 10 h darkness in a growth chamber at a controlled temperature (25 ± 2°C). Mean Germination Time (MGT) was also calculated (Hartmann and Kester, 1983).

Micropropagation studies

To develop a micropropagation protocol, seeds were germinated under aseptic conditions on Murashige and Skoog's (1962) basal medium supplemented with 3% sucrose and 0.8% agar-agar. Epicotyl segments sectioned from 15-20 days old *in-vitro* germinated seedlings were used as an explant. A single explant was placed in each 25 x 150 mm culture tuber containing 15 ml medium supplemented with Kin at 0.5, 1.0, 2.0 and 4.0 μM in combination with NAA at 0.0, 0.25 or 0.50 μm concentrations.

At initial stages explants were cultured individually for 30 days. Thereafter they were periodically sub-cultured every 15 day (three times) in the same media to obtain high multiplication rates.

Organogenesis protocol

Leaf explant (15-18 mm^2) was isolated from four months old *in vitro* shoots. Callus induction and shoot regeneration was carried out on MS basal medium supplemented with BA (1.0, 2.5, 5.0, 10.0 μM) in combination to NAA (0.0, 0.1, 1.0 and 10.0 μM).

Induced callus was transferred after 30 days in the same media for growth and shoot regeneration. About 1 cm long regenerated shoots were separated from the callus and transferred to MS medium supplemented with a combination of 1.0 μM Kinetin and 0.25 μM NAA for elongation and multiplication for another 30 days.

In vitro rooting and acclimatization

Rooting of shoots was done on half strength MS medium supplemented with different concentrations of IBA (0.0, 1.0, 2.5 or 5.0 μM). Rooted shoots were removed from rooting medium and washed thoroughly in running tap water to remove all traces of medium attached to the roots. They were then transplanted in plastic pots containing a mixture of autoclaved soil + peat moss + soilrite (2:1:1 v/v). Pots were covered with potythene bags to maintain high humidity.

All experiments were carried out in complete randomized block design. Mean data on percentage were transformed using arcsin transformation, while mean data on number was transformed using logarithmic transformation before analysis, and then converted back for presentation. Fisher's Least Significance Difference (F – LSD < 0.05) was applied to show statistical significance of difference among the means (Wilkinson, 1986).

RESULTS

Seed as a propagule

Results indicate high seed production (>500 seeds per inflorescence), good initial seed viability (90-100%) and high germination potential (>70%) all across popu-lations. However variations were recorded in some characters among populations i.e. seed weight, length and germination responses (Table 1). Seeds collected from P3 population showed the best response in germination (95.6%) and MGT (3.2 days).

Table 1. Comparative seed morphology in selected populations of *S. obvallata*

Parameters	Populations			LSD (P<0.05)	F-Ratio
	P1	P2	P3		
Number of seeds	517a±319	568a ±88	580a ±190	231.79	0.13^{ns}
Per inflorescence	2.30a±.05	1.90b ±0.13	2.30a ±0.12	0.21	11.68^{*}
Av. Weight/ seed (mg)	5.70b±0.10	4.90c ±0.10	6.20a ±0.10	0.09	387.00^{***}
seed length (mm)	1.30a±0.10	1.30a ±0.00	1.30a ±0.30	0.40	0.00^{ns}
seed diameter (mm)	71.1b±3.8	88.9a ±3.8	95.5a ±3.8	11.19	12.19^{**}
germination (%)	4.5a±0.4	4.3a ±0.4	3.2b ±0.4	0.84	6.25^{*}
MGT (days)					

Mean values followed by different lower-case within a row are significantly different (P<0.05). *P<0.05, **P<0.01, ***P<0.001, nsNon significant

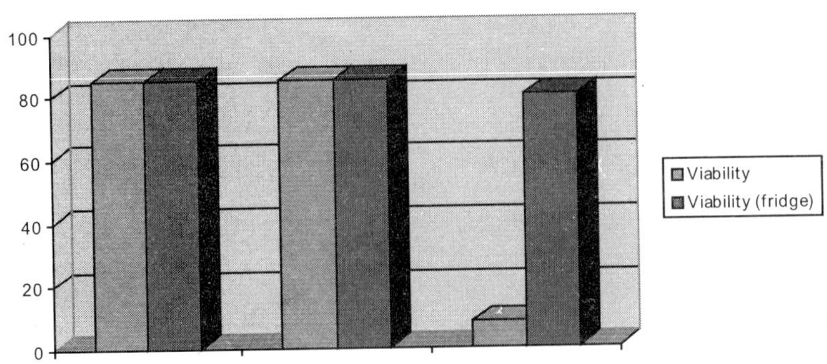

(Storage Time Period (Months)

Figure 1. Changes in viability and germination patterns of seed during storage of *S. obvallata*

Seed storage treatments and time showed significant ($P<0.05$) effect upon seed viability and percent germination (Fig. 1) of *S. obvallata*. Freshly harvested seeds of the species exhibit high degree of initial viability (93%) and with the passage of time they lost their viability. It was observed that initially all the viable seeds could germinate (93%). These values however significantly decreased according to storage period. Results indicate that seed storage time period was negatively correlated with percent viability ($r = -0.963$; $P<0.01$ and germination ($r = -0.960$; $P<0.01$) of seeds. Whereas, there was a positive correlation between viability and germination ($r = 0.995$; $P<0.01$) irrespective of storage temperature. The high initial viability of seeds was maintained upto six months storage period under control condition (room environment). Thereafter, a sudden drop in final percent germination (10%) was recorded which corresponded to the viability loss (13%). In comparison to control this viability period was extended upto 18 months and seeds stored within 4°C remained relatively stable. In general, seed storage at low temperature (4°C) proved to be the best.

Micropropagation

The explants on medium without growth regulators (control) initiated shoot growth, but failed to multiply (Table 2). However, multiplication rate and shoot length was significantly higher in cultures supplemented with plant growth regulators. Kinetin alone was less favourable for shoot multiplication. Addition of an auxin (NAA) showed significant improvement. Among all the concentrations tested, the best response (5 shoots per explant; shoot length 6 cm) was achieved in Kin (1.0 µm) in combination of NAA (0.25 µM) at second subculture. Serial transfer thus favoured shoot length and number for most of the combinations.

Table 2. Effect of different PGR combinations and progressive subcultures on shoot length and number in shoot cultures of *S. obvallata*

PGR's (µm)		Shoot Induction		Sub-culture I		Sub-culture II		Sub-culture III	
		Shoot							
KIN	NAA	Number	Length (cm)	Number	Length (cm)	Number	Length (cm)	Number	Length (cm)
-	-	0.4	0.4	0.0	0.0	0.0	0.0	0.0	0.0
0.5	-	1.5	4.3	4.4	2.7	1.9	2.6	1.2	3.3
0.5	0.25	1.2	3.8	2.2	4.8	2.0	3.7	1.2	3.8
0.5	0.50	1.1	3.6	1.8	3.5	2.3	3.8	1.4	3.3
1.0	-	1.4	4.0	1.8	3.3	3.1	3.3	1.3	3.5
1.0	**0.25**	1.8	4.6	2.2	4.8	**5.0**	**6.0**	3.0	5.2
1.0	0.50	2.2	5.4	3.4	4.9	2.3	4.2	3.8	4.2
2.0	-	1.6	2.0	2.1	2.1	2.8	2.3	1.2	3.1
2.0	0.25	3.1	3.2	3.3	3.7	3.6	3.0	1.7	4.3
2.0	0.50	1.8	4.5	1.8	3.7	2.2	4.0	1.6	4.5
4.0	-	1.2	1.4	1.0	3.9	0.0	0.0	0.0	0.0
4.0	0.25	1.0	1.5	1.2	1.8	1.7	2.3	1.8	2.3

4.0	0.50	0.8	1.2	1.6	1.3	0.0	0.0	0.0	0.0
LSD ($p<0.05$)		0.83	1.40	1.02	1.52	1.02	1.14	1.23	1.34

Data recorded after 15 days of transfer to MS medium after first subculture.
After: Joshi and Dhar, 2003.

Organogenesis

Formation of light green and friable callus was evident within three week s on the cut surfaces of leaves after initiation of the treatments. No callus was observed on hormone free medium. Addition of BA and NAA was found suitable both for callus induction and shoot differentiation (Table 3). BA (2.5µM) plus NAA (10.0µM) induced the highest percentage (100%) of callus with maximum fresh weight (0.88g). Results after first transfer of callus showed that both BA and NAA were necessary for viable callus induction and even if callus is formed in absence of NAA, it turned brown or black after 30 days and showed no further growth. After first subculture, 100% callus differentiated and each clump produced 12$\pm$2.6 shoots in MS medium supplemented with 5.0µM BA and 1.0 µM NAA. These shoots elongated to 2.1 cm in length. Each shoot after separation and transfer to Kin (1.0 µM) and NAA (0.25 µM) further elongated to 5.6 cm and multiplied (5 shoots / explant).

Table 3. Effect of different PGR combinations on callus induction and differentiation of *S. obvallata*[a]

PGR'S (µM)		Callus induction (%)	Average fresh weight (g)	Shoot regeneration (%)	Number of shoots/explant	Average shoot length (cm)
BAP	NAA					
0.0	0.0	-	-	-	-	-
1.0	0.0	-	-	-	-	-
1.0	0.1	41.6(25.0)	*	-	-	-
1.0	1.0	50.0(0.0)	0.17(0.08)	-	-	-
1.0	10.0	83.3(14.4)	0.39(0.34)	-	-	-
2.5	0.0	-	-	-	-	-
2.5	0.1	25.0(0.0)	*	-	-	-
2.5	1.0	50.0(25.0)	0.14(0.09)	41.6(25.0)	1.6(0.5)	0.3(0.1)
2.5	10.0	100.0(0.0)	0.88(0.21)	-	-	-
5.5	0.0	21.0(6.0)	*	-	-	-
5.0	0.1	25.0(0.0)	*	-	-	-
5.0	1.0	75.0(0.0)	*	100.0(0.0)	12.0(2.6)	2.1(0.2)
5.0	10.0	91.6(12.5)	0.46(0.19)	-	-	-
10.0	0.0	13.0(0.0)	*	-	-	-
10.0	0.1	17.0(6.0)	*	-	-	-
10.0	1.0	50.0(0.0)	0.35(0.29)	-	-	-

10.0	10.0	66.6(12.5)	0.10(0.03)	-	-	-
LSD P<0.05)		16.9	0.34	19.9	0.48	0.31
F-ratio		17.4	5.8	61.0	93.2	152.7

Data are mean of three replicates each with eight explants, recorded after 30 days of transfer to MS medium; values in parenthesis are the ± SE; - explants dead/ not responding; * callus that survived only for 30 days. After: Dhar and Joshi, 2004.

Rooting and ex vitro transplantation

All treatments applied in the present study, induced rooting (Table 4). In most of the treatments root initiation was observed from the seventh day after transferring shoots to rooting medium. In absence of growth regulator, relatively lesser number of small roots was induced in 80% of shoots. 100% rooting was achieved with 2.5μM IBA in half strength medium and further increase to 5.0μM decreased rooting percent.

Table 4. The effect of various IBA treatments on root induction in shoots of *obvallata*. Data recorded after four weeks of transfer to rooting medium.

IBA (μM)	Rooting (%)	Number of roots per shoot	Root length (cm)
0.0	80.00	17.5	4.1
1.0	86.66	21.6	4.6
2.5	100.00	14.2	1.8
5.0	73.33	15.6	2.6

LSD P<(0.05) for rooting % 36.98; root number 7.7; root length 2.28

Approximately 60% *ex vitro* transplantation success was achieved. Initially the growth performance was slow immediately after transplanting (day 0-12), thereafter it improved or became static comparatively. However intensive research is required in the direction.

DISCUSSION

Seed as a propagule

The target species inhabits zones exposed to severe cold. Under such harsh environmental conditions plants, at time, are said to develop certain tolerance survival strategies through which energy investment is utilized in the production of large amount of high quality seed to overcome morality risks (Angevine and Chabot, 1979). Perhaps this could be one of the reasons that the target species produces large number of viable seeds all across the populations. At the same time high germination capacity suggests that just like some other Asteraceae members in the Himalaya, seed of *S. obvallata* do not possess true dormancy and when transferred to favourable temperature and light regimes germinate very well (Bhatt *et al.,* 1985). However, similar to some other alpine herbs there is a general trend of intraspecific variation in seed germination responses (Joshi and Dhar, 2003; Majkhola *et al.*, 2003). The small differences in seed viability among populations can be attributed to the micro-climatic conditions (Karssen, 1970).

Experiments on seed storage indicate that viability largely remains unaffected when stored in low temperature (4°C). These results indicate the non-recalcitrant nature of the seeds which are also apparently not sensitive to chilling (Roberts, 1973). Maintenance of viability in low temperature is in conformity with the habitat requirement of the species. Seed maturation of *S. obvallata* takes place in October. Thereafter, seeds get buried in snow and this is the natural way of retaining the viability for next active growing season. This is also in tune with the results of earlier studies, which show that seed stored at low temperature can help in maintaining viability for a longer time period in alpine plant species (Nautiyal *et al.*, 1985; Pandey *et al.*, 2000). Further investigations on seedling survival indicated that it was very difficult under the present experimental conditions (data not shown). Thus high mortality and slow growth of germination indicate habitat specific requirements for seedling establishment.

In vitro propagation

Keeping in view the endangered status of the species, vegetative propagation protocol development seems not to be feasible, as the use of rhizome can leads to destruction / decomposition of the donor plant is due course of time. At the same time propagation through seed germination gets limited because of specific growth requirements of the seedlings. Thus, considering the potential of the target taxa and limitations of conventional propagation protocol, role of tissue culture becomes important.

Following the procedures without harvesting or damaging the source material, approximately 20 to 41 well-rooted plantlets can be developed from a single explant, within 12-21 weeks. These protocols can be very useful to multiply an elite stock of the target species in limited time. Significant improvement in shoot formation over control has previously been achieved with addition of cytokinins like BA and Kin in many Composites (Conchou *et al.*, 1992; Le, 1994; Nin *et al.*, 2000; Mandal *et al.*, 2000).

Using the above mentioned protocols, various objectives can be fulfilled. Where micropropagation protocol can provide a successful and rapid technique that can be used, for mass multiplication of *S. obvallata* aiming at *ex situ* conservation. The callus regeneration system may assume importance for further advance studies in genetic improvement, it also has a considerable potential in the future as an alternative for the production of known and new secondary metabolites.

CONCLUSIONS

The present study attempts to provide options for: (i) domestication of the species; (ii) industrial exploration and utilization and (iii) mass multiplication and thereby ensuring the conservation of the species. Taking a lead from the study a systematic approach, can be developed for exploration, maintenance, conservation and sustainable utilization of the endangered state flower of Uttaranchal.

REFERENCES

Agrawal, P.K. *et al.* (1973). *Seed Res.* **1:** 83

Angevine, R. and B. F. Chabot (1979). In: *Plant Population Biology* (Eds. O.T. Solbrig, S. Jain, G.B. Johnson and P.H. Raven). Columbia Press, New York. pp. 188-206.

Bhatt, R.M., Nautiyal, S. and A.N. Purohit (1985). *Seed Res.* **13:**1.

Conchou, O., Nichterlein, K. and A. Vomel (1992). *Pla. Med.* **58:** 73.

Cuenca, S., Amo-Marco, J.B. and R. Parra (1999). *Plant. Cell Rep.* **18:** 674.

Dhar, U and M. Joshi, (2004). *Plant Cell Reports,* (in press).

Echeverrigaray, S. *et al.* (2000). *Plant Cell Tis. Org. Cul.* **60:** 1.

Fauconnier, M.L, *et al.* (1996). *Agriculture and Forestry: Medical and Aromatic Plants IX, Vol.* 37. (Ed. Y.P.S. Bajaj). PP. 16.

Hartmann, H.T. and D.E Kester (1983). *Plant Propagation: Principles and Practices.* Prentice Hall Pvt. Ltd., New Delhi.

Joshi, M. and U. Dhar (2003). *Seed Sci. Tec.* **31:** 737.

Joshi, M. and U. Dhar (2003). *Plant cell Report.* 21: 933.

Karssen, G. M. (1970). *Act. Bot. Neo.* **19:** 81.

Kirtikar, K.R. and B.D Basu (1984). *Indian Medicinal Plants, Vol. 1-5.* Bishen Singh Mahendra Pal Singh, Dehradun.

Le, C.L. (1994). *Rev. Sui. Vit. Arb. Hor.* **26:** 391.

Mandal, A.K., Chakraborty, A.D. and S.K. Datta (2000). *Pla. Cell Tis. Org. Cul.* **60:**33.

Manjkhola, S., U. Dhar and R.S. Rawal, (2003). *Seed Sci. Tec.* **31:** 571.

Murashige, T. and F. Skoog, (1962). *Phy. Pla.* **15:** 473.

Nautiyal M.C., Rawat, A.S. and J.K. Tiwari (1999). *Seed Res.* **14:**133.

Negi, K.S., Gaur, R.D and J.K. Tiwari (1999). *Ethnobotany.* **11:** 9.

Nin, S., Schiff, S., Bennici, A.. and R. Magherini (1994). *Adv. Hor. Sci.* **8:** 145.

Pandey, H. *et al.* (2000). *Seed Sci. Tec.* **28:** 39S.

Roberts, E.H. (1973). *Seed Sci. Tec.* **1:** 499.

Sastry, A.R.K. and S. Chatterjee (2000). In *Setting Biodiversity Conservation Priorities for India* (Eds. S. Singh, A.R.K. Sastry, R. Mehta and V. Uppal). WWF India, New Delhi. PP. 467.

Wildi, E., W. Schaffner and K. Berger-Buter, (1998). *Plant Cell Rep.* **18:** 336.

Wilkinson, L. (1986). *SYSTAT: The System for Statistics.* Evaston II, Systat. Inc.

8

MEDICINAL MUSHROOMS FOR HEALTHY LIFE

Arun Arya and Chitra Arya

Botany Department, Faculty of Science,
The Maharaja Sayajirao University of Baroda, Vadodara - 390 002

ABSTRACT

Mushrooms find their place in traditional folk medicines throughout the world since ancient times. Mushrooms are nutritionally classified as functional foods or Nutraceuticals. Mushrooms are a rich source of protein. According to leading nutrition experts, "*It will be a choicest food supplement for calorie conscious younger generation.*" Mushrooms are appreciated for their flavour and texture or their tonic like attributes. The Greeks regarded mushrooms as "*providing strength to soldiers in war.*" Romans considered them as "*Food of Gods*" and Chinese treated them as "*Elixir of life*". The paper describes important medicinal mushrooms, their uses and active compounds. *Ganoderma, Lentinus, Agaricus blazei, A. campestris, Agrocybe aegertia, Amanita muscaria, Coprinus comatus, Coriolus versicolor* and others are described in detail.

INTRODUCTION

Famous French philosopher Voltair once said, "The Vedas are the most precious gift for which the West has ever been indebted to the East." Atharva Veda is full of herbal plants and products. There are 5987 mantras (rucha) in this holy text. The 16th mantra of 7th sukta in 8th chapter of Atharva Veda says,

Mumuchana ayoshadhayo agne vairshwa naradadhi.
Bhumi santanwatiriti yasan raja vanaspatih.

Herbs produced from earth relieve us from alimentary diseases and keep us healthy. Soma is the king of these herbs. The extract of this soma ras medicine should keep our body and mind fit.

Use of *Ficus religiosa* (Peepal), Kusha (a grass) and Soma is life giver has been said in mantra (8:7:20) of Atharva Veda. Mention of Soma is also there in (1:30:4) mantra of RigVeda. However, the exact plant or mushroom mentioned as soma is not yet clear. Scientist believe either it was a fermented product or hallucinogenic mushroom like *Amanita muscaria*. Fascinating book *"Soma-Divine Mushroom of Immortality"* written by Wasson (1968) mentions this mushroom as real soma.

The variety of mushroom species discovered is vast. Researches have identified up to 270 mushrooms (Wright, 2004; Wasser, 2002) with medicinal benefits, however, that may be just a small fraction of the actual mushrooms existing in the wild that have yet to be found. Some of the medicinal mushrooms include reishi, maitake, shiitake, *Cordyceps*, turkey tail, lions mane, *Psilocybe, Amanita muscaria, Flamulina velutipes,* ergot and giant puffball (*Calvacin*). Cultivation of these can give you more money. Mushrooms are immuno modulators. Reishi has compounds that most other mushrooms do not, including triterpenes, which are found in many longevity herbs.

Scientifically any mushroom is a fruiting body of the fungus; a group of non chlorophyllous eukaryotic organisms earlier classified along with plants. These acquire nutrition by absorption from their surroundings. The mycelium is hidden in ground or in wood and these bodies are produced in different shapes and sizes. Any fungus body which can be seen by naked eyes is a mushroom. It includes members of Ascomycotina [(like Truffle (*Tuber*) and Morel (*Morchella*)] and Basidiomycotina i.e. members of Agaricales and Aphyllophorales. The poisonous mushrooms are known as toadstool and Aphyllophorales include hard or tough fruiting bodies and bracket fungi. Each basidia produces 4 basidiospores while each ascus produces 8 ascospores in a sac like structure. In a normal gill fungi the spore fall rate is of the order of about half a million spores per minute. Spores are violently discharged.

SOME IMPORTANT MEDICINAL MUSHROOMS

Ganoderma lucidum (Reishi mushroom)

This polypore fungus is an ingredient of several Chinese medicinal preparations. The species name *lucidum* comes from laccate, red surface of the fruiting body. It has a short stalk (Plate 1). The stalk is absent in *G. applanatum*. In China it is known as Lin Zhi Cao. It is reputed to possess hypocholesterolemic activity. It is rich in oxygenated lanosterol derived triterpenoids and more than 90 terpenoid compounds have been isolated from this fungus. Several of these triterpenoids were inhibitors of cholesterol bio-synthesis. An immuno-suppressant glycoprotein has been isolated from mycelia of *G. lucidum*. This glycol protein is useful in treatment of allergies Polysaccharide Ganodermin shows antitumour activity through immunopotentiation and not through direct cytotoxicity against tumor. From cultured mycelium of *G. lucidum* ganoderic acids X, Y, Z have been isolated, which were cytotoxic to hepatoma cells *in vitro* (Toth *et al.,* 1983, Wright 2004).

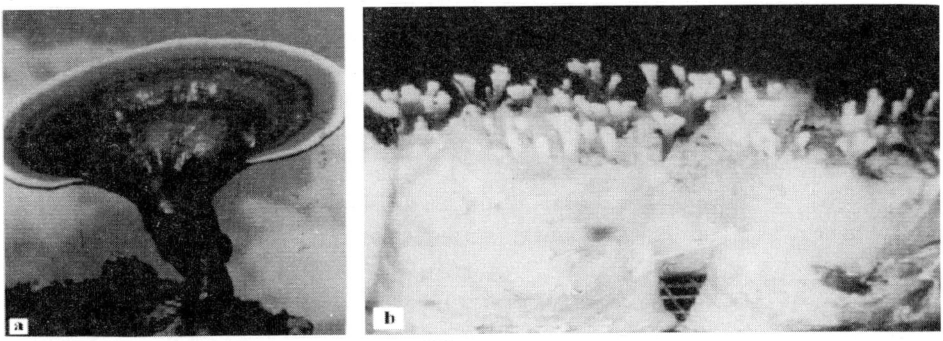

Plate - 1

Lentinus edodes (Shiitake mushroom)

The mushroom has stalk and a pileus (Plate 2, Figure b). The pileus is white or off white in colour with recurved peripheral surface. It has been credited with antitumour, antiviral, hypolipidemic, hypocholesterolemic, and antibiotic properties. It is one of the most valuable medicinal mushroom known to have antiviral (against HIV aids) properties. The antitumour properties are attributed to the polysaccharides-Lentinan and emitanin. Lentinan is now used as an antitumour drug. Hypolipidemic property is due to presence of eritadenine and antibiotic activity is due to the presence of Lenthionine-a pentathiepene.

The regular consumption of this mushroom reduces the cholesterol content of blood. A phenolic extract taken from this mushroom repress the influenza virus of mice. A watery extract of this mushroom prevents the multiplication of poliovirus. Lentinan has an intensive anti-cancerous effect. Lentinan sulphate is used for clinical trials against AIDS (Gareth, 1990). Powder from fruiting bodies and hot water extract from mycelium is used to inhibit HIV infection of cultured human T-cells.

Agaricus blazei (Himematsutake mushroom)

Name *Agaricus* is derived from Agarica district of Sermatica province in Russia from where this mushroom was first reported. The mushroom is fleshy with numerous gills present on lower surface of pileus. The mushroom has glycoprotein which shows antitumour activity. The extract of this fungus improves liver functions. Three ergosterol derivatives I, II and III showed antitumour properties (Mizuno *et al.,* 1989). A high antitumour activity was reported from xyloglucan-protein complex (Mizuno *et al.*, 1990).

Agaricus campestris

The polysaccharides of this mushroom consists of mannose (42%) glucose, (25.5%) xylose (16.6%) and protein moiety composed of 17 amino acids. The hot water extract contains a glycoprotein fraction showing antitumour activity against Sarcoma-180 in 1 CR mice (George *et al.*, 1997).

Plate - 2

Agrocybe aegertia

The extract of this mushroom is bactericidal against many pathogenic bacteria and yeast. The filtered medium (0.1 ml) had antibacterial and antifungal activity corresponding to 100 IU of streptomycin and 500 IU of fungicidin respectively.

Amanita muscaria (fly agaric)

It is brilliant red in colour with remains of universal veil on pileus. It has a volva around the bulbous base. Among the medicinal mushrooms, poisonous and some inedible ones it had a very prominent place in early times. Use of these poisonous mushrooms as medicines has been considered in China "as using of poison as an antidote for poison." *Amanita muscaria* is a deadly poisonous one; it has been therapeutically used as a powder, tincture for swollen glands, nervous troubles, and epilepsy etc. for ailments of heart, inflammations of eyes etc. This mushroom is now used in homoeopathic medicines under the name 'agar'. Muscinol and ibotenic acid can cure the malfunctioning of "GABA" system of brain (Kogagaard, 1981)

Coprinus comatus (Inky cap mushroom)

This is found on dead wood or dung and litter. Black spored species has inaequihymeni-ferous gills that dissolve into black, inky liquid. This mushroom lowers blood glucose concentration and is beneficial for treatment of diabetes mellitus. In mice fed with fruit bodies (3.6 g / kg body wt.) of *C. comatus* lowering of plasma glucose was observed after 10 hour. The mushroom shows anti fungal activity.

Coriolus versicolor

The fruit bodies are hard, few cms in size with grayish or brownish rings on upper surface. This mushroom belongs to polypore group of fungi.

Macromolecules isolated from the fruit bodies and mycelia of *C. vesicolor* were found to possess antitumour and immuno-stimulant properties.

Pleurotus ostreatus (Oyster mushroom)

P. ostreatus is an edible fungus and is cultivated in large quantities. In production it ranks second in India. The basidiocarp is fan or shell shaped with short stalk on one side (may be absent). A polysaccharide isolated from this fungus was found to lower serum and liver lipids in Syrian hamsters. It has antioxidants. An antitumour glucan has been isolated from this mushroom.

Grifola frondosa (Maitake)

Maitake means "dancing mushroom" in Japanese. It has been observed that when ten week aged rats with well established high blood pressure were fed with 'maitake' for weeks showed reductions in blood pressure. The extract can inhibit the growth of tumours and stimulate the immune system in cancerous mice. Its consumption may cause side effects like stomach upset and soft bowels.

Psilocybe mexicana

About 140 species of *Psilocybe* are known of which about 80 are hallucinogenic. Psilocybin extracted from this mushroom has the structure N, N-dimethyl-H-phosphoryloxytryptamine. There is resemblance in structure of lysergic acid (LSD) and psilocybin. This mushroom is used in psychotherapy and neuro-physiological research.

Calvatia gigantea (Giant puffball)

It is a member of Gasteromycetes. Development takes inside ground, when mature the bodies come out from leaf litter. The fruiting body is large spherical containing Gleba inside. The outer covering is called peridium. It was found that people of a village in Kashmir regularly consuming this mushroom had no history of any cancer. Studies have revealed the presence of Calvacin in giant puff balls which is anticancerous.

FUTURE THRUST

India has taken up mushroom cultivation as a cottage industry. Himachal Pradesh and Kashmir Valley are most suitable for cultivation of button mushroom. Authors developed a new method to cultivate *Pleurotus* mushroom in earthen pots (Plate 2, Figure a). India is exporting about 100,000 kg dried mushrooms annually earning a considerable amount of foreign exchange (about 4 million rupees from black morels alone in *Morchella* spp.). These morels are exported to Switzerland, France, Germany, Belgium, and USA. Chang a leading mushroom biologist told that industrial cultivation of medicinal mushroom should be started. Estimated production cost of Oyster mushroom is Rs. 15, Button mushroom is 25, Reishi mushroom and Shiitake Rs. 50 per kg each.

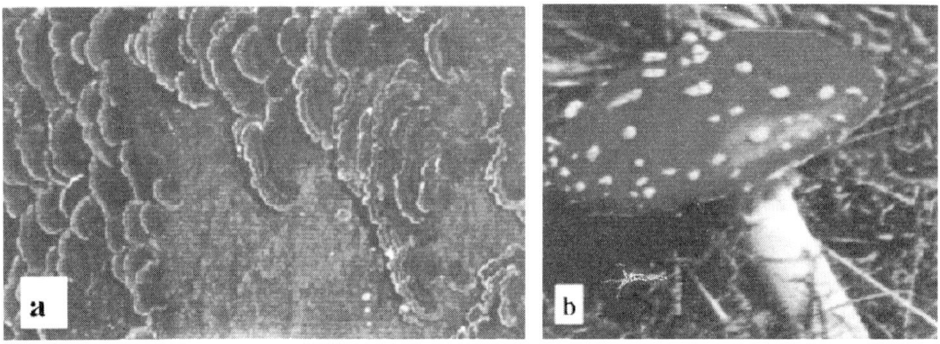

Plate - 3

According to an estimate (Purkayastha, 1981), 620 kg. of button mushroom can be cultivated in a square meter of land. This is based on 5 crops/year and a yield of 25kg/m^2 of bed per crop. No light is required for growth as in most crops. Mushroom requires only 4-6 weeks for a single crop. The cultivation of white button mushroom the most popular species began in vicinity of Paris in 1650. The rat ear fungus *Auricularia auricula* was first to be cultivated by Chinese people. Efforts are on to find out solutions to cultivate some of the most delicious mushrooms. These mushrooms are having symbiotic relationships with higher plants and form ectomycorrhizae with them. Cultivation of Reishi mushroom and Shiitake on saw dust and wood logs is known and efforts are on to grow them on industrial scale to meet the increasing demands of pharmaceutical industries.

REFERENCES

Gareth, J.E.B. (1990). *Mycologist* **4:** 119.

George, V., Singh, O.V. and Koshy, A.S. (1997). In: *Supplement to cultivation and utilization of medicinal plants* (eds. Atal C.K. and B.M. Kapoor) RRL, Jammu Tawi, p. 793.

Kogagaard, L. (1981). *Science News* **119**: 932

Mizuno, T. *et al.* (1989). *Jpn Kokai Tokkyo Koho*, JP **01 66**: 127

Mizuno, T. *et al.* (1990). *Jpn Kokai Tokkyo Koho, JP* **02 78** : 630

Potron, M. (1956). *Concurs Medical,* **36:** 3795

Purkayastha, R.P. (1981). *Everyman's Science,* **26:** 132-138

Wasser, S.P. (2002). *Herbal Gram* **56** : 28-33

Wasson, R.G. (1968). *Soma Divine Mushroom immortality,* Harcourt Brace Jovanovich, New York, 320 pp

Wright, T. (2004). *Nutraceutical World* **7(4) :** 26-30

Toth, J.O., Luu, B. and Ourisson, G. (1983). *J. Chem. Res.* (M) 2719.

9

CULTIVATION OF REISHI MUSHROOM [*GANODERMA LUCIDUM* (LEYES) KARST.]

Chitra Arya

Department of Botany, Faculty of Science,
The Maharaja Sayajirao University of Baroda, Vadodara-390 002. Gujarat (India)

ABSTRACT

The wood rotting fungus, *Ganoderma lucidum* (Leyes.) Karst., is called lingzhi in China and Reishi in Japan and used for the treatment of diseases like cardiovascular, anti-inflammatory, human leukocyte cancer management, longevity, vitality, hepatitis B, diabetes, hypertension etc. This fungus is rich in oxygenated lanosterol derived triterpenoids and more than 90 terpenoid compounds. This mushroom was cultivated on saw dust substrate supplemented with corn meal (10%), pearl millet (10%), gypsum and chalk powder for maintaining pH of 5-5.5, and moisture level 65-70%. Polypropylene bags were filled with 2 kg of moist substrate. Bags were sterilized at 15 psi for 2 hour for two successive days and then inoculated with saw dust and pearl millet spawn (at 5% dry weight basis). Bags were incubated in dark at room temperature 25-30^0 C for 45 days and then top was opened. The fruiting was initiated with exposure to light (500-800 lux) and humidity 95%. Three flushes of fully mature red Reishi were obtained in 3 months cycle with 15% B.E.

INTRODUCTION

In this world, every human being is a God blessed one. If he is healthy, he can celebrate his happy life. To retain the body structure, to avoid and rectify the diseases, the human started to consume medicines, which is composition of powerful chemicals. Our body suffers with several diseases. Even the developed countries agree that the only solution to the diseases should not be medicine. 'Make the food as medicine, and kill the disease before it kills us'.

Mushrooms are nutritionally functional food and a source of physiologically beneficial and non toxic medicines (Wasser 2002). 'Soma' the divine drink is supposed to be obtained from the mushrooms Rig Veda says '*Apamiva bhavatu rakshasa sah* 9:85, which means with the help of soma we may be free from diseases and demons.

Mushrooms are valuable health food –low in calories, high in vegetable proteins, zinc, chitin, chiton fiber, vitamins, and minerals. Mushrooms also have a long history of use in traditional Chinese medicine. Their legendary effects on promoting good health and vitality and increasing the body's adaptive abilities have been supported by recent studies. These studies suggest that mushrooms are prebiotic they help our body strengthen it self and fight of illness by maintaining physiological homeostasis restoring our bodies health balance natural resistance to disease. The compounds they contain have been classified as host defense potentiators, which can have immune system enhancement properties. Their legendary effects on promoting good health and vitality and increasing the body's adaptive abilities have been supported by recent studies.

Chemicals present in Reishi Mushroom and their biological activities

For 4,000 years, the Chinese and Japanese have called upon Reishi to treat liver disorders, hypertension, arthritis, and other ailments. Certain studies have demonstrated antiallergenic, anti-inflammatory, antibacterial, and antioxidant effects (Kino *et al.*, 1989).

This fungus, *G. lucidum* with woody texture, is also used in several medicinal preparations by the tribal of Madhya Pradesh in India (George, 1997). It is reputed to possess hypocholesterolemic activity. This fungus is rich in oxygenated lanosterol derived triterpenoids and more than 90 terpenoid compounds have been isolated from this fungus. Many of the triterpenoids isolated from *G. lucidum* are C-3 epimers and C-3/C-15 positional isomers in pairs.

Several of these triterpenoids were inhibitors of cholesterol biosynthesis. Bioassay of hypolipidemic activity was based on the inhibition of (2-MC) acetate incorporation into cholesterol by rat liver slices (Shiao *et al.*1989). Biological activity was observed in the glycoproteins and polysaccharides obtained from the fruit bodies and mycelia of *G. lucidum (*Lee *et al.*1990). Alkali soluble glycoproteins isolated from the fungus showed anti-complementary activity and it was also shown to increase hemolytic plate forming cells of the spleen in ICR mice (Lee *et al.*, 1990). A beta- D-glucan isolated from this fungus showed immunostimulative antitumor activity (Mizuno *et at.*, 1988). An immunosuppressant glycoprotein having m.wt. 16000-18000, free of human hemoglutination ability is prepared from mycelium of *G. lucidum.* This glycoprotein is useful in the treatment of allergies (Tsunoo *et al.*, 1988). Moderate hypoglycemic activity was reported in the heteroglycan fraction obtained from this fungus (Hikino & Mizuno, 1989).

A novel protein with mutagenic activity *in vitro* and immuno modulatory activity *in vitro* has been isolated from the mycelial extracts of this medicinal fungus – Ling Zhi (*G. lucidum*) and has been named LZ – 8. Isolated glycoprotein fraction GL showed 81% inhibition of tumor growth in mice. This fraction had mol. wt. – 47KD and analysis showed the presence of 82% polysaccharide, 8% protein and 0.9% hexosamine. The polysaccharide moiety consisted of glucose (63%), galactose (27%), mannose (7%) and fructose (3%). GL exhibited the antitumor activity through immunopotentiator. From the cultivated mycelium of *G. lucidum* ganodermic acids were isolated. Ganodermic acids Z, Y, X, W, V, and U were cytotoxic to hepatoma cells *in vitro* (Toth *et al.,* 1983).

The fungus is being cultivated in China, Japan, Korea, Taiwan, Thailand, Malaysia, Indonesia and USA; In India literature shows that Sharma and Rai (2002) cultivated it on saw dust, Tiwari *et al.* (2004) cultivated *Ganoderma* on wood log and wood chips and Rai (2003) on saw dust. In the present studies results of the cultivation of *G. lucidum* on mixed saw dust of various forest tree species is reported.

MATERIALS AND METHODS

Two different cultures were used in the studies one mother culture of G. lucidum was procured from the National Mushroom Gene Bank of the NRCM and the other was isolated from wild *G. lucidum.*

Studies were carried out for 7 days in petriplates (Radial diameter) and conical flasks (visual) on pearl millet supplemented with mixed saw dust, optimum temperature, pH, rate of supplementation and moisture content for mycelial growth of *G lucidum.* It was observed that addition of 10% corn meal and 10% pearl millet, 70% moisture, 30°C temperature and pH 5.5 was optimum for mycelial growth and specially fruiting of *G lucidum* in 500ml. flasks by the two strains.

Table 1. Substrate selection of *Ganoderma lucidum* study in polypropylene bags

No.	Substrate	Mycelial Colonization	Days for Pins after opening	No. of Primordia in 1st flesh	Healthy fruiting body
1	RS	++	-	-	-
2	RS + RB	+++	-	-	-
3	SD	+++	15 days	4	1
4	SD + RB	+++	10 days	4	2
5	SD + RB + CM	++++	9 days	5	3
6	SD + RB + PM	+++++	7 days	7	4
7	SD + RB + 10% CM + 10% PM	+++++	7 days	9	6

RS = Rice Straw; RB= Rice Bran; SD= Saw Dust; CM= Corn Meal; PM= Pear Millet

Substrate was supplemented with $CaSO_4$ and $CaCO_3$ to adjust the pH 5.5. The mixture (100g.) was filled in small polypropylene bags cotton plugged and sterilized at 15 psi. for 45 min. Each bag was inoculated with about 5g of spawn. Bags were incubated at $30^0 \pm 2^0C$ in BOD incubator for 20 days. Observations on the rate of mycelial colonization were recorded. After completer colonization cotton plug was removed and substrate was transferred to environmentally controlled growing rooms. The fluorescent light of 500 lux was maintained 8 hours daily, introducing fresh air to obtain CO_2 at about 1200 $\pm$ 100 ppm maintaining temperature of 30 ± 2°C and relative humidity at 95%. After pinhead formation, humidity was reduced to 85-90%. Observations on pinhead formation, cap formation and growth were recorded in various substrates.

Cultivation trials

It was observed from the substrate selection experimentation that though mycelial growth was slightly slower in saw dust but fruiting was best in mixed saw dust and rice bran (for aeration) supplemented with corn meal and pearl millet. Cultivation trials were therefore laid on this substrate. Saw dust fortified with 10% corn meal and 10% pearl millet (dry weight basis) and $CaSO_4$ and $CaCO_3$ were mixed in appropriate quantities to give pH of above 5-5.5. Saw dust of different forest trees like *Tectona grandis* (Verbenaceae), *Acacia arabica* (Mimosaceae) and *Terminalia catappa* (Combretaceae) were used. The substrate was moistened to provide moisture content of 65%. Substrate was filled in polypropylene bag and cotton plugged and sterilized @ 15 p.s.i. for 2 hours then aseptically inoculated with pearl millet saw dust spawn @ 5% on dry weight basis. The bags were kept in growing rooms and incubated in dark 30 ± 2°C and at 80% humidity. After complete colonization the bags were opened form upper surfaces. After maintaining at 95% humidity 30°C temperature darkness, and high CO_2 (5000 ppm) for 3 days. Fresh air was introduced to give 1000-1200 ppm CO_2 and light was switched on to provide about 500 lux brightness. Humidity was maintained 90-95% RH. After pinhead formation was observed the humidity was reduced to 85%. Above conditions continued throughout the growing period except lowering of humidity to 70% for gap thickening (after white margin on the cap disappeared) observations on no of fruit bodies and yield was recorded in all three flushes appeared in 3 months. Experiment was conducted in 10 bags.

G. lucidum showed typical fruiting behaviour, mycelium aggregated to form pinheads of about 0.6-0.9 cm diameter, white in colour, pinhead appeared after 6-8 days after opening the bags and continued to appear for 4-5 days. Approximately 40-50% pinheads aborted and rest grew to form long red stems with white tips. After lowering of humidity by about 10% the tip of the stem started flattering to form the cap (pileus). In the beginning stem was red and cap was yellow with white margin and the margin growth continued with colour of the pileus attached to the stem become brownish red. Margin of the pileus stopped growing but red coloration continued. Mature fruit bodies were brownish red with no white margin. Humidity was further lowered (65%) which resulted in thickening of pileus and shading of the brown spores. At this stages fruit bodies were harvested and weighed. Mushrooms were dried in oven at 50°C and on an average gave 40% dry weight. 2nd and 3rd flushes appeared after about 20-25 days intervals. Cropping was continue upto 3 months after opening the of the bags. Total operation took 5 months on an average produced 15% B.E. (150 gram fresh mushrooms in 1 kg dry substrate). The experiment with wild strain did not produced fruiting bodies.

The biological efficiency was calculated by following formula:

$$\text{B. E.} = \frac{\text{fresh wt of mushroom produced X 100}}{\text{Total wt of substrate used}}$$

RESULTS AND DISCUSSION

Growth of fungal contaminants like *Rhizopus stolonifer* and *Trichoderma viride* was observed in bags without Bavistin. After 60 days the polypropylene bags were opened and solid mass covered with white mycelium was placed in a chamber on a raised platform. Humidity was maintained 80-85%. C : N ratio which is different in various plant tissues is known to be important factor in influencing the rate of decomposition. On '0' day the C:N ratio of the substrates was 300 : 1 on 75 day the ratio was 190 : 1 and it became 80 : 1 after 60 days. There was a significant drop in the carbon content. Biodegradation of lignocellulosic substrate by fungi is studied by numerous investigators. Recycling of nitrogen by the fungi through autolysis and reuse and action of diazotrophs has been suggested (Levi *et al.*, 1968, Balaji *et al.*, 1999).

On fluffy mycelium warty out growth were seen in presence of natural light, after 68-70 days of spawning. After 75-80 days young buttons were observed which were reddish brown in colour, out of these many young buttons have aborted and about 25% were enlarged further. At this time the moisture level reduced till 65-70%. After 90 days of spawning, these young buttons enlarged to form orange reddish crust on the pilius. At the centre they showed different brown coloured line patterns and at margin of basidiocarp white border was present (Plate I, II). When mature spores shading and bodies became hard, stipe became dark brown and hard, this is the correct time of harvesting. Fifteen g fresh *Ganoderma* fruiting bodies were obtained in 1 kg. dried substrate, 15 BE obtained at room temperature on saw dust without any supplement indicates that this species can also be exploited commercially in our country in colder months. More studies like nutrient supplementation and improved cultivation practices are however needed to further improve the yield on industrial scale.

Fig. a. Sawdust substrate for cultivation ganoderma lucidum

Fig. b. Reddusg growth primordial of basidiocarp seen on saw dust substrate

Plate - I

Cost of production of *Ganoderma* was about 220/- per kg. The expenditure on building, labour, and electricity were main expenses besides cost of saw dust for synthetic logs, fungicides, and polypropylene bag. Being very important medicinal the selling cost is estimated to be Rs. 400-500/- approximately. The production cost is inversely proportional to the size of unit. In other words the amount of mushroom production will increase the production cost per kg of mushroom will decrease. The above data clearly reveal that investment in *Ganoderma* mushroom production unit is economically viable.

Fig. a. Fruiting bodies of *Ganoderma lucidum* (Leyes.) Karst. produced on tree trunks

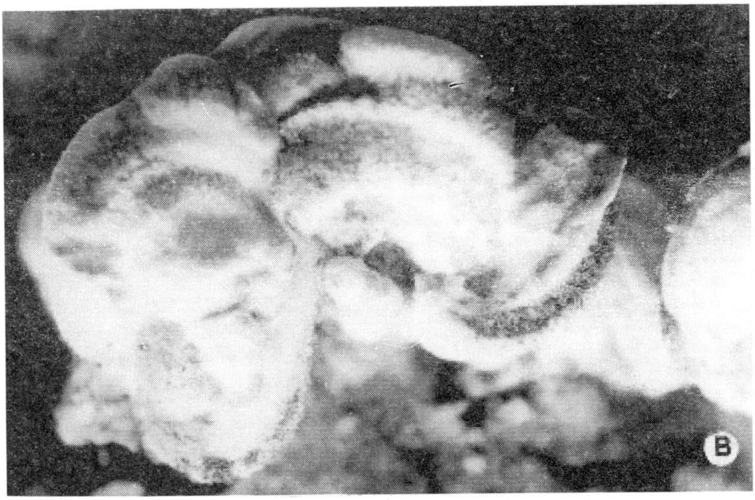

Fig. b. Fruiting bodies of *Ganoderma lucidum* (Leyes.) Karst. cultivated in lab.

Plate – II

REFERENCES

Balaji S., Sujatha, A. and I. Kalyanasundaran (1999). *Mushroom Research.*, 8(1): 31.

George, V., Singh, O.V. and A.S. Koshy 1997. In: *Suppl. To Cultivation and Utilization of Medicinal Plants* (Eds. Atal C.K. & Kapoor B.M). RRL., Jammu Tavi pp. 793.

Hikino, M. and T. Mizuno (1989). *Planta Med.*, 55: 385.

Kino, K., *et al.* (1989). *J. Biol. Chem.*, 264: 472.

Lee, J.W., Chung, H.C., Jeong, H. & K.H. Lee (1990) *Hanguk Kyunhakhoechi*, 18: 137.

Levi, M.P., Merrill, W. and E.B. Cowling (1968) *Phytopathol*, 58:626.

Rai, R.D. (2003) *Mushroom Research*, 12(2): 87-91.

Sharma, J.R. (1995) *Hymenochaetaceae of India,* Botanical Survey of India, Calcutta pp. 219.

Sharma, S.R. and R.D. Rai. (2002) *J. Non Timb. For. Prod.* 9(1/2): 16.

Shiao, M.S. *et al.* (1989) *Chung Yang Yen Chiu Yuan Chiu So Chuang Koan*, 9: 235.

Tiwari, C.K., Meshram, P.B. and A.K. Patra (2004) *Indian For.*, September, 1057

Toth, J. O. *et al.* (1983) *Chem. Res.* (M) 2719.

Tsunoo, H., Kino, K. and A. Yamasita (1988) *Eur. Pat. Appl, EP* 288: 959.

Wasser, S.P. (2002) *Herbal Gram,* 56: 28-33.

10

MICROMORPHOLOGICAL CHARACTERS AS BIOMARKERS FOR SOME OF THE MEDICINAL PLANTS OF GUJARAT

Dipa. Gohil, Vinay M. Raole and M. Daniel

Dept. of Botany, Faculty of Science, The Maharaja Sayajirao University of Baroda, Vadodara - 390 002

ABSTRACT

Twelve medicinal plants of Gujarat have been screened for their micro-morphological characters such as epidermal cell complex, stomatal complex, and trichomes. Trichomes were found to be very distinctive and useful in identifying each taxon. Therefore, a key for identification of these plants have been prepared. The trichomes thus form a valid biomarker.

INTRODUCTION

Micromorphological studies of any taxa include characteristic features of epidermal cell complex, stomatal complex, trichome complex, or any other epidermal inclusions. Importance of micromorphological characters; specifically those of trichomes are widely recognized for taxonomic considerations of Angiosperms (Banerjee, 2002). A majority of the flowering plants can readily be identified with as much easy by their vegetative characters as by their floral characters (Davis and Heywood, 1973; Praveen, 2000). These characters aid as biomarkers for easy identification of plants even in vegetative state.

In the present study, the emphasis is given on trichome studies. Trichomes are epidermal appendages of diverse form, structure and functions. They may occur on all parts of the plants. Trichomes show wide variations within families and the smaller plant groups and even in the same plant. On the other hand, one may observe considerable uniformity in trichomes within a plant group also.

Structurally, trichomes are classified as unicellular and multicellular. The unicellular may be unbranched or branched. Multicellular trichomes may consist of a single row (uniseriate) or of several layers (multiseriate). Some multicellular trichomes are branched in dendroid manner; others have the branches oriented largely in one plane.

In the present work 12 medicinal plants found in Baroda have been studied for their micromorphological characters especially trichomes and stomatal features.

MATERIALS AND METHODS

All the plants were collected from the campus of M.S.U. The voucher specimens were deposited in the herbarium of M.S. University of Baroda. For trichome studies, the leaves were boiled in 10% KOH for 5-10 min. and after decanting the KOH, these leaf samples were washed in water. These treated leaf samples were then stained with saffranin and mounted in glycerin and observed under compound microscope.

Observations

The trichomes observed in the plants studied are described below and data on trichome frequency and trichome index are provided in Table 2. The data on epidermal cells and stomata such as epidermal frequency, stomatal index and their frequency are presented in Table 1. The details of trichomes of each plant studied are presented below.

1. *Aerva lanata* Juss. Only eglandular trichome is present here. It is multicellular and uniseriate. The surface shows wartiness due to small protuberances. The walls of joints of the trichome show many finger like projections on either side. The basal cell is swollen.

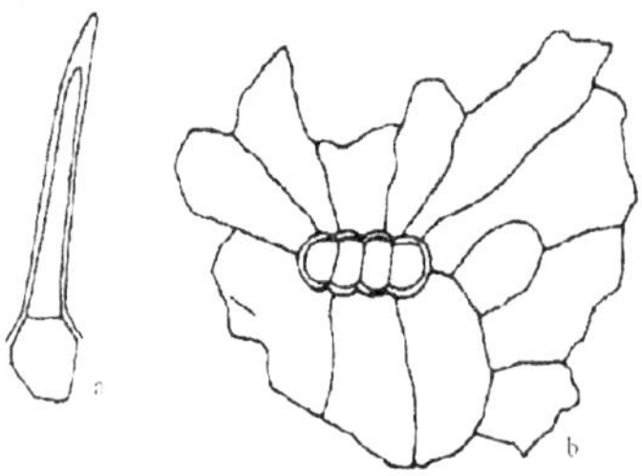

2. *Antigonon leptopus* H. & A. This plant also has both glandular and eglandular trichomes. The glandular trichome shows 3-4 celled sessile head and the polar end of the head show hemispheric cell. The eglandular trichome is unicellular and very short in length. Tip of trichome is conical.

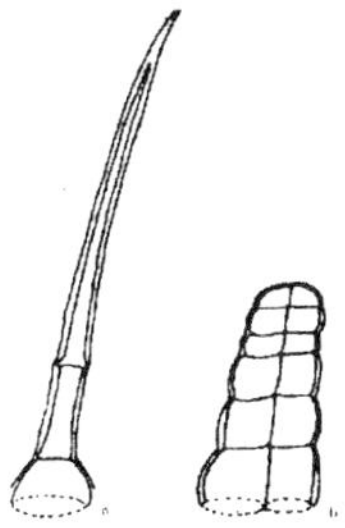

3. ***Blumea membrancea* DC.** Here also both glandular and eglandular trichomes are present. The eglandular trichome (a) shows swollen basal cell. Basal cell is somewhat hemispheric. Eglandular trichome is multicellular, uniseriate and the joints show swellings as those of *Jatropha*. Glandular trichome (b), is multicellular, biseriate and the head is not well differentiated and 3-4 celled.

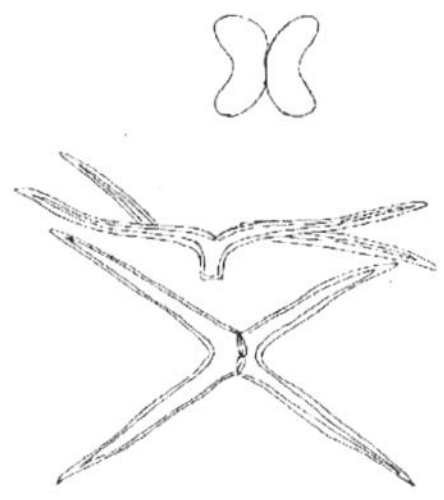

4. ***Buddleia asiatica* Lour.** Trichomes are of two types, glandular and eglandular. The eglandular trichome is two celled joined at the base and each cell is further dichotomously branched. Hence, it shows four arms and each arm is spread in the form of a cross. This trichome shows a small stalk formed of the bending of the base of two cells. The glandular trichome is 2-celled with kidney shaped cells joined at the back.

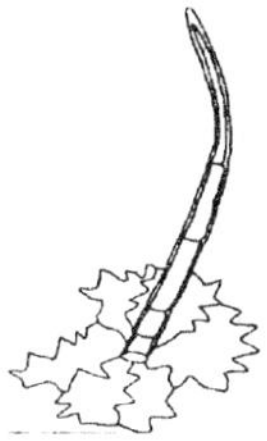

5. ***Euphorbia hirta* L.** The trichomes observed are eglandular uniseriate unbranched filaments. Apical cell is pointed.

6. ***Glinus lotoides* Loefl.** This plant shows presence of 3-5 armed stellate hairs. Each arm of the stellate hair is dichotomously branched. Sometimes the arm shows presence of cross septa.

7. ***Hygrophila serpyllum* Anders.** Here also, both eglandular and glandular types of trichomes are present. Eglandular trichomes are of two types, viz. unicellular (a) and multicellular uniseriate trichome (b). Unicellular trichome show pitted surface. Multicellular uniseriate filamentous trichome show swollen base. The base is hemispheric and formed of 3 swollen cells. The glandular trichomes are of two types one is with single cell head (c) and another one with 3-celled head (d). Both trichomes are stalked (multicellular uniseriate stalk).

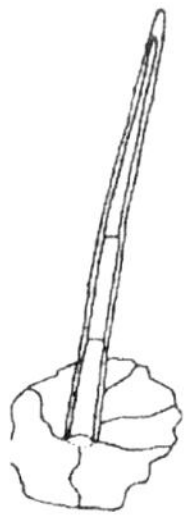

8. *Jatropha gossypifolia* L. Trichome observed here is multicellular, uniseriate, and eglandular. The apical cell of the trichome is pointed. Surface of the trichome is smooth.

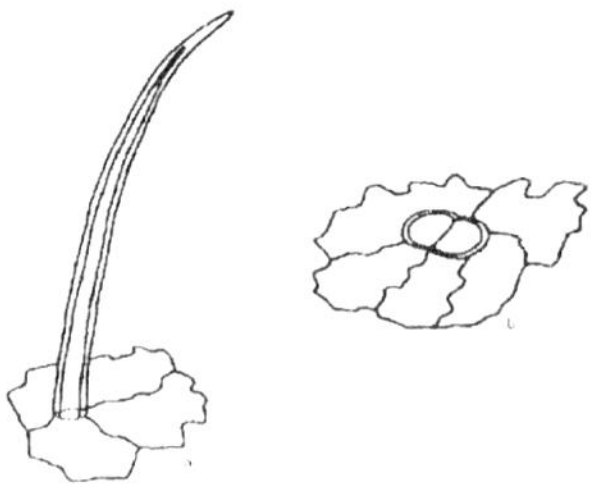

9. *Lantana camara* L. Both glandular and eglandular trichomes are present. Eglandular trichome is unicellular with slight curve at its tip. The glandular trichome is sessile and 2-celled. Each cell is hemispheric.

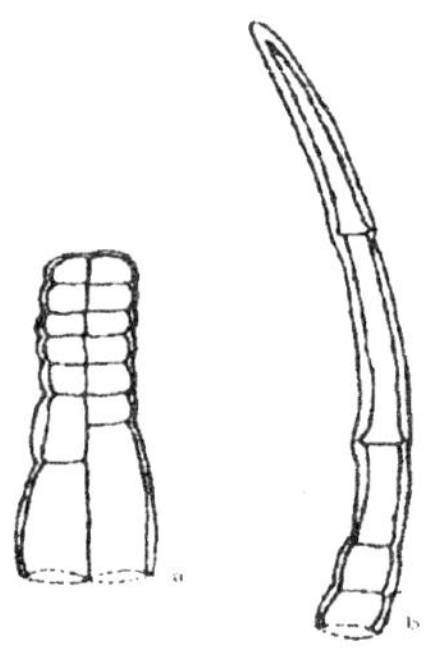

10. *Sphaeranthus indicus* L. Here two types of trichomes viz. glandular and eglandular are present. The glandular trichome is multi-cellular filamentous and biseriate. The trichome shows an undifferentiated head (4-5 celled). Stalk is absent and basal cells are swollen.

Eglandular trichome is multicellular filamentous uniseriate. The basal 2-3 cells are swollen. The joints are swollen similar to the joints of bones.

11. *Terminalia arjuna* W. & A. Here very long unicellular glandular trichomes are present. It shows a very narrow lumen at the base only.

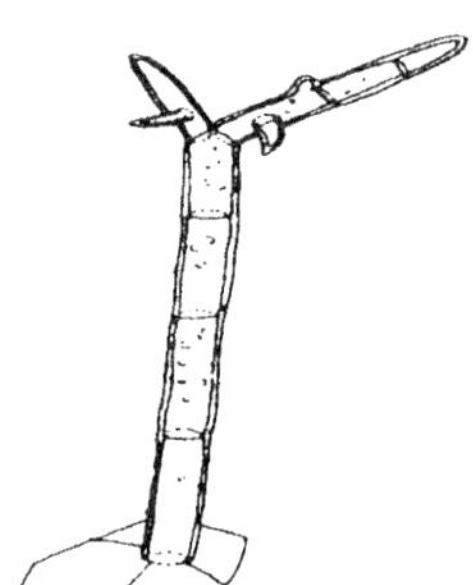

12. *Withania somnifera* Dun. Only eglandular trichome is present here. It is multicellular, uniseriate and branched. Many a times these trichomes are vigorously branched. The branches appear like horns. Surface of the trichome is warty.

The general description and types of trichomes along with stomata of the investigated taxa are described below.

Epidermal cell complex

Epidermal cells were four to many sided. All the taxa investigated show epidermal cells with irregular cells with wavy and curved walls except *Withania somnifera*, Dun. and *Antigonon leptopus*, H. & A. which showed cells with straight walls. Epidermal cell frequency ranged from 18 to 61.2 per mm^2.

Stomatal complex

Stomata observed were of different types such as anisocytic, anomocytic, diacytic, tetracytic, polycytic, pentacytic and paracytic. Stomatal frequency ranged from 7 to 12 per mm^2 and stomatal index ranged from 11.56 to 26.67 per mm^2.

Trichome complex

Both glandular and eglandular types of trichomes were present. The trichome were unicellular/multicellular uniseriate or multicellular multiseriate.

All the investigated taxa studied showed unbranched trichomes except for *Withania somnifera*, Dun. where branched trichomes were observed. The data on stomata and trichomes in the investigated taxa are presented in Tables 1 & 2.

Table 1. Stomatal and Epidermal characters in some Medicinal Plants of Baroda

Taxa	Epidermal Cell Shape	Epidermal Cell Frequency (per mm^2)	Stomatal Type	Stomatal Frequency (per mm^2)	Stomatal Index (per mm^2)
Aerva lanata Juss.	Irregular	41.2	Anisocytic Anomocytic	12	22.56
Antigonon leptopus H. & A.	Irregular	61.2	Anisocytic Anomocytic Tetracytic Paracytic	8	11.56
Blumea membranacea DC.	Irregular	21.8	Anomocytic	7	24.30
Buddleia asiatica Lour.	Irregular	63.2	Anomocytic	21.4	25.29
Euphorbia hirta L.	Irregular	30	Anisocytic Anomocytic	8	21.05
Glinus lotoides Loefl.	Irregular	81.2	Anomocytic Polycytic	10.8	25.71
Hygrophila serpyllum Anders.	Irregular	25.6	Diacytic	8.6	25.14

Jatropha gossypifolia L.	Irregular	31.6	Tetracytic	9.8	23.67
Lantana camara L.	Irregular	28.6	Diacytic Anisocytic Anomocytic	10.4	26.67
Sphaeranthus indicus L.	Irregular	18	Anisocytic Anomocytic	6.2	25.61
Terminalia arjuna W. & A.	Irregular	23.4	Anomocytic Anisocytic Pentacytic	10.2	30.35
Withania somnifera Dun.	Irregular	24.4	Anisocytic	6.4	20.78

Table 2. Trichomes in some medicinal plants of Baroda.

Taxa	Trichome Type	Trichome Frequency (per mm^2)		Trichome Index (per mm^2)	
		Eglandular	Glandular	Eglandular	Glandular
Aerva lanata Juss.	Multicellular uniseriate eglandular	2.4	---	5.50	---
Antigonon leptopus H. & A.	Both multicellular glandular & unicellular eglandular	1	1	1.6	1.6
Blumea membranacea DC.	Both multicellular glandular and eglandular	1	1	4.38	4.38
Buddleia asiatica Lour.	Both eglandular and glandular	14.8	2	18.97	3.16
Euphorbia hirta L.	Multicellular uniseriate eglandular	1	---	3.22	---
Glinus lotoides Loef.	Stellate hair	1.2	---	3.70	---
Hygrophila serpyllum Anders.	Both eglandular and glandular	10.6 1.2	2.2 1.8	29.28 3.31	7.91 6.56
Jatropha gossypifolia L.	Multicellular uniseriate eglandular	8.5	---	9.71	---
Lantana camara L.	Both multicellular glandular and unicellular eglandular	1.2	1.2	4.02	4.02
Sphaeranthus indicus L.	Both eglandular and glandular	1	1.6	5.26	8.16
Terminalia arjuna W. & A.	Unicellular glandular	1	---	1.6	---
Withania somnifera Dun.	Multicellular uniseriate branched eglanudular	1.6	---	6.15	---

DISCUSSION

The present study of micromorphological characters of the leaf epidermis revealed that these characters are useful for the identification of these investigated taxa. These features are now considered important in taxonomic studies and they can be considered as biomarkers. In this study, the trichomes of the plants screened are very much distinct from each other. The nature, branching and surface of the trichomes are variable and hence, they appear very sound biomarkers.

A dichotomous key for identification of these investigated taxa is given below

1. Trichomes eglandular only:
    2. Trichomes branched:
        3. Trichomes dichotomously branched like Stellate hairs .... ***Glinus lotoides*** Loefl.
        3. Trichome multicellular uniseriate randomly branched .... ***Withania somnifera*** Dun.
    2. Trichome unbranched:
        4. Trichome multicellular uniseriate with Smooth surface:
            5. Stomata only tetracytic .... ***Jatropha gossypifolia*** L.
            5. Stomata both anisocytic and anomocytic .... ***Euphorbia hirta*** L.
        4. Trichome multicellular uniseriate with spiny surface .... ***Aerva lanata*** Juss.
1. Trichomes both eglandular and glandular or only glandular:
    6. Eglandular trichome absent, glandular unicellular trichome .... ***Terminalia arjuna*** W. & A.
    6. Eglandular trichome present with glandular trichomes.
        7. Eglandular unicellular trichomes:
            8. Multicellular glandular trichome with two celled head.. ***Lantana camara*** L.
            8. Multicellular glandular trichome with 3-4 celled head .... ***Antigonon leptopus*** H. & A.
        7. Eglandular multicellular trichomes:
            9. Eglandular trichome with dichotomous branching .... ***Buddleia asiatica*** Lour.
            9. Eglandular trichome unbranched:
                10. Eglandular both unicellular and multicellular uniseriate .... ***Hygrophila serpyllum*** Anders.
                10. Eglandular but only multicellular:
                    11. Stomata only anomocytic... .... ***Blumea membranacea*** DC.
                10. Stomata both anisocytic and anomocytic....... ***Sphaeranthus indicus*** L.

REFERENCES

Banerjee, A, Chowdhury, H.R., Karr, K. and S. Mandal (2002). *Phytomorphology,* **52 :** 223.

Davis, P.H. and V.H. Heywood (1973). *Principles of Angiosperm Taxonomy.* R.E. Krieger, New York, USA.

Pandey, C.N., Raval, B.R. and S. Mavli (2005). *Medicinal Plants of Gujarat.* GEER Foundation, Gandhinagar.

Parveen, N.S., Murty, K.S.R. and T. Pullaiah (2000). *Phytomorphology,* **50:**205.

Sivarajan, V.V. and I. Balachandran (1994). *Ayurvedic Drugs and Their Plant sources,*Oxford & IBH Publishing Co. Pvt. Ltd. New Delhi.

11

PHARMACOGNOSTIC AND PHYTOCHEMICAL STUDIES ON *FICUS BENGHALENSIS* LINN. AND *FICUS RELIGIOSA* LINN.

Richa Pandey and M. Daniel

Phytochemistry Lab., Dept. Botany, M.S. University of Baroda, Vadodara – 390 002

ABSTRACT

The bark of *F. benghalensis* is found to contain, well developed cork, isodiametric cells of sclereids, compound acentric starch grains in thick walled cells, crystals of calcium oxalate, and a characteristic type of sclereids having striated walls. In *F. religiosa,* bark is divided into three regions. The outer most is three to four layers of cork and few layers of sclereids, followed by middle thick walled cells (but without sclereids) having characteristic fibres, and inner most zone of rectangular patches of elongated sclereids which contain thick walled cells, calcium oxalate crystals, and starch grains. The starch grains are of two types, one larger with concentric striations (found in inner zones) and other smaller ones. The smaller are compound type and are acentric.

The leaf of *F. benghalensis* is characterized by presence of hypodermis, cystolith, double layers of palisade, covering trichome with silicified wall, and anomocytic stomata having thick walled guard calls in lower epidermis. *F. religiosa* leaf contains single layer of epidermis, trichomes rare, single layer palisade, and have 2 groups of V.B. which are surrounded and separated by stone cells. *F. benghalensis* bark shows 5-6 sesquiterpenes, vanillic, syringic, melilotic, p-coumaric acids, procyanidin and quinones. These compounds also found to present in latex and leaves. *F. religiosa* bark contain 5 sesquiterpenes, 5-methoxyl quercetin, vanillic, syringic, melilotic, o- and p-coumaric acids and *cis* and *trans* ferulic acids. These compounds are also present in leaves.

INTRODUCTION

Ficus benghalensis (Nyagrodhah) and *F. religiosa* (Aswattah) are two well-known medicinal plants belonging to the Moraceae, a family characterized by milky latex, two stigma and a single pendulous ovule (Lawrence, 1951). All the parts of both the plants, are being used for treatment of various diseases since ancient times. The bark of *F. benghalensis* is used for urinogenital disorders, nervous disorders, chronic blood dysentery, skin diseases, rheumatism and diabetes; while the aerial roots are employed for gonorrhea and dysentery. The leaves of this plant are vermicidal, and the latex is used against rheumatism. The bark of *F. religiosa* is hypoglycemic and is used in ulcer and skin diseases whereas its leaves are employed for menorrhagia, hemorrhage, bleeding piles and blood dysentery (Anon., 2000).

Although these are very important medicinal plants, very little work is done on the pharmacognostic and phytochemical biomarkers of these plants. Only the T.S. of the bark of both the plants have been studied by Iyer and Kolammal (1960). Benghalenoside, phytosterol and procyanidin from the bark while quinones from leaves are the compounds reported from *F. benghalensis*. α-Sitosterol and stigmasterol are found to be present in the bark of *F. religiosa* (Anon., 2000).

In the present study, bark and leaf of both the plants have been screened for phytochemicals like flavonoids, quinones, phenolic acids, alkaloids, steroids etc. Pharmacognostic studies also have been conducted to generate biomarkers, which can be effectively used for the quality control analysis of drug in the powder form or in the extract form.

MATERIALS AND METHODS

The various plant parts were collected from campus near Sanskrit Mahavidyalaya of M.S. University, Baroda. They were properly identified and dried. Standard procedures were followed for analysis of various compounds (Harborne, 1984; Daniel, 1991). The anatomical and histological characters were analyzed by taking hand sections and by using differential staining techniques (Wallis, 1955).

RESULTS AND DISCUSSION

a. Pharmacognosy

Ficus benghalensis

The young and old bark and leaves are studied. **The young bark** shows well developed cork layers, appearing yellowish – green; cells are more than twice in length as compared to width. The number of layers depends upon the age of bark. Cork is followed by rectangular, compactly arranged layer of cambium. After cambium the inner cortex is present which is divided in three parts. The outer most is consisting of parenchyma, sclereids, laticifers and thick walled cells having crystals of calcium oxalate. Some sclereids contain more than one crystal. The laticifers appeared as elongated tubes with granular fluorescent inclusions. The middle portion consists of square to isodiametric sclereids. These cells are compactly arranged and they do not have any inclusions, crystals etc. The inner most portion consisted of elongated sclereids, stone cells, laticifers and cells with starch grains. This position of cortex has characteristic thick walled cells which appeared like crumbled cloth; and are spherical in shape (Fig. 1).

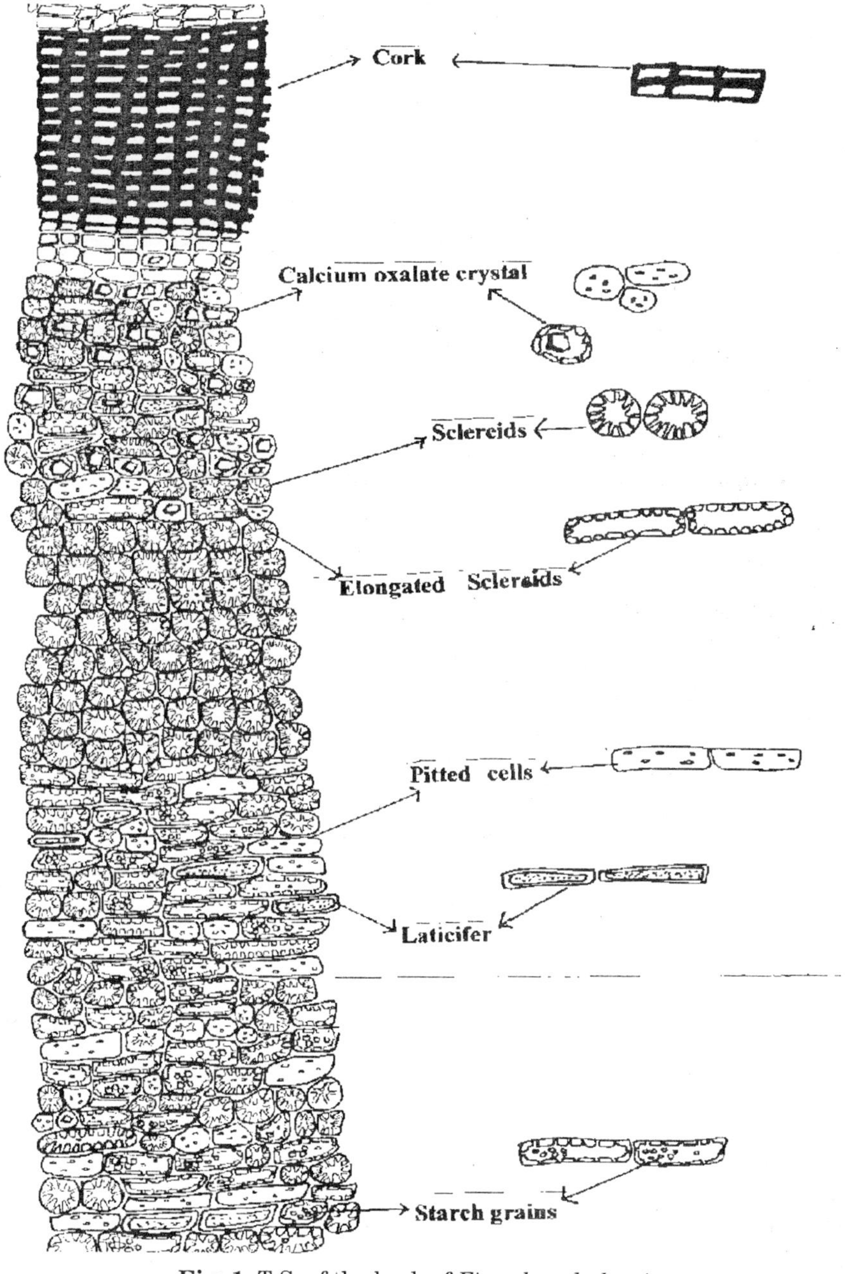

Fig. 1. T.S. of the bark of *Ficus benghalensis*

The old bark showed the outermost cork layer having lenticels. The cork is followed by cork cambium leading to flattened sclerenchymatous cells. Below this region barrel shaped alternating patches of thick walled cells are there. This region is followed by homogenous thick walled cells containing laticifers at intervals.

The leaf was analyzed by T.S. (Fig. 2), powder study and micro-morphological evaluation. The T.S. of leaf shows a single layer epidermis followed by two to three layers of hypodermis. The epidermis is covered by a thick layer of cuticle. Covering trichomes are frequently present. The inner wall of trichomes is silicified (it gives light pink colour with phenol). The base of few trichomes shows hair like appendages. Hypodermal cells are angular and 4 to 5 times larger as compared to epidermal cell. In midrib region hypodermis is collenchymatous. The mesophyll is divided into palisade and spongy parenchyma. Palisade is double layered. The cells of upper layer are bigger as compared to lower one. Spongy parenchyma have finger like projections. Vascular bundle is covered by stone cells. The laticifers are present in pith and phloem region. They are anastomosing and branched. Sphaeraphides and other crystals are frequently present near fibres.

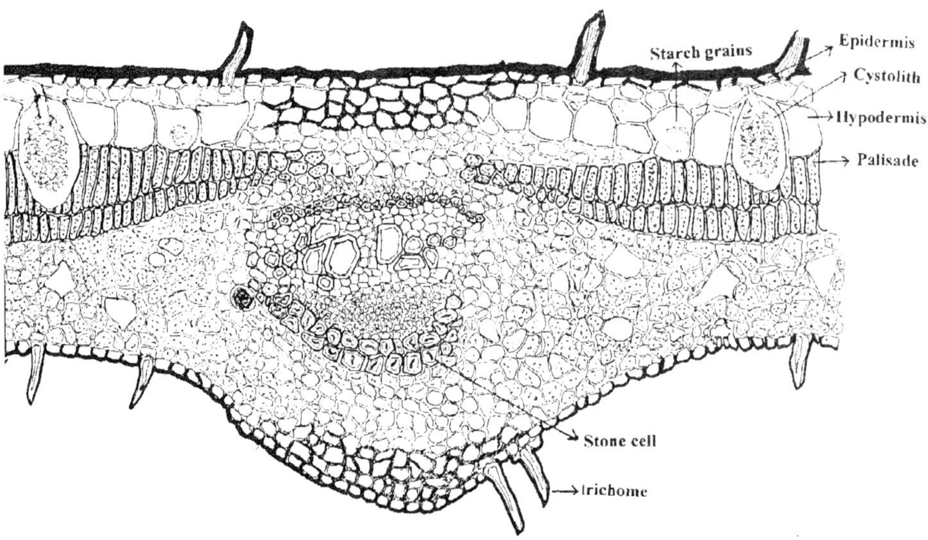

Fig. 2. T.S. of the leaf of *Ficus benghalensis*

Powder study : The powder is light dull green, odorless, bitter in taste. It shows the presences of anomocytic stomata with thick walled guard cells, cystolyths, double layered palisade and fragments of palisade with hypodermis, silicified covering trichomes, cystolyth with prominent stalk, compound and simple starch grains with concentric striations, tracheids with hexagonal area of thickening having slit like opening, as well as tracheids with circular bordered pits, vessels with reticulate and annular thickening and the prominent stone cells. There stone cells have double layers of thickening (Fig. 3).

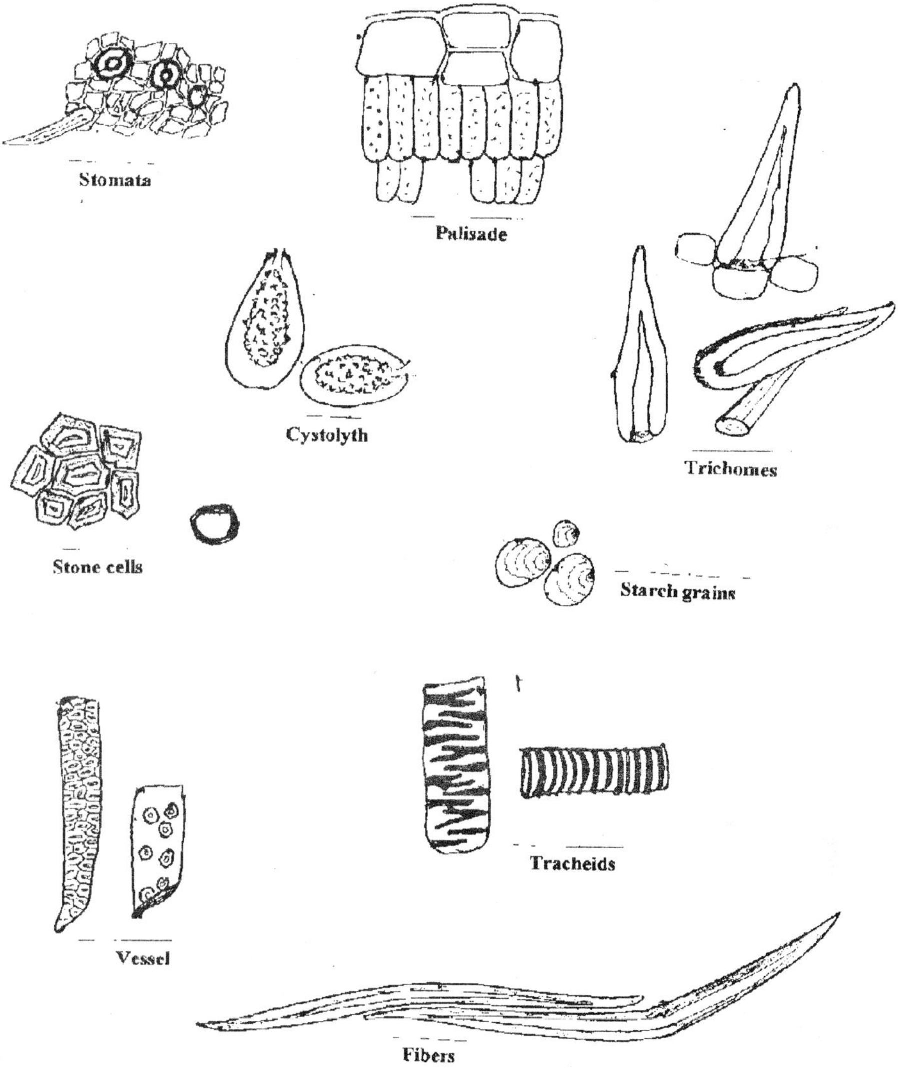

Fig. 3. Powder study of the leaf of *Ficus benghalensis*

Ficus religiosa

The bark is light gray and having small fractures. The T.S. of bark (Fig. 4) shows 2-4 layers of cork; followed by cork cambium. The cortex is divided in three zones. The outer most zone consists of spherical to angular sclereids, sclerenchyma, crystals of calcium oxalate and small compound, centric starch grains in thick walled cells.

Middle portion of cortex has sclerenchyma, parenchyma, laticifers, few sclereids and 5-6 layers of prominent fibres. There fibres are characteristic since they are doubly thickened. The lumen is very small. The innermost zone of cortex consists of rectangular patches. There patches are formed by vertical and horizontal row of elongated sclereids. In these patches sclerenchyma, sclereids, fibres, thick walled cells with starch grains; and with crystals of calcium oxalate are present. Sometimes more than one crystal are present in a cell. The starch grains are of two types in this zone, the bigger one having concentric striations. It may be simple or compound. The smaller one is compound and acentric. The fibres which are found in this zone have thin layer of lignin and thick inner deposition of gelatin, the lumen appeared as a slit.

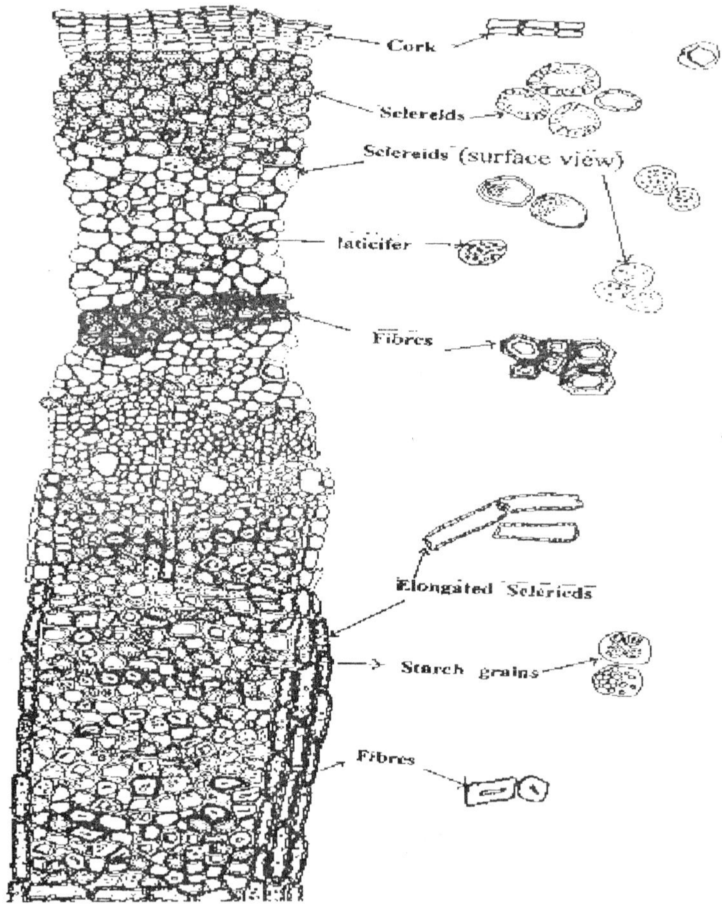

Fig. 4. T.S. of the bark of *Ficus religiosa*

The **T.S. of leaf** showed single layered epidermis covered with cuticle, trichomes were rare. The mesophyll was divided into palisade and spongy parenchyma. Palisade is single layered. The first layer of spongy parenchyma is consisting of loosely arranged elongated cells. The vascular bundle is covered by endodermis. The inner and radial walls of endodermis was thickened. Vascular bundle is covered by stone cells. In vascular region laticifers are present. The lower epidermis contain sunken stomata: and few sac-like structures having stalks but no depositions (Fig. 5).

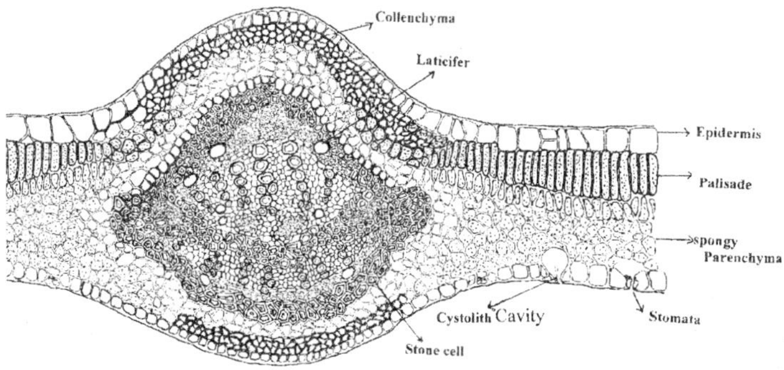

Fig. 5. T.S. of the leaf of *Ficus religiosa*

Powder study: The powder is bright green, odorless and bitter in taste. This shows presence of epidermis, palisade, stone cells, fibres, laticifers, and tracheids with bordered pits and vessels with reticulate and annular thickening (Fig. 6).

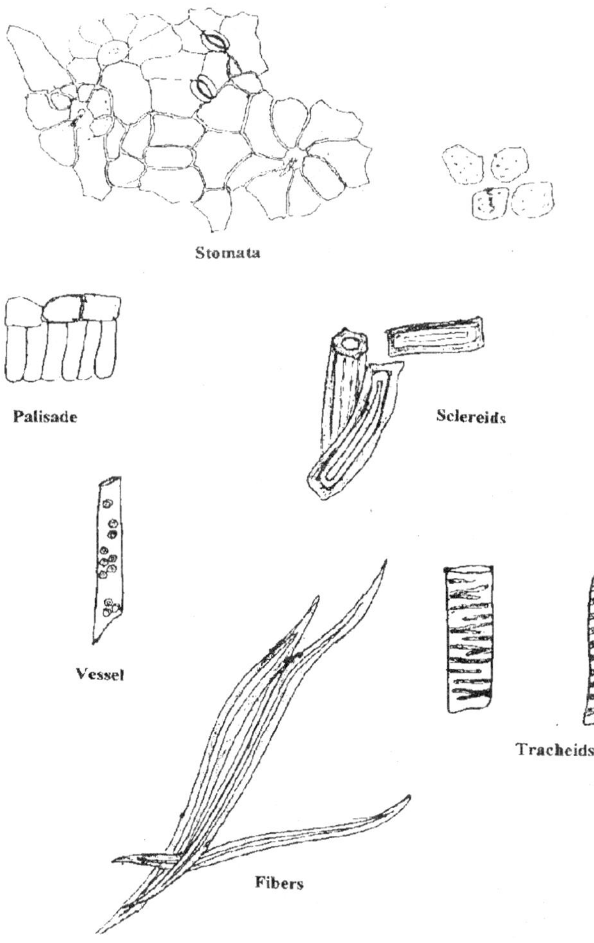

Fig. 6. Powder study of the leaf of *Ficus religiosa*

Phytochemical analysis: The bark and leaf of both the plants are screened for flavonoids, phenolic acids, steroids, quinones, and volatile oil. For steroids, quinones and volatile oil, plant material is extracted with solvent ether while for flavonoids and phenolics acids, the extraction was with methanol. Flavonoids and phenolic acid analysis have been done by P.C. while for others T.L.C. was adopted (Harborne, 1964). The results are presented in Table 1.

Table 1. Phytochemical Analysis

Plant name	Plant Part	5-OMe Quercetin	Phenolic Acids							Alkal-oids	Stero-ids	Quin-ones	Volatile oil	Sapo-nins
			1	2	3	4	5	6	7				Sesquit-erpenes	
Ficus benghalensis	Bark	-	+	+	+	+	-	-	-	-	+	+	+	-
	Leaf	-	+	-	+	-	-	+	+	-	+	+	+	-
	Latex		+	+	+	+	-	-	-	-				
F. religiosa	Bark	+	+	+	+	+	+	-	-	-	+	+	+	-
	Leaf	+	+	+	+	+	+	+	+	-	+	+	+	-

1 – Vanillic acid, 2. Melilotic acid, 3. Syringic acid, 4. *p*-coumaric acid, 5. *o*-Coumaric acid, 6. *cis*-Ferulic acid, 7. *trans*-Ferulic acid

F. benghalensis bark shows 5-6 sesquiterpenes, vanillic, syringic, melilotic, p-coumaric acids, procyanidin and quinones. Flavonols are absent here. These compounds are also found to be present in latex and leaves. *F. religiosa* bark contain 5 sesquiterpenes, a flavonol 5-methoxyl quercetin and phenolic acids such as vanillic, syringic, melilotic, o- and p- coumaric, and *cis* and *trans* ferulic acids. Leaves also contain the very same compounds.

The present study unearths a number of distinctive phytochemical and pharmacognostic characters which will be helpful in validating the genuineness of the powders and extracts of these drugs available in the market and also some of the active compounds.

REFERENCES

Anonymous, (2000). *Supplement to Wealth of India* K.G. CSIR, New Delhi.

Daniel, M. (1991). *Methods in Plant Chemistry & Economic Botany*, Kalyani Publishers, Delhi.

Harborne, J.B. (1984). *Phytochemical Methods,* 2 ed. Chapman & Hall London.

Iyer, N.K and M. Kolammal (1960). *Pharmacognosy of Ayurvedic Drugs*, Univ. Kerala, Trivandrum, **1(3):** 87

Lawrence, G.H.M. (1951). *Taxonomy of Vascular Plants,* Macmillan, New York.

Wallis, T.E. (1955). *Practical Pharmacognosy*, J. & A. Churchill, London.

12

CHEMICAL AND PHARMACOGNOSTIC CHARACTERS OF *LAGERSTROEMIA FLOS-REGINAE* AND *WOODFORDIA FLORIBUNDA*

Elizabeth M. Daniel, Susy Albert and M. Daniel

Phytochemistry Laboratory, Department of Botany, The Maharaja Sayajirao University of Baroda, Vadodara, 390 002, Gujarat, India

ABSTRACT

Woodfordia floribunda and *Lagerstroemia flos-reginae,* two important medicinal plants of India are studied for their pharmacognostic and phytochemical characters. *Woodfordia* is characterized by the unicellular, conical trichomes and spherical glandular hairs in leaves and patches of stone cells, intraxylary phloem and sphaeraphides in stem. The leaves are found to contain flavonoids such as quercetin and 3' OMe quercetin and phenolic acids like syringic, vanillic, gallic and *p*- hydroxy benzoic acids. The stem of this plant contained only acids such as syringic, vanillic and *p*-hydroxybenzoic acids. In *Lagerstroemia,* leaves possessed two layers of palisade and sphaeraphides along the veins while the stem contained mucilage cells, tannin cells, stone cells in phloem and intraxylary phloem. Chemically this plant contained quercetin and 4'OMe quercetin as well as syringic, gallic, vanillic, *p*-coumaric and *p*-hydroxybenzoic acids in both leaves and stem.

INTRODUCTION

The family Lythraceae comprises of 11 genera classified under two tribes Ammanieae and Lythreae. In Gujarat this family is represented by only four genera *Woodfordia*, *Lawsonia*, *Lagerstroemia* and *Ammania*. Members of this family are of great medicinal importance.

Woodfordia floribunda is a much branched shrub with long spreading branches and numerous brilliant red flowers in dense axillary paniculate cymose clusters. The dried flowers of this plant are used extensively in Ayurvedic arishtas, for tonic, in disorders of mucous membrane, hemorrhoids and derangements of liver. Extract of the plant is used for tooth ache and gum trouble, and it also stimulates the contraction of intestinal loop. Fresh leaves are used as an excellent remedy for snake bite and among Sandals it is used as a remedy in excessive bleeding during pregnancy, cholera, fever, muscular pain, sores and spleen complaints (Jain and Tarafdar, 1970). Leaves are known to contain 12-20% tannin and 'lawsone', the naphthaquinone of henna. 'Woodforticosin' isolated from the dried leaves are useful as an inhibitor of topoisomerase and as neoplasm inhibitors. It also serves as an antitumor agent.

Lagerstroemia flos-reginae Retz. is a medium sized large deciduous tree with a rounded crown and with mauve to purple flowers borne in long panicles. Leaves of this plant are purgative, deobstruent and diuretic. Decoction of leaves and dried fruits prepared like tea are used for diabetes mellitus in Phillipines. Roots, stem and leaves are found to contain hydrocyanic acid. Ellagitannins isolated from the plant parts are activators of glucose transport in fat cells.

The chemistry and pharmacognosy of these plants which are prerequisites of the quality control procedures are poorly known with only the pharmacognosy of *Woodfordia* flowers been conducted so far. Therefore, the present study has been conducted to analyze the phytochemicals and the characteristic microscopical features of the leaves and stem of *Woodfordia floribunda* and *Lagerstroemia flos-reginae*.

MATERIAL AND METHODS

Plant parts stem and leaves used for the present study were collected locally from regions in and around Vadodara, Gujarat. The voucher specimens are deposited in the Herbarium (BARO) of the M.S. University of Baroda.

(a) Pharmacognosy

Pieces of stem and leaves were fixed on the spot in FAA (Johansen, 1940), processed to obtain paraffin embedded sections. Transverse sections of leaves and stem were obtained at 8-12 μm on a rotary microtome, deparaffinized, stained in toluidine blue and observed to study the internal features Epidermal preparations were made by using the epidermal peels obtained directly or by heating in 10% KOH and then stripping of the epidermal layers gently from the mesophyll tissue with the help of pointed needles and forceps. Epidermal peels obtained were then washed in water, stained with aqueous safranin or toluidine blue and mounted on a clean microscopic slide. The slides were examined under microscope and camera lucida diagrams were drawn. The quantitative and qualitative data comprising of stomatal index, trichome index, type of stomata and epidermal cells etc. were based on the average of 20 readings.

(b) Phytochemical studies

The leaves and stems of both these plants have been analysed for their phytochemicals such as alkaloids, flavonoids, phenolics, saponins, tannins etc. using standard procedures (Harborne, 1984; Daniel, 1991).

RESULTS AND DISCUSSION

Micro morphological features

Leaves of both *Woodfordia* and *Lagerstroemia* are hypostomatic with stomata present only on the lower/abaxial surface of the leaf. Epidermal cells are mostly polygonal and isodiametric. There is no definite pattern of arrangement of epidermal cells. The epidermal cell walls are straight to curved or sinuate. The surfaces of the cells are smooth without any striations.

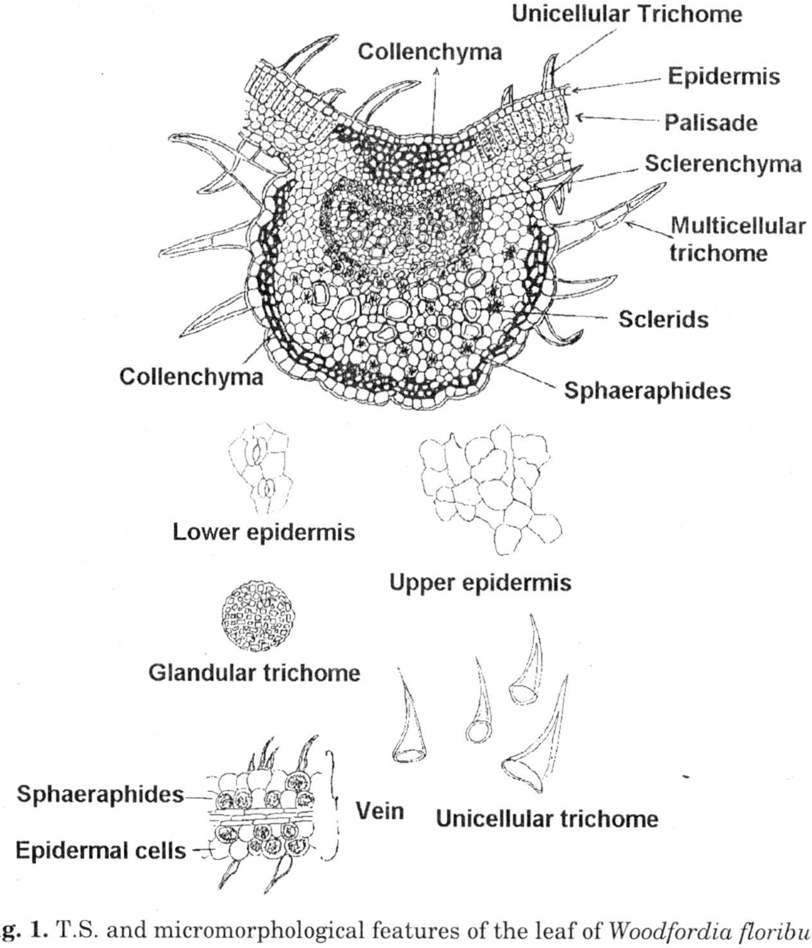

Fig. 1. T.S. and micromorphological features of the leaf of *Woodfordia floribunda*

Stomata are of Ranunculaceous type in both *Woodfordia* and *Lagerstroemia*. They are distributed evenly over the entire lower surface of the lamina in between the veins, but generally not over the finer veins and main veins. Epidermal cells widely separate them. Each stoma is surrounded by 5-6 subsidiary cells which are indistinguishable in size, shape or form from the other epidermal cells. The stomatal index of *Woodfordia* is 18-19 while that of *Lagerstroemia* is 18-20. On either side of the main and finer veins in the leaf of *Woodfordia* the epidermal cells show presence of a single large sphaeraphide within each cell.

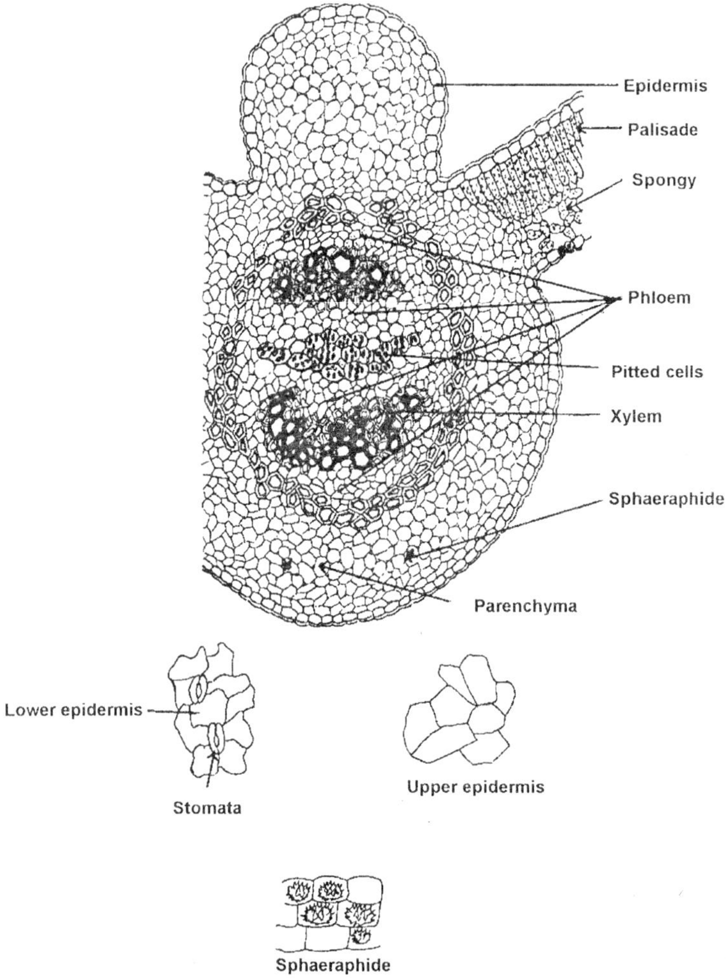

Fig. 2. T.S. micromorphological features of the leaf of *Lagerstroemia flosreginae*

Trichomes are completely absent in *Lagerstroemia* while in *Woodfordia* both abaxial and adaxial surface of the leaf is densely covered with covering trichomes. On the upper surface, unicellular conical covering trichomes with prominent glandular base are evenly distributed. On the lower surface of the leaf the veins are covered with long unicellular trichomes while the lamina surface is densely covered with tufts of shaggy small unicellular hooked/

curved trichomes. Glandular spherical/spheroidal trichomes appear sunken in pits filled with secretion which makes the lower surface sticky. Each glandular trichome consist of a one celled stalk and and a glandular head of 20-40 or more radially elongated cells more or less separated. Length of the trichome is 63.84 μm. The trichome index is 48-50 (Figs. 1 and 2).

Anatomical attributes of leaves

Leaves in *Woodfordia* and *Lagerstroemia* are dorsiventral with a distinct differentiation of the mesophyll tissue into palisade and spongy layer protected on either side by upper and lower epidermis respectively. In *Woodfordia* palisade tissue is made up of a single layer of short palisade cells which are compactly arranged, while in *Lagerstroemia* palisade layer constitutes of two layers of elongated palisade cells which are loosely arranged. Both layers of palisade cells are of the same height. Spongy layer is composed of rounded or elongated cells with large prominent intercellular spaces.

In *Woodfordia* the midrib is continuous with few layers of collenchyma, within which compactly arranged thin walled polygonal cells of parenchyma are present embedded in which a semicircle fibro vascular bundle is present. A large number of sclereids and sphaeraphides are evenly distributed lining the vascular bundle. In *Lagerstroemia* the midrib contains two large crescentiform collateral vascular bundles placed opposite to each other and are separated by thick walled pitted parenchyma cells. Parenchymatous cells surrounding the vasculature show sphaeraphides. Cortical region of midrib is heterogenous consisting of few layers of collenchyma followed by parenchyma (Figs. 1 & 2).

The results obtained from this investigation revealed a number of interesting features which may be useful in classifying the two plants. Variability in epidermal cells architecture among different groups of plants is an attribute that makes cuticular feature a reliable tool in taxonomic analysis and diagnosis (Olowokudejo, 1990, Metcalfe and Chalk, 1950). Leaf of *Woodfordia* is distinctly differentiated by the presence of trichomes. The upper and lower leaf surface is also apparently different with hooked covering trichomes and multicellular glandular trichomes on the abaxial surface and only conical covering trichome on the upper / adaxial surface.

Anatomical attributes of stem

Epidermis in the stem of *Woodfordia* and *Lagerstroemia flos-reginae* are single layered made up of barrel shaped cells which appears to be crushed due to the enlarging and addition of internal tissues by the secondary growth. In *Woodfordia* the epidermis is followed by few layers of hypodermis comprising of collenchyma. Cortex is composed of sclerenchyma cells interspersed with mucilage cells followed by large parenchyma cells intermingled with islands of stone cells. In *Lagerstroemia flos-reginae,*it is made up of 8-10 layers of large polygonal parenchyma cells. Vascular region comprises of solid cylinder of xylem with phloem on both sides in *Woodfordia* while in *Lagerstroemia flos-reginae* phloem is present only towards outside.

Isolated cells of phloem and pith contain sphaeraphides in *Woodfordia*. Uniseriate medullary rays are distinctly present in the xylem cylinder of *Woodfordia* while in *Lagerstroemia flos-reginae* medullary rays are completely absent (Figs. 3 & 4).

Phytochemical studies

Lagerstroemia flos-reginae

The results of screening leaves and stem of this plant is presented in Table 1. Both the stem and leaves were found to be very rich sources of flavonoids such as quercetin and 4'-OMe quercetin. The phenolic acids present were syringic, vanillic, gallic, *p*-coumaric and *p*-hydroxy benzoic acids. In addition quinenes, catechins and steroids also were found to be present. Alkaloids were found to be absent.

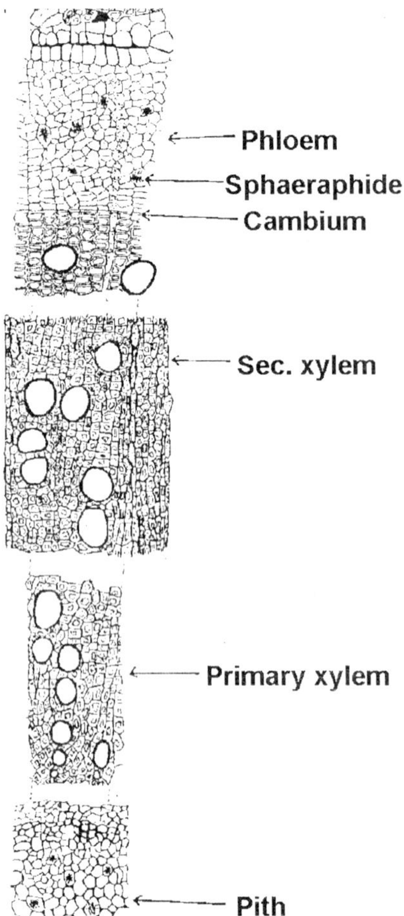

Fig. 3. T.S. of stem of *Woodfordia floribunda*

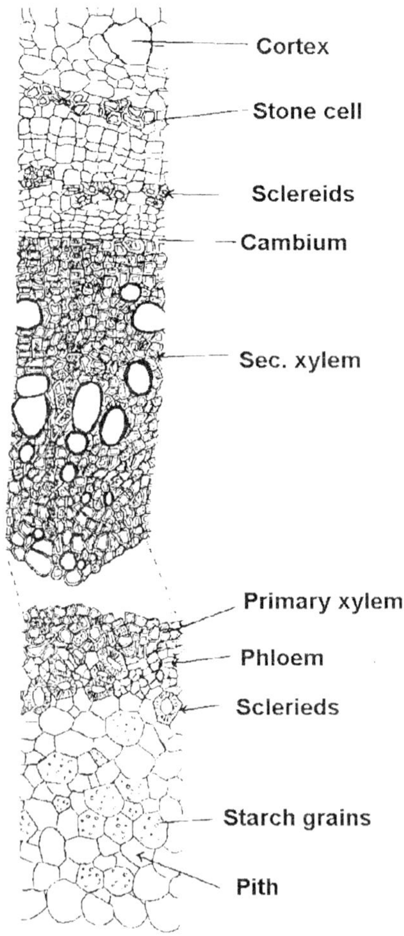

Fig. 4. T.S. of stem of *Lagerstroemia flos-reginae*

Woodfordia floribunda

The results of screening this plant is also presented in Table 1. Stem was found to be devoid of flavonoids while the leaves were found to possess quercetin and 3'-OMe quercetin. Except for gallic acid, both stem and leaves contain the same phenolic acids such as vanillic, syringic and *p*-hydroxybenzoic acid. Lawsone and steroids were seen in both the parts. Alkaloids were absent.

Table 1. Distribution of phytochemicals in the leaves and stems of *Lagerstroemia flos-reginae* and *Woodfordia floribunda*

S. No	Name of the plant	1	2	3	4	5	6	7	8	9	10	11	12
1.	*Lagerstroemia flos-reginae* (leaf)	+	-	+	+	+	+	+	+	-	-	+	+
	(stem)	+	-	+	+	+	+	+	+	-	-	+	+
2.	*Woodfordia floribunda* (leaf)	+	+	-	+	+	+	-	+	-	-	+	+
	(stem)	-	-	-	+	+	-	-	+	-	-	+	+

(1) Quercetin, (2) 3'OMe quercetin, (3) 4'OMe quercetin, (4) Syringic acid, (5) Vanillic acid, (6) Gallic acid, (7) *p*-Coumaric acid, (8) *p*-OH Benzoic acid, (9) Glycoflavones, (10) Alkaloids, (11) Quinones, (12) Steroids.

REFERENCES

Daniel, M. (1991). *Methods in Plant Chemistry and Economic Botany*, Kalyani Publishers, New Delhi.

Harborne, J.B. (1984). *Phytochemical Methods*, (2nd Ed.), Chapman and Hall London.

Jain, S.K. and Taraphdar, C.R. (1970). *Econ. Bot.* 24: 241-278.

Johansen, D.A. (1940). *Plant Microtechnique*, (1st Ed.). McGraw Hill Book Company, New York.

Metcalfe, C.P. and L. Chalk (1950). *Anatomy of the Dicotyledons*, **1:** 649-655, Oxford at the Clarendon Press.

Olowokudejo, J.D. (1990). *Phytomorphology* 40 : 407-422.

13

STUDIES ON THE PHARMACOGNOSY AND PHYTOCHEMISTRY OF ONE OF THE BALAS, *SIDA VERONICAEFOLIA* LAM.

Anamika Kumari and M. Daniel

Phytochemistry laboratory, Department of Botany,
The M.S University of Baroda, Vadodara - 390 002, India.

ABSTRACT

The root, stem and leaves of *Sida veronicaefolia* Lam., one of the plants used as **Bala** an ayurvedic medicine, has been studied for their phytochemical and pharmacognostic biomarkers. Alkaloids and phenolic acids like gallic acid were seen in leaves while stem contains quinones and no gallic acid. Roots contained ephedrine and other phenolic acids of stems. Pharmacognostically, the roots were peculiar in having large parenchymatous cells within the xylem and also in xylem rays. The leaves possessed 2-layered palisade, unicellular non-glandular trichome; one celled glandular trichome, poorly branched stellate hairs and loosely arranged spongy tissue.

INTRODUCTION

Sida veronicaefolia Lam. (Syn. *S. humilis* Cav., *S. cordata* Borss.) is one of the plants, the roots of which are used as **Bala** in Ayurvedic medicines. Bala is a very important rasayana drug known for its demulcent, aphrodisiac and tonic properties. It is also useful in neurological disorders, general debility, urinogenital diseases, tuberculosis and diabetes. A number of species of *Sida, Abutilon, Urena, Pavonia* and *Grewia* are used as Bala in various parts of the country. *S. veronicaefolia* is designated as Nagabala

(Bhumibala) while *S. cordifolia* and *S. acuta* are the accepted Bala. *S. spinosa* is Atibala (Nagabala) and *S. rhombifolia* is the Mahabala (Sahadevi).

Though the acclaimed members of "Bala" group are four to five, the survey of literature shows more than fifteen plants are used thus. In the absence of biomarkers (both the phytochemical and pharmacognostic) it is almost impossible to identify any of these plants in its drug form in commercial samples. Only roots of very few plants such as *S. acuta, S. cordifolia, S. rhombifolia, Abutilon indicum* have been studied for their phytochemistry (that too very few reports). Anatomy of the roots also is very poorly studied. The data available pertains to *S. acuta, S. cordifolia, S. rhombifolia* and *S. veronicaefolia* only, but a detailed phytochemical or pharmacognostic study was carried out only on *S. cordifolia* (Handa *et al.*, 1999). The studies conducted on these plants so far were concentrated on the roots which are the official drug Bala, but it is seen that almost all parts of the plant have been used in medicines for treating various ailments. None of the other parts (other than roots) of any plant has been screened for their active principles or for pharmacognostic data. This being a large vacuum, an attempt is made here to study the phytochemistry and pharmacognosy of all the parts of *S. veronicaefolia,* so that the biomarkers and active principles of all the parts are identified and classified. Absolutely no work is carried out on the chemistry of this plant and pharmacognosy is known only by the T.S of root (Iyer and Kolammal, 1960).

S. veronicaefolia is a trailing hispid procumbent herb with roots sometimes found at the nodes. Leaves are cordate to ovate. Flowers are pale yellow axillary, solitary or in pairs or in small cymes. Fruit is a schizocarp enclosed within the calyx. The whole plant is said to have the medicinal properties attributed to Bala. Leaves are eaten as a pot-herb by Santals. A poultice of them is applied to cuts and bruises. They are reported to be given in diarrhoea in pregnant women. The plant enters into a compound given to cattle suffering from the rinderpest disease.

MATERIALS AND METHODS

The plant materials were collected from the Science faculty campus of The M.S University, Baroda. The voucher specimen of the plant is deposited in Herbarium, Department of Botany (BARO), M.S University of Baroda. Phytochemical analysis of roots, stem and leaves of all the plants for their secondary metabolites such a alkaloids, flavonoids, steroids and other compounds was done by standard methods of chromatography and spectrophotometry (Harborne, 1984; Daniel, 1991). Standard methods (Wallis, 1953) were followed for the pharmacognostic studies (including micromorphological studies).

RESULTS AND DISCUSSION

1. Phytochemistry

The distribution of various phytochemicals in different parts of *S. veronicaefolia* is presented in Table 1. Roots were found to contain alkaloids, flavonoids, phenolic acids and mucilage. The alkaloid present was ephedrine (in fairly large amounts). The flavonoids were in traces. Melilotic, *cis*-ferulic, *trans*-ferulic, vanillic and syringic acids were the phenolic acids identified.

Table 1. The distribution of phytochemicals in different parts of *Sida veronicaefolia*

Sr. No.	Plant part	1	2	3	4	5	6	7	8
1.	Leaves	_	+	+	+	+	+	+	_
2.	Stem	_	+	+	+	+	_	+	+
3.	Roots	+	_	+	+	+	_	+	_

1. Ephedrine, 2. Other alkaloids, 3.Vanillic acid, 4. Syringic acid, 5. *cis* and *trans* Ferulic acid, 6. Gallic acid, 7. Melilotic acid, 8. Quinones

Almost all the compounds found in roots, except for ephedrine, were present in stem also. The flavonoids and phenolics were found to be the same as that of root, but quinones (two spots) were the new compounds located here. There was one alkaloid in stem (which is yet to be identified). In leaves gallic acid was isolated alongwith the other phenolic acids located from roots and stem. The alkaloids present in leaves were two in number, one is the same as that of stem, which are yet to be identified.

2. Pharmacognosy

In the **T.S of the leaf** (Fig. 1), midrib portion is characterized by a conical bulge on the upper side and a collenchymatous hypodermis on both upper and lower regions. The vascular bundles were collateral with a sclerenchymatous (single layered) patch on the lower side and very few scattered fibers on the upper side. The mesophyll consisted of palisade and spongy tissues. The palisade was double layered and was finely packed with chloroplasts. The spongy tissues contained loosely arranged parenchyma with large intercellular spaces. Both in the lower and upper epidermis there were unicellular elongated trichomes and glandular (unicellular) trichomes. Stomata are seen only in the lower epidermis. The powder of leaves exhibited unicellular glandular and non-glandular trichomes, double layered palisade, spongy parenchyma, phloem ray cells containing sphaeraphides, xylem parenchyma cells also with sphaeraphides and three armed stellate trichomes (Fig. 2).

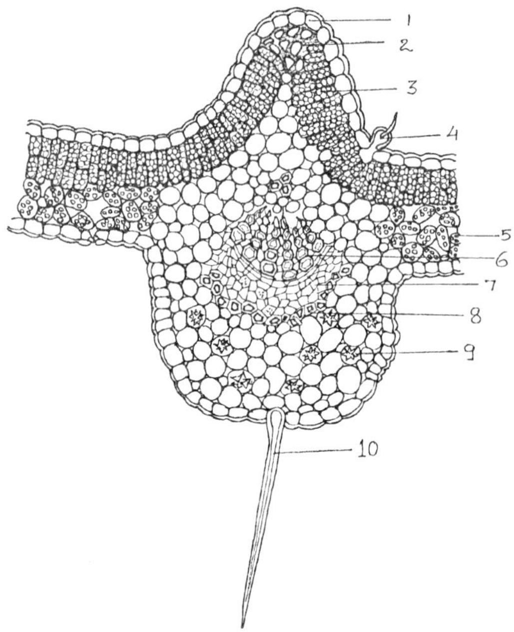

Fig. 1. T.S. of leaf: *Sida veronicaefolia*

1. Epidermis, 2. Collenchyma, 3. Palisade tissue, 4. Glandular trichome, 5. Spongy cells, 6. Xylem, 7. Phloem, 8. Sclerenchyma, 9. Sphaeraphides, 10. Unicellular trichome.

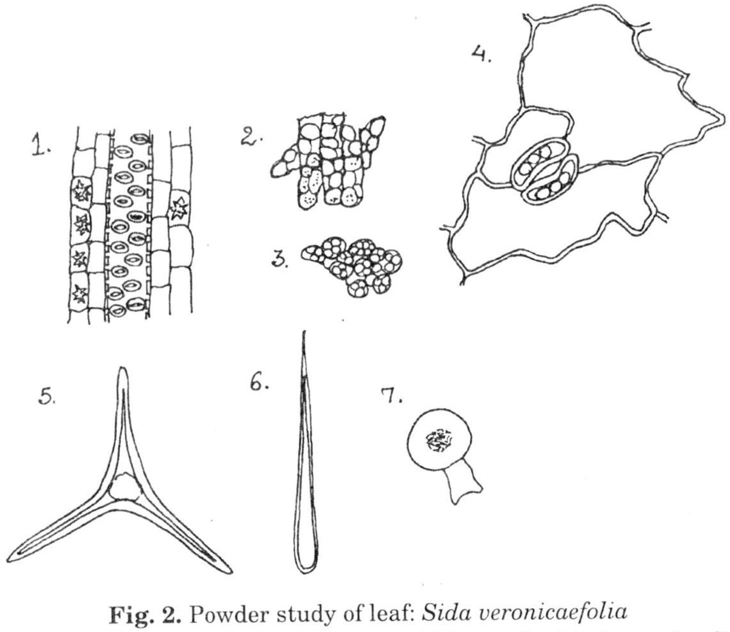

Fig. 2. Powder study of leaf: *Sida veronicaefolia*

1. Xylem parenchymna cells with sphaeraphides, 2. Double layered palisade, 3. Spongy parenchyma, 4. Stomata, 5. Three armed stellate trichomes, 6. Unicellular non-glandular trichomes, 7. Unicellular glandular trichome.

In the **T.S of the stem** (Fig. 3), hypodermis consisted of 2-3 layered chlorenchyma. Cortex contained an outer 2-3 layered collenchyma (very thick walled) layer and an inner1-2 layered parenchyma. Phloem consisted of tangential layers of bast fibers alternating with sieve tubes (stratified phloem). Bast fibers contain both thick walled (small lumen) and thin walled (big lumen) sclerenchyma. The rays were comparatively very frequent.

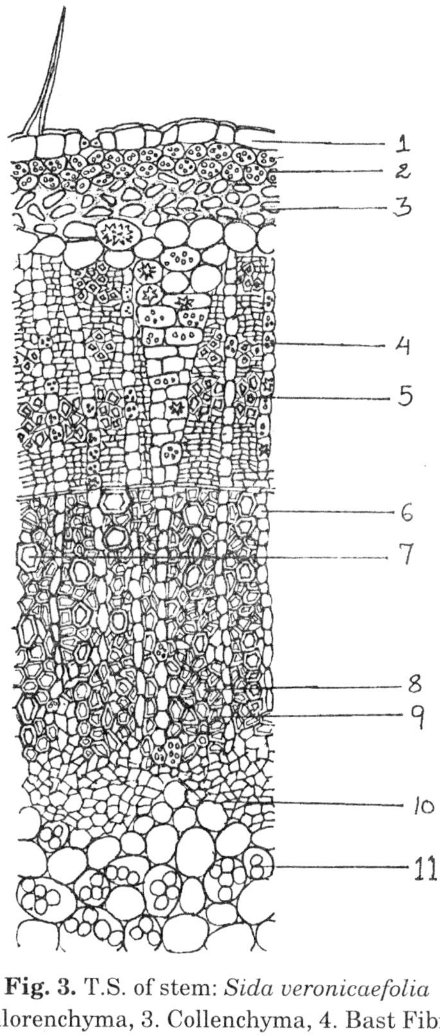

Fig. 3. T.S. of stem: *Sida veronicaefolia*
1. Epidermis, 2. Chlorenchyma, 3. Collenchyma, 4. Bast Fibres, 5. Phloem ray, 6. Xylem ray, 7. Metaxylem, 8. Protoxylem, 9. Tracheids, 10. Xylem parenchyma, 11. Pith cells with starch grains.

The structural features found in both **R.L.S** (Fig. 4) and **T.L.S** (Fig. 5) of stem were more or less similar to those of root. The phloem rays were rich in calcium oxalate containing cells and xylem containing both sclerified and parenchymatous cells. The sclerified rays were biseriate and contain starch grains, while the parenchymatous rays were large and rectangular containing abundant starch grains. The powder of stem revealed xylem parenchyma cells as a line adjoining the vessels, cork cells alongwith parenchyma cells containing sphaeraphides, phloem fibers, xylem vessels with bordered pits, rectangular thick walled pitted xylem ray cells containing starch grains, unicellular glandular and non-glandular trichomes, three armed stellate hairs, angular starch grains, large rectangular crystals of calcium oxalate (Fig. 6).

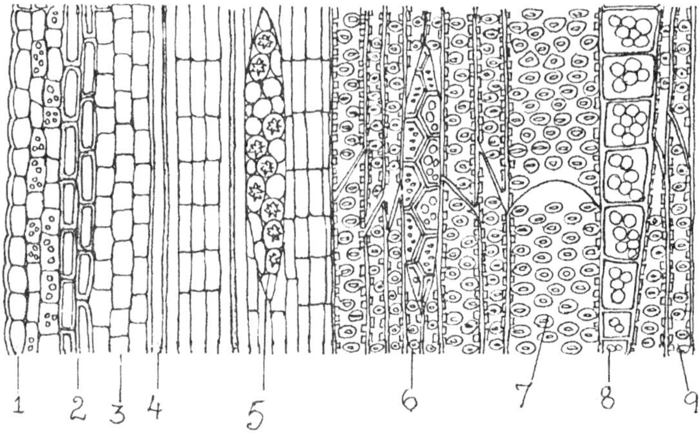

Fig. 4. T.L.S. of stem: *Sida veronicaefolia*
1. Epidermis, 2. Collenchyma, 3. Cortical parenchyma, 4. Bast fibres, 5. Phloem ray with Ca-oxalate crystals, 6. Biseriate medullary rays with thick walls containing starch grains, 7. Vessel, 8. Uniseriate parenchymatous xylem ray, 9. Protoxylem-Annular thickening.

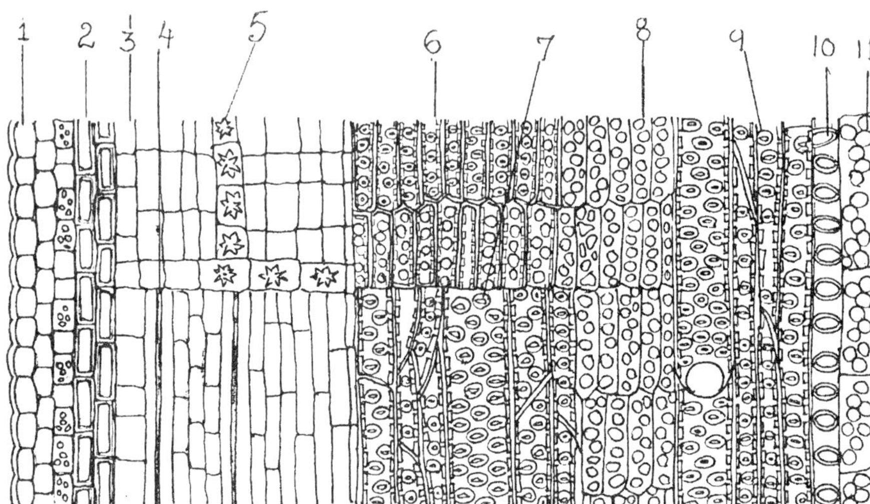

Fig. 5. R.L.S. of stem: *Sida veronicaefolia*
1. Epidermis, 2. Collenchyma, 3. Cortical parenchyma, 4. Bast fibres, 5. Phloem ray with Ca-oxalate crystals, 6. Biseriate medullary rays with thick walls containing starch grains, 7. Vessel, 8. Uniseriate parenchymatous xylem ray, 9. Protoxylem-Annular thickening, 10. Pith cells containing starch grains.

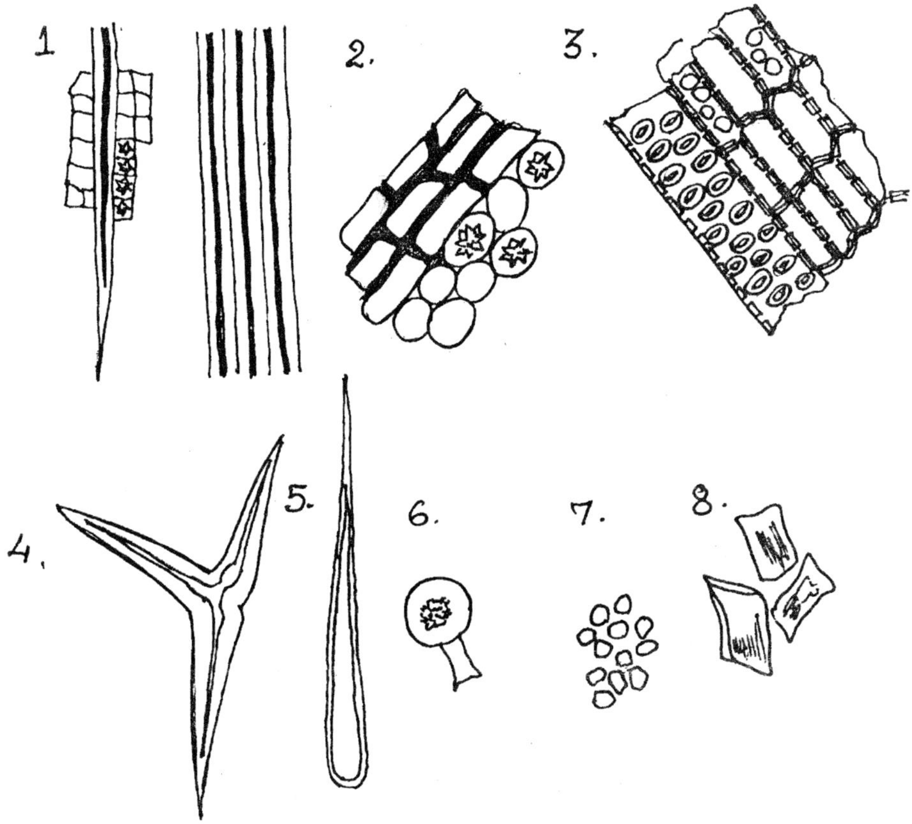

Fig. 6. Powder study of stem: *Sida veronicaefolia*
1. Phloem fibres with adjoining parenchyma with sphaeraphides, 2. Cork cells alongwith parenchyma cells containing sphaeraphides, 3. Xylem vessels with bordered pits with adjoining rectangular thick walled pitted xylem ray cells containing starch grains, 4. Three armed stellate hairs, 5. Unicellular non-glandular trichome, 6. Unicellular glandular trichome, 7. Angular starch grains, 8. Large rectangular crystals of calcium oxalate.

In the **T.S of root** (Fig. 7), the phellem or cork zone was wavy, very thin and consisted of 4-6 rows of narrow tangentially elongated cells, of which the outermost one or two rows of cells were light brown in colour. Cortex is of 10-12 layers of isodiametric parenchymatous cells, many of them containing sphaeraphides. Phloem was stratified with groups of bast fibers .The outer layers of bast fibers were comparatively small, having smaller lumen and the inner layers are larger in size possessing bigger lumens. Phloem rays were found to contain both sphaeraphides and starch grains. Secondary xylem was found to contain large patches of parenchyma mostly as tangential layers. These cells were fairly large, 3-4 times larger than the nearby tracheids and contain abundant starch grains. Adjacent to this parenchyma on either side were patches of sclerenchyma (fibers) which were slightly bigger than the tracheids. Vessels were comparatively few and were comparatively smaller in size. Xylem rays were two layers thick and contained starch gains only. Primary Xylem was tetrarch and contained large clusters of sclerenchyma alternating with the xylem groups.

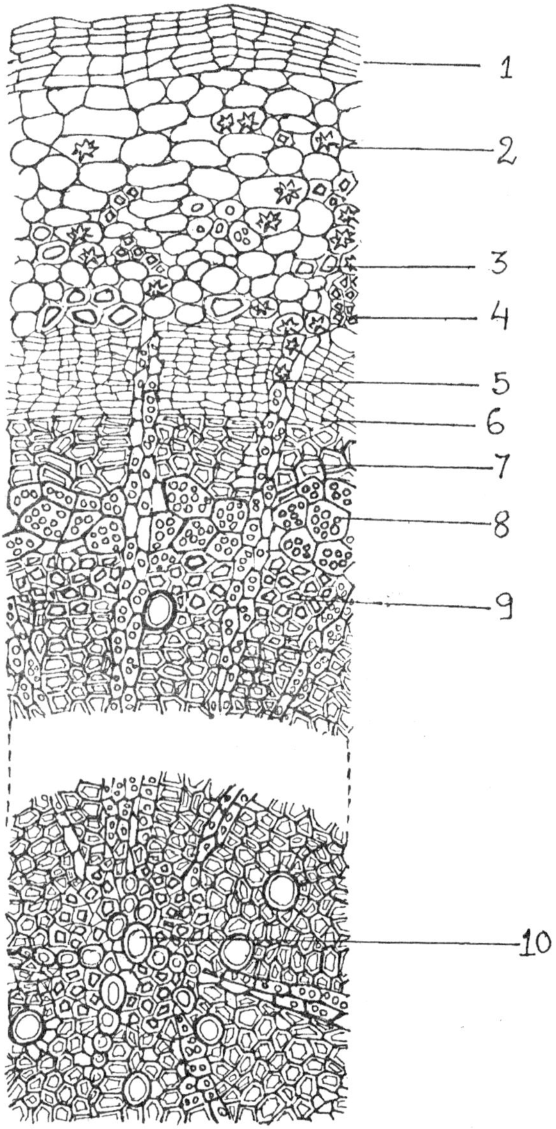

Fig. 7. T.S. of root: *Sida veronicaefolia*

1. Cork, 2. Cortex containing sphaeraphides, 3. Bast fibres with large lumen, 4. Bast fibres with smaller lumen, 5. Phloem rays containing sphaeraphides, 6. Phloem, 7. Trachied, 8. Patches of Xylem parenchyma, 9. Fibres, 10. Protoxylem.

From the **T.L.S** of the root (Fig. 8), it is clearly evident that phloem rays consist of vertical rows of sphaeraphides-containing cells. Xylem rays were of two types - (1) Thick walled biseriate rays. (2) Thin walled multiseriate rays.

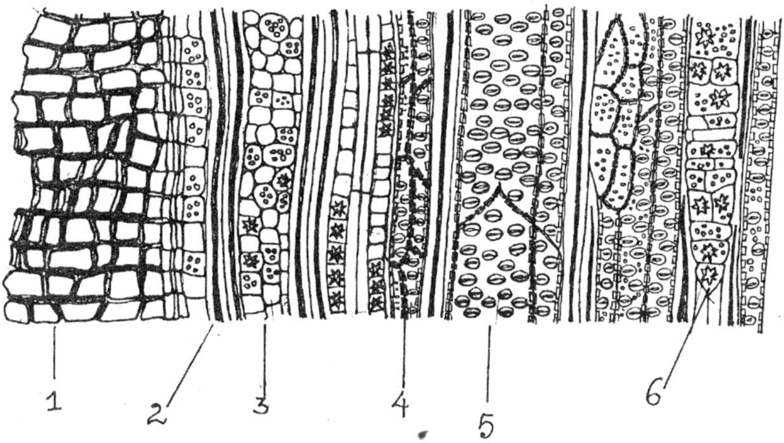

Fig. 8. T.L.S. of root: *Sida veronicaefolia*
1. Cork, 2. Bast fiber, 3. Phloem rays with starch grains and sphaeraphides, 4. Thick walled xylem rays, 5. Vessels, 6. Thin walled xylem rays.

The former type contained ovate or elongated cells with bordered pits and having starch grains or sphaeraphides. These cells appear rectangular in **R.L.S.** (Fig. 9). The later type of rays consist of multiseriate, 3-4 layers thick, very large and distinctive parenchymatous cells with abundant starch grains. The sclerenchyma present in xylem are heavily thickened. The vessels are also bordered pitted with 4-5 rows of bordered pits in broader vessels. The cross walls of vessels were oblique. Tracheids also contain bordered pits. The powder of the root is distinguished by thin-walled xylem parenchyma (adjoining vessels), thick walled pitted xylem rays, cork cells, phloem ray cells containing sphaeraphides (Fig. 10).

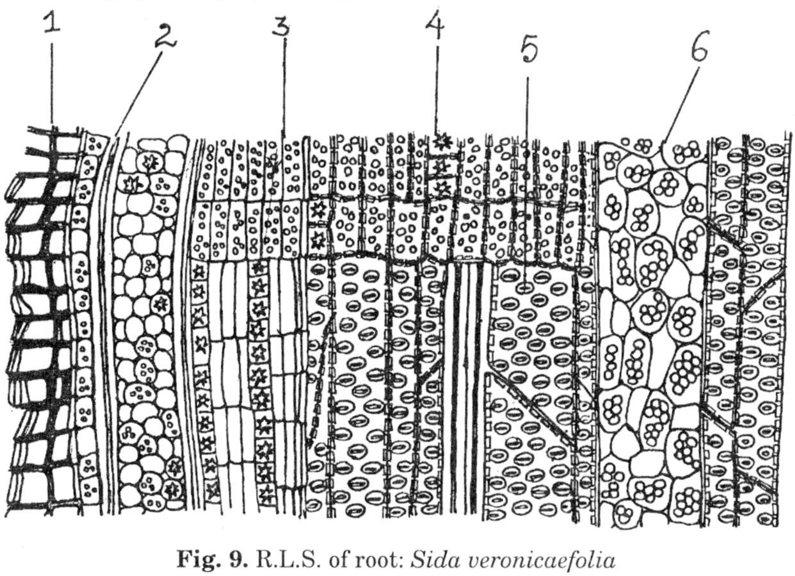

Fig. 9. R.L.S. of root: *Sida veronicaefolia*
1. Cork, 2. Bast fiber, 3. Phloem rays with starch grains and sphaeraphides, 4. Thick walled xylem rays, 5. Vessels, 6. Thin walled xylem rays.

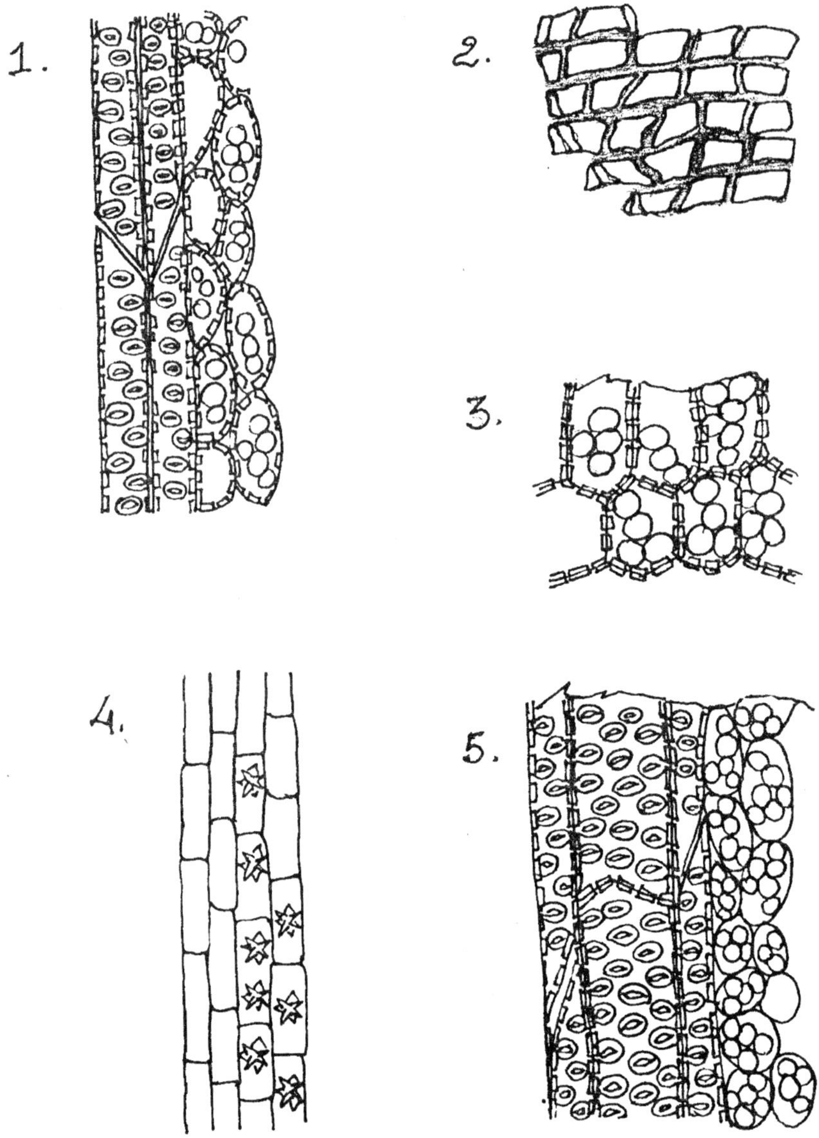

Fig. 10. Powder study: root of *Sida veronicaefolia*
1. Thick walled pitted xylem rays adjoining tracheids, 2. Cork cells, 3. Xylem ray, 4. Phloem ray cells containing sphaeraphides, 5. Thin-walled xylem parenchyma adjoining vessels.

CONCLUSIONS

Sida veronicaefolia is peculiar in having a little or no flavonoids, ephedrine only in roots, gallic acid in the leaves and quinones in the stem. Pharmacognostically the leaves exhibit poorly branched stellate hairs (branches mostly 2-3), presence of unicellular trichomes (both elongated, non granular and spherical glandular one celled) and double layered palisade. Both stem and roots possess distinct large xylem parenchymatous cells and abundant xylem fibres.

REFERENCES

Daniel, M., (1991). *Methods in Plant Chemistry and Economic Botany*, Kalyani Publishers, Delhi.

Handa, S.S. *et al.* (1999). *Indian Herbal Pharmacopoeia,* RRL, Jammu & IDMA, Mumbai, **2 :** 129.

Harborne, J.B. (1984). *Phytochemical Methods,* 2nd ed. Chapman & Hall London.

Iyer, N.K and M. Kolammal (1960). *Pharmacognosy of Ayurvedic Drugs,* Univ. Kerala, Trivandrum, **1(5):** 100.

Wallis, T.E. (1953). *Practical Pharmocognosy*, J. & A. Churchill Ltd., Gloucester Place, W.I. pp. 187.

14

PHARMACOGNOSTIC AND PHYTOCHEMICAL STUDIES ON *EUPHORBIA THYMIFOLIA* LINN AND *EUPHORBIA HIRTA* LINN.

Sallykutty Thomas and M. Daniel

Phytochemistry Laboratory, Department of Botany,
The Maharaja Sayajirao University of Baroda, Vadodara - 390 002.

ABSTRACT

Pharmacognostic and phytochemical studies on *E. thymifolia,* a well-known ayurvedic drug **Dugdhika,** is carried out and the results were compared with that of *E. hirta*. Leaves, stem and roots were subjected to detailed pharmacognostic and phytochemical studies. Leaves showed non-articulate laticifers ramified into the mesophyll cells and vascular bundles surrounded by a single layered parenchymatous bundle sheath. Stomata were predominantly anomocytic in both the species; anomalous types such as contiguous stomata and stomata with cytoplasmic connections were recorded only on *E. thymifolia*. Four types of trichomes were noticed on the leaf lamina and margins *viz:* - uniseriate warty covering long and short hair, uniseriate curved and multiseriate capitate hair. The histochemical studies were conducted by standard methods. Starch grains, alkaloids, fats, tannins, proteins and saponins were located in sections. Powder analysis showed starch grains, fragments of trichomes, tracheids and fibres. *E. thymifolia* was found to contain both flavones and flavonols while only flavonols were located in *E. hirta*. *trans-* Ferulic and *O*-coumaric acids were present only in the latter plant.

INTRODUCTION

The drug **Dugdhika** is considered acrid, bitter, hot in action, diuretic and aphrodisiac. It purifies blood, cures skin diseases, poisonous affections, cough asthma and other respiratory disorders. The plant juice is a good galactogogue. The whole plant is medicinal. Many authors, including Sivarajan and Balachandran (1994), recognize two different varieties of this drug; the smaller variety and a bigger variety, the former one is identified as *Euphorbia thymifolia* and the latter one as *E. hirta*. In ayurvedic texts of North India, *E. thymifolia (Chamaesyce thymifolia* Millsp.) is considered as the actual source of *dugdhika*, but in some parts of India *E. hirta* is used as the drug. The former plant was found to be effective in the treatment of patients of bronchial asthma (Sharma and Sharma, 1971; 1972; Sharma *et al.*, 1982). The extract and pure compounds such as rugosin B isolated from the plant were reported to possess antioxidant and antiviral activities (Lin *et al.,* 2002). The extract of the plant showed antibacterial and antifungal activities (Khan, *et al.* 1988; Satyanarayana and Kawdikar, 1978; Kawdikar *et al.*, 1976). Lee *et al.* (1990) reported hydrolysable tannins from *E. thymifolia*. The leaves and stem were found to contain 5, 7, 4'-trihydroxy flavone-7glycoside, 12-deoxyphorbol, 4β-hydroxy phorbol and alkaloids.

E. hirta (Syn. *Chamaesyce hirta* Millsp, *Euphorbia pilulifera* Auct.non.L, *Euphorbia capitata* Lam.) also is used in traditional medicine for the treatment of several ailments like diarrhoea, dysentery, intestinal parasitosis, asthma, bronchitis, hay fever and conjunctivitis. Besides these, hypotensive, tonic and hypoglycemic activities also are reported (Anon, 2002). The aqueous extract of the herb exhibits sedative, analgesic, antipyretic and anti-inflammatory activities (Lanhers *et al.*, 1991). A flavonoid quercetin, with antidiarrhoeial activity, was isolated from this plant (Galvez *et al.*, 1993). The herb is well known for its anticancer property. Leaves are reported to have antifungal and antibacterial activities. The root exudates exhibited nematicidal activity. No detailed study has been reported on the pharmacognosy and phytochemistry of these two plants. Therefore the detailed pharmacognostic and phytochemical characters have been conducted for both *E. thymifolia* and *E. hirta* that will be helpful to identify the former from the latter.

MATERIALS AND METHODS

The plant materials for the present study were collected from different localities of the campus of the M.S. University of Baroda Gujarat. The voucher specimens of both these plants are submitted in BARO, the herbarium of Department of Botany, The Maharaja Sayajirao University of Baroda, Vadodara. Plant materials were washed, shade dried and completely dried by keeping in an oven at 60°C. The dried materials were powdered and stored in airtight plastic bags. This powder was used for the analysis of all the chemical constituents. Micromorphological studies were carried out on fresh materials. Epidermal peels were taken and stained with Toluidine blue (0.5%) and mounted in glycerine. Leaf constants such as stomatal index, stomatal frequency and size of guard cell, vein islet number, vein termination number and palisade ratio were calculated. Transections of leaf, stem and root were taken by free hand sections, and were stained in Toluidine blue and mounted in glycerine (Johansen, 1940).

For the study of vessel elements, fragments of stem and root were macerated using a mixture of nitric acid (10%) and potassium dichromate (10%) solution in equal proportions. The vessel elements were stained with Toluidine blue and mounted in glycerin and the margins of the cover slips were sealed with DPX. The slides were examined under the microscope and Camera Lucida sketches were drawn and the cells measured using an ocular micrometer. The quantitative data are based on the average of 20 readings. Histochemical tests were conducted on fresh materials and were located in sections according to Johansen (1940); Gibbs (1974) and Jensen (1962). The finely powdered drug was scanned under 400x magnifications for recording the cell elements. For Phytochemical analysis dried powdered materials were analysed for flavonoids, phenolic acids, steroids, alkaloids and proanthocyanidins using standard procedures (Mabry *et al.*, 1970; Harborne, 1984; Daniel, 1991). The identification of compounds were done by colour reactions and spectral data and confirmed by co-chromatography with authentic samples.

RESULTS

a. Pharmacognostic characters

Stomatal complex: Leaves were amphistomatic bearing anomocytic stomata (Fig. 2). Anomalous types such as contiguous stomata and stomata with cytoplasmic connections (Fig. 1a-b) were also seen in *E. thymifolia*. Leaf constants such as stomatal index/mm^2, stomatal frequency/mm^2, size of guard cell μm, vein islet number/mm^2, vein termination number/mm^2 and palisade ratio were enumerated and are presented in Table 1 and 2. Stomatal index in both the surfaces (26.8/mm^2, 31/mm^2) vein islet number (10.7/mm^2) and vein termination number (15.7/mm^2)) were highest in this plant.

Table 1. Leaf Constants

		Adaxial epidermis					
Sr. No.	Name of Plant	S.I/ mm^2	S. F/ mm^2	S.G.C (μ m) L B	V. I. N/ mm^2	V. T. N/ mm^2	P. R.
1.	*E. thymifolia*	26.8	10.8	13.3 12.3	10.7	15.7	6.8
2.	*E. hirta*	17.6	11.2	17.5 11.6	3.6	7.9	8.4

Table 2. Leaf Constants

		Abaxial epidermis		
Sr. No.	Name of Plant	S. I/ mm^2	S. F/ mm^2	S. G. C (μ m) L B
1.	*E. thymifolia*	31	10.2	14 15.8
2.	*E. hirta*	23.2	25.4	21 13.1

S.I.-Stomatal Index; S.F.-Stomatal Frequency; S.G.C.-Size of Guard Cell; V.I.N - Vein Islet Number; V.T.N. - Vein Termination Number; P.R. - Palisade Ratio; L - Length; B - Breadth.

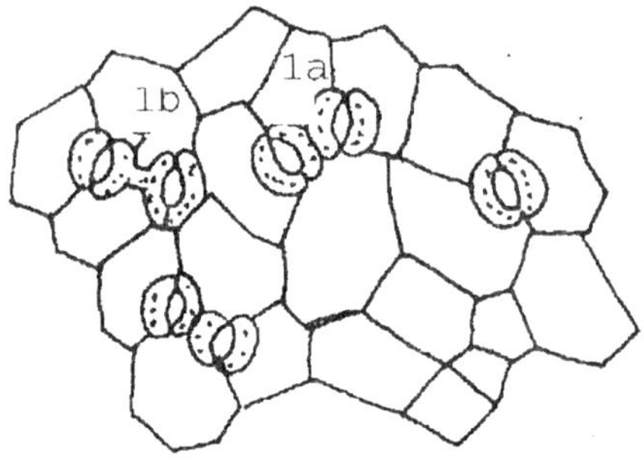

Fig. 1a-b. Surface view of adaxial epidermis of *E. thymifolia* leaf lamina 1a. Contiguous stomata, 1b. Stomata with cytoplasmic connection

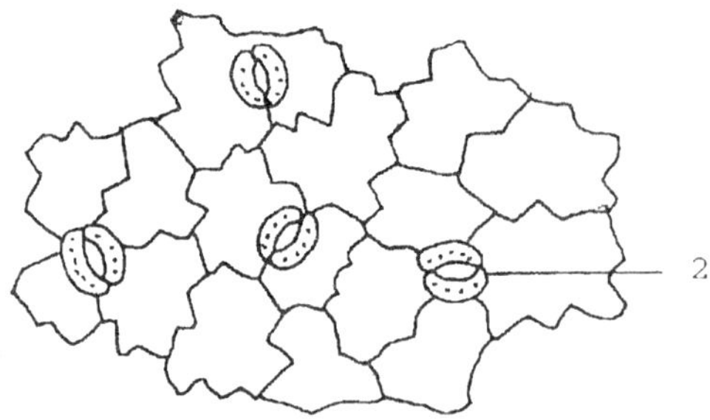

Fig. 2. Surface view of abaxial epidermis of *E. thymifolia* leaf lamina (Anomocytic stomata).

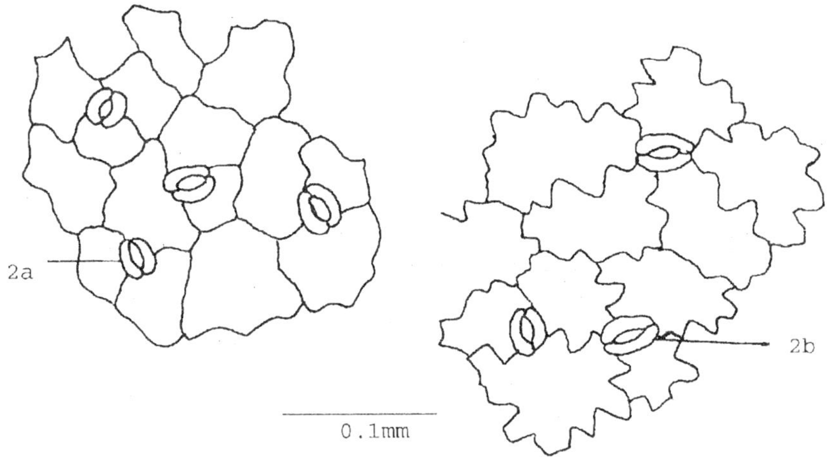

Fig 2a. Surface view of adaxial epidermis of *E. hirta* leaf lamina (anisocytic stomata)
Fig 2b. Surface view of abaxial epidermis of *E. hirta* leaf lamina (anomocytic stomata)

***Trichome complex* (Fig. 3a)**

1. Uniseriate warty covering short and long trichomes 3-4 cells thick with blunt end were confined to the abaxial and marginal surfaces of the leaves.
2. Multicellular multiseriate trichomes were found in the margins of the leaves.

Anatomy (Fig. 5a-5b). The epidermis of the leaf consisted of a single layer of isodiametric parenchyma cells with thick cuticle. The abaxial epidermal cells were smaller and thin walled. The mesophyll was differentiated into single layered palisade and few layered spongy cells. Non-articulate branched laticifers were found ramified into the mesophyll cells. Leaves were found to possess simple and compound starch grains. Leaves possessed a sheath of large rectangular cells surrounding each vascular bundle (Fig. 4). T.S through midrib showed an arc of median vascular bundle and few collenchyma on the abaxial side.

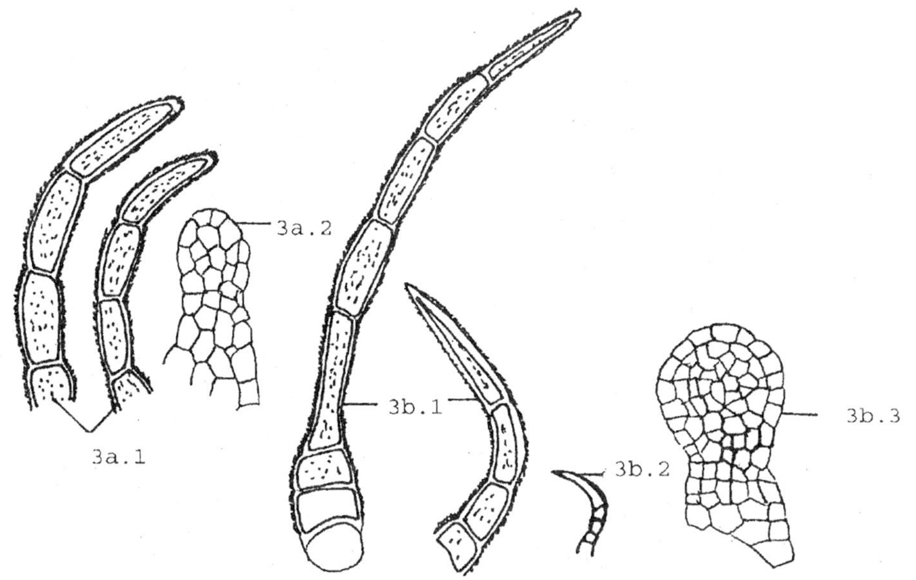

Fig. 3a. Trichome complex of *E. thymifolia*; **3a.1.**Warty covering trichome; **3a.2.** Multiseriate multicellular hair.
Fig. 3b. Trichome complex of *E. hirta*; **3b.1.** Warty covering trichome with pointed end, **3b.2.** Short curved hair **3b.3.** capitate hair.

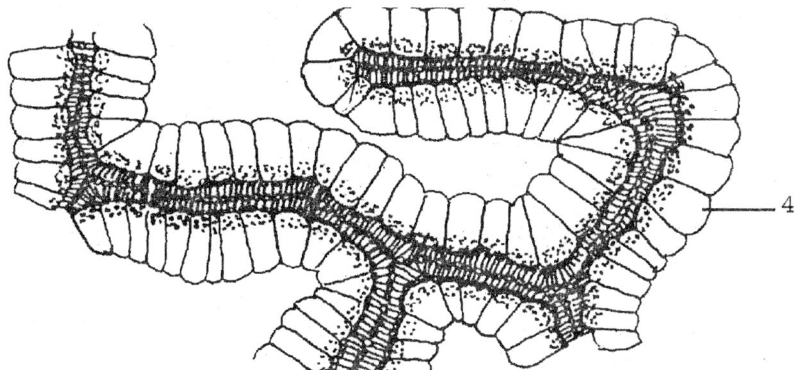

Fig. 4. Sheath of large rectangular cell surrounding the vein

T.S. of stem (Fig. 6a-6b) was circular in outline. Cells of the epidermis were barrel shaped with thick cuticle and were filled with dark pink contents. Cortex was differentiated into thick walled parénchyma and chlorenchyma. Simple and semi compound starch grains and latex cells were found in the cortex and pith region. Groups of radially arranged fibres were also found in the cortex. Xylem composed of radially arranged small xylem groups or solitary associated with fibres. Medullary rays were biseriate. Phloem cells were narrow parenchymatous and contained latex canals. Pith consisted of parenchyma cells with intercellular spaces.

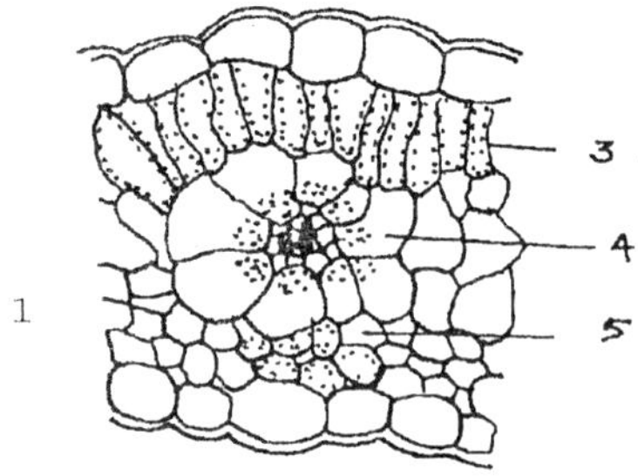

1. A portion of the T.S. of leaf lamina

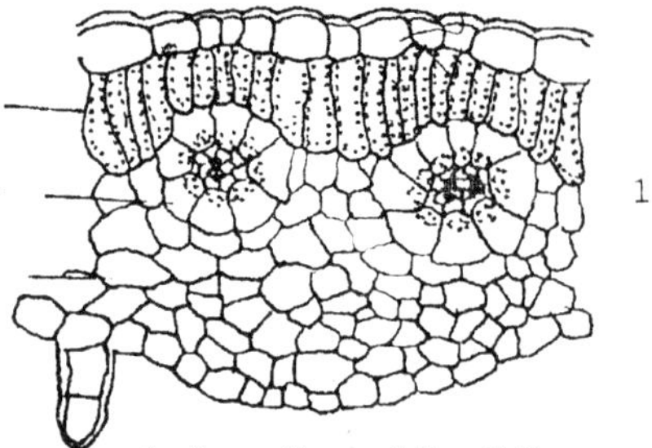

1. A portion of the T.S. of leaf lamina

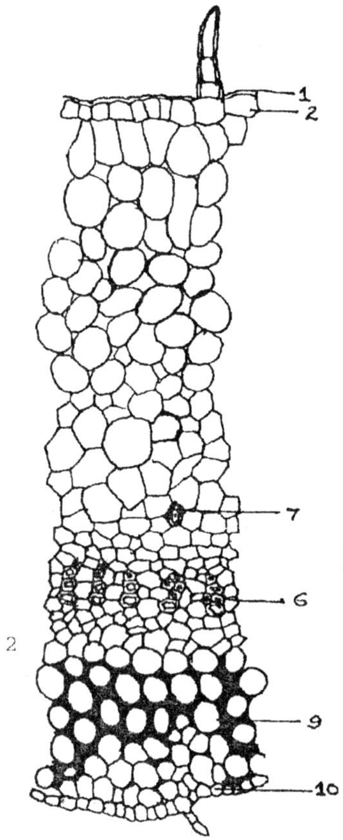

Fig. 5a. Histology og E. thymifolia leaf
2. A portion of the T.S. of midrib

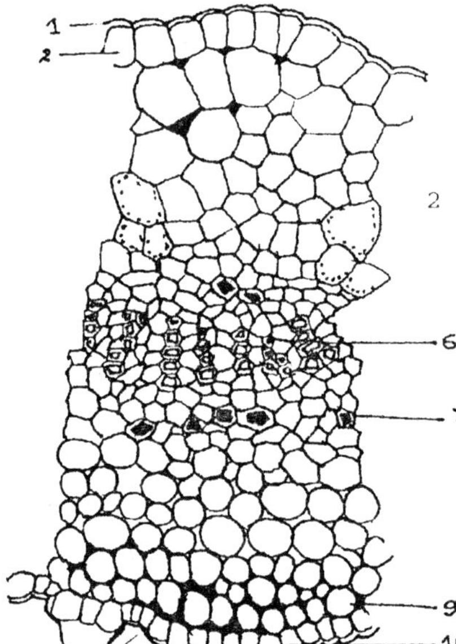

Fig. 5b. Histology E. hirta leaf
2. A portion of the T.S. of midrik

1. Cuticle 2. Upper epidermis
3. Palisade tissue 4. Bundle sheath like cell 5. Sporty tissue
6. Vascular bundle 7. Laticiferous cell, 8. Starch grain, 9. Collenchyma
10. Lower epidermis

0.1mm

In T.S root (Figs. 7a-7b) was differentiated into cork, cortex and vascular cylinder. Cork composed of tangentially elongated thin walled cells. Some of the cells in the peripheral rows contained dark brown contents. Cortex formed of several rows of thin walled parenchyma cells with very small intercellular spaces traversed by tangentially running laticiferous canals, but those towards the phloem were narrow. Most of these cells were filled with starch grains. Xylem consisted of radially arranged xylem vessels associated with fibres and tracheids; medullary rays were biseriate in continuation with that of phloem. Phloem was very narrow, parenchymatous, traversed by laticiferous canals.

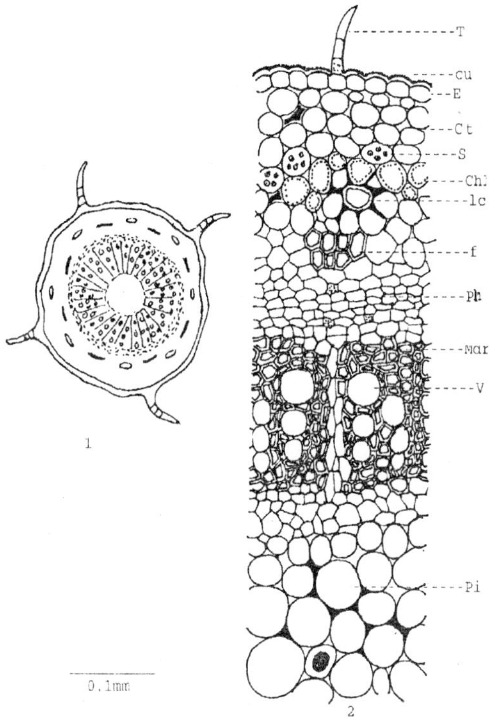

Fig. 6a. Histology of stem of *E. thymifolia* Linn. 1. Diagramatic sketch of the T.S. of the stem, 2. A portion of the T.S. of stem enlarged T. Trichome; C, cuticle; E, epidermis; Ct, cortex; S, starch grain; Chl, chlorenchyma; lc, latex canal, f, fibres; ph, phloem; Mdr, medullary ray; V, vessel, Pi, pith

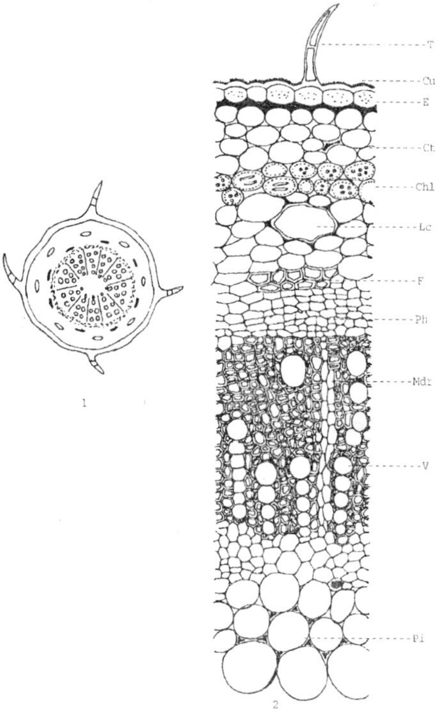

Fig. 6b. Histology of stem of *E. hirta* 1. Diagramatic sketch of the stem; 2. A portion of the T.S. of the stem enlarged T. Trichome; C, cuticle; E, epidermis; Ct, cortex; S, starch grain; Chl, chlorenchyma; lc, latex canal, f, fibres; ph, phloem; Mdr, medullary ray; V, vessel, Pi, pith

When macerated, the vessel elements were of different shapes and sizes. Some of them were cylindrical having peg like outgrowths at the ends. Tracheids were longer than vessels and were having pointed ends. Vessel elements of the stem were long cylindrical, their length ranging from 150-225 μm and 15-30 μm in diameter. They were very small to moderately large. Perforation plate was simple, oval placed at the terminal ends. Simple and scalariform pits were arranged along the lateral walls of the element (Fig. 8a). Vessel elements in root were long cylindrical their length ranging from 165-225 μm and 30-45 μm in diameter. They were very small to moderately large. Perforation plate was simple oval to circular placed at the terminal or sub terminal end of the element. Scalariform and reticulate pits were arranged along the lateral walls (Fig. 8b). Tail was short with pointed end.

Secondary metabolites like starch, protein, fat, tannin, saponin and alkaloid in leaves and stem were looked into for their presence and localization in certain organs or tissues. Starch was located in the cortex, pith and vascular tissues. (Table 3).

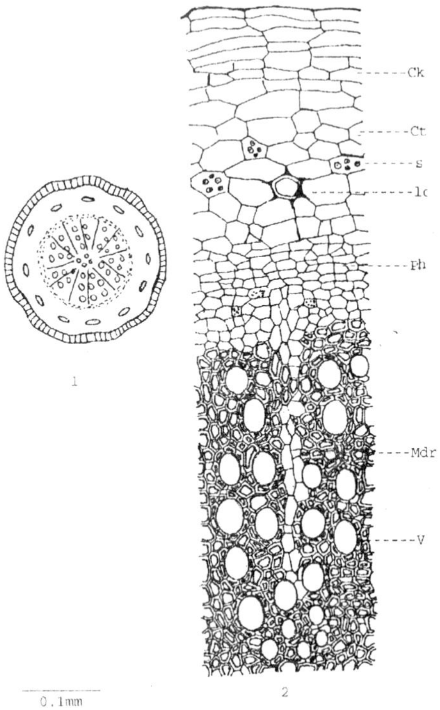

Fig. 7a. Histology of root of *E. thymifolia* Linn.
1. Diagramatic sketch of the T.S. of the root; 2. A portion of the T.S. of root enlarged
Ck, cork, Ct, cortex; S, starch grain; lc, latex canal; Ph, phloem, Mdr, medullary ray; V, vessel

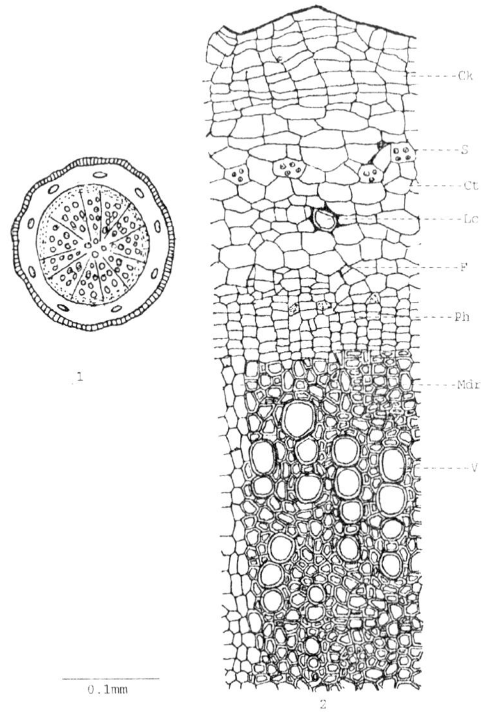

Fig. 7b. Histology of *E. hirta* root
1. Diagramatic sketch of the T.S. of the root, 2. A portion of the T.S. of the root
Ck, cork; S, starch grain; Ct, cortex; Lc, latex cell; F, fibres; Ph, phloem; Mdr, medullary ray; V, vessel

Table 3. Histochemical Tests

S.No.	Test	*E. thymifolia*		*E. hirta*	
		Leaf	Stem	Leaf	Stem
1.	Starch	+	+	+	+
2.	Protein	—	—	+	—
3.	Tannin	+	+	+	+
4.	Saponin	—	—	+	+
5.	Fat	+	+	+	+
6.	Gums & Mucilage	—	—	+	+
7.	Alkaloid	+	+	+	+

Powdered drug of the whole plant contained fragments of trichomes, simple and semi-compound starch grains, latex cells, broken parts of tracheids, fibres and vessel elements with spiral, annular, scalariform and helical thickenings. (Fig. 9a).

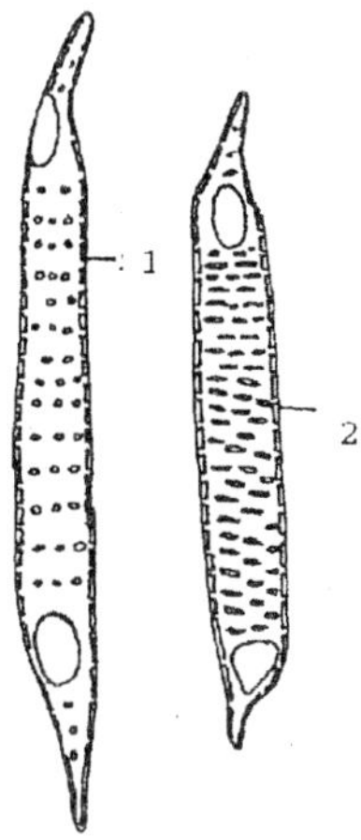

Fig. 8a. Vessel elements of *E. thymifolia* stem 1. simple pit, 2. scalariform pit

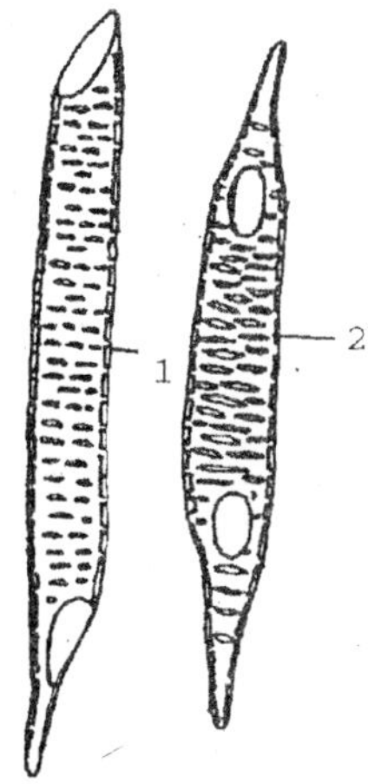

Fig. 8b. Vessel elements of *E. thymifolia* root; 1. scalariform, 2. reticulate

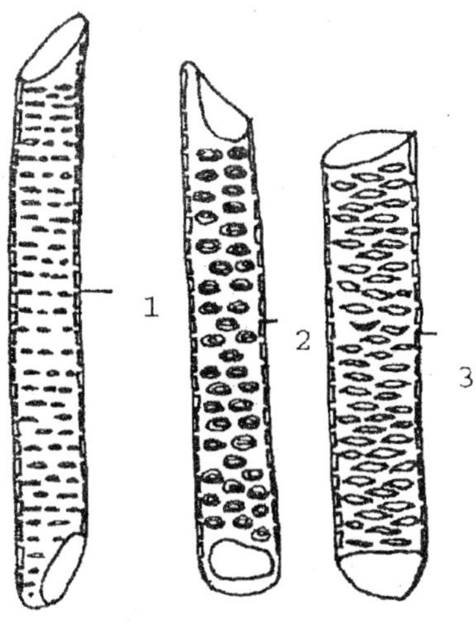

Fig. 8c. Vessel elements of *E. hirta* stem 1. scalariform; 2. bordered pit, 3. reticulate

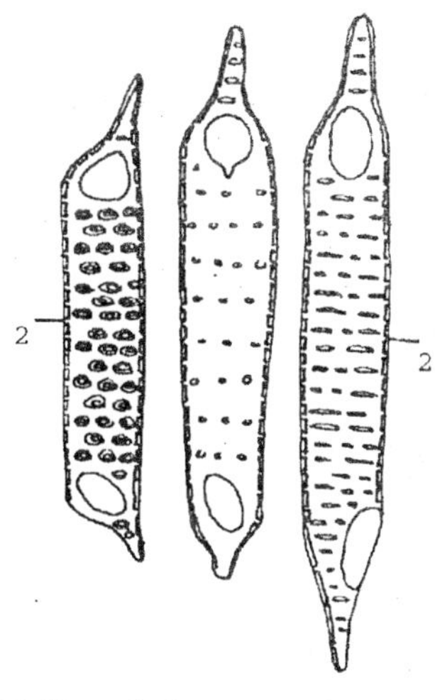

Fig. 8d. Vessel elements of *E. hirta* root

E. hirta was different in having adaxial anisocytic stomata (Fig. 2a) and short curved trichomes (Fig. 3b.2), warty covering trichome 3-8 celled with pointed end (Fig. 3b.1), multicellular multiseriate capitate hair (Fig. 3b.3) and an arc of median vascular bundle (Fig. 5b). Yellowish and short curved hairs and rod shaped starch grains were also found in the stem. Vessel elements were cylindrical; perforation plate was simple and was placed at the terminal ends. In stem the length of the vessel elements ranging from 705 - 750 µm and 30-50 µm in diameter (Fig. 8c). They were very small to extremely large. Perforation plate was simple oval placed at the terminal ends of the element. Scalariform, bordered pits and reticulate thickenings were seen on the lateral walls of the element. End wall was either transverse or oblique. Tail was short. In root, vessel elements were ranging from 150-270μm and 45-60 μm in diameter. They were moderately small to moderately large. Perforation plate was simple circular to oval placed at the terminal end of the element. Simple, bordered and scalariform pits were seen on the lateral walls of the element (Fig. 8d). Tail was short with pointed end.

Powdered drug of the whole plant consisted of fragments of yellowish and short curved trichomes, warty covering trichomes with pointed ends, rod shaped starch grains and vessel elements with round bordered pits (Fig. 9b).

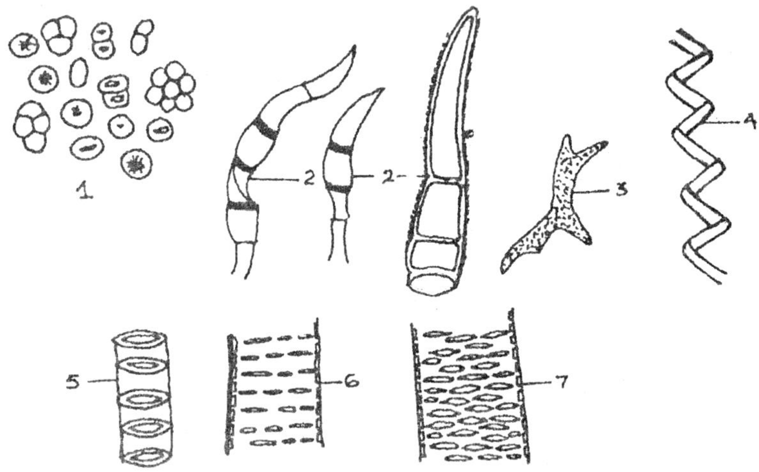

Fig. 9a. Powder characters of *E. thymifolia*

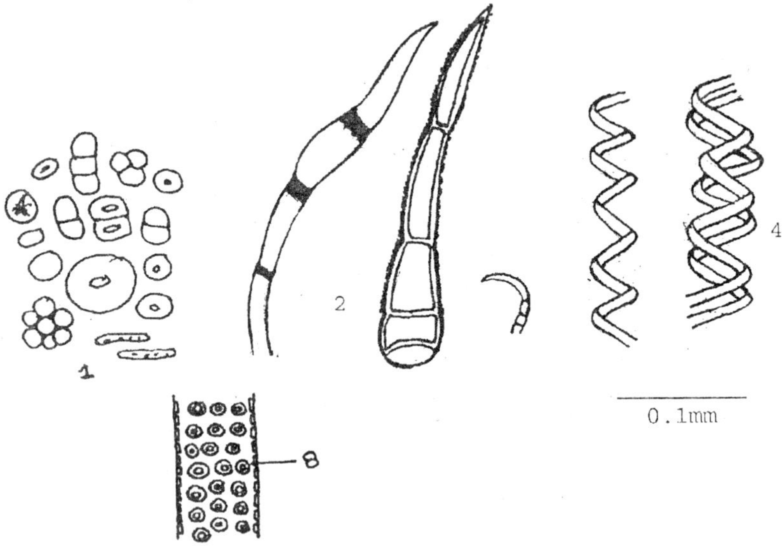

Fig. 9b. Powder characters of *E. hirta*

1. Simple, compound and rodshaped starch grains, 2. trichomes, 3. latex cell, 4. spiral, 5. annular, 6. scalariform, 7. reticulate, 8. bordered pits

Phytochemical analysis

The leaves and stem were then taken up for phytochemical analysis. Flavonoids, phenolic acids, alkaloids, steroids, gallic acid and proanthocyanidins were found to be present. Among the flavonoids located flavones such as apigenin, 3'-OMe luteolin and flavonol like kaempferol along with vanillic, syringic, *p*-OH Benzoic acid, melilotic, *cis*-ferulic, *p*-Coumaric acid and alkaloids were seen in leaves whereas the stem possessed apigenin along with vanillic, syringic, melilotic, *cis*-ferulic acid, alkaloid and steroids.

The leaves of *E. hirta* were found to contain only flavonols Quercetin, 3'-OMe quercetin, 3'-4'-diOMe quercetin, 7'- OMe quercetin, 7'3'-di OMe quercetin, 7'3'4'-tri OMe quercetin and no flavones. Similarly the stem was found to contain 3'-OMe quercetin and 3'-4'-diOMe quercetin. *trans*-Ferulic and *o*-coumaric acids were found in addition to the phenolic acids present in the former plant. Gallic acid and proanthocyanidins were located only in the leaves of both the plants (Table 4).

Table 4. The Distribution of various phytochemicals in leaves / stem of *E. thymifolia* and *E. hirta*

S.N.	Name of Plant	1	2	3	4	5	6	7	8	9	10	11	12	13	14	15	16	17	18	19	20	21
1.	*E. thymifolia* (leaves)	+	+	+	-	-	-	-	-	-	+	+	+	+	+	-	+	-	+	+	+	+
	Stem	+	-	-	-	-	-	-	-	-	+	+	-	+	+	-	-	-	-	+	+	-
2.	*E. hirta* (leaves)	-	-	-	+	+	+	+	+	+	+	+	+	+	+	+	+	+	+	+	+	+
	Stem	-	-	-	-	+	+	-	-	-	+	+	-	+	+	-	-	+	-	+	+	-

(1) Apigenin, (2) 3'- OMe luteolin, (3) Kaempferol, (4) Quercetin, (5) 3'-OMe Quercetin, (6) 3'4' –di OMe Quercetin, (7) 7' – OMe Quercetin, (8) 7'3' – di OMe Quercetin, (9) 7'3'4' – tri OMe Quercetin, (10) Vanillic acid, (11) Syringic acid, (12) *p*-OH Benzoic acid, (13) Melilotic acid, (14) *cis*-ferulic acid, (15) *trans*-ferulic acid, (16) *p*-Coumaric acid, (17) *o*-Coumaric acid, (18) Gallic acid, (19) Alkaloids, (20) Steroids, (21) Proanthocyanidins.

DISCUSSION

The drug *dugdhika* (*E. thymifolia* and *E. hirta*) has been used in ayurvedic medicine for the treatment of bronchial asthma. The study was planned to locate both pharmacognostic and phytochemical markers useful in the identification of both the plants. Based on the detailed pharmacognostic characters, both of the plants were found to be different *i.e.*, *E. thymifolia* is characterised by the absence of adaxial curved trichome, warty covering trichomes with blunt end and presence of contiguous stomata and stomata with cytoplasmic connections. *E. hirta* was characterised by adaxial curved trichome, warty covering trichomes with pointed end, anomocytic stomata, and rod shaped starch grains. Similarly the root anatomy too is different as large intercellular spaces in the cortex were noted in the former one while they were absent in the latter.

The phytochemical investigation revealed that *E. thymifolia* was rich in flavones and flavonols where as *E. hirta* was found to contain only flavonols quercetin and its methoxy derivatives and also possess *trans-* ferulic and *o-* coumaric acids. The present results have shown that the two taxa are distinct from each other and this distinction was more phytochemically than pharmacognostically.

REFERENCES

Anonymous (2002). *Wealth of India: Raw materials,* **III**. C.S.I.R. New Delhi.

Daniel, M. (1991). *Methods in plant chemistry and Economic botany*. Kalyani Publishers, Delhi.

Galvez, J., Zarzuel A., Crespo M.E., Lorente M.D., Oceto M.A. and J. Jimenez (1993). *Planta Med.* **59**: 333.

Gibbs, R.D. (1974). *Chemotaxonomy of Flowering Plants.* Mc Gill Queens Montreal, London.

Harborne, J.B. (1984). *Phytochemical Methods* (2 ed), Chapman and Hall, 104-Gloucester, London.

Jensen, W.A. (1962) *Botanical Histochemistry: Principles and Practice*, Freeman and Company U.S.A.

Johansen, D.A. (1940). *Plant Microtechnique*, Tata Mc Graw-Hill publishing company Ltd. New Delhi.

Kawdikar, S.R., Kazmi, S.M. and V.B.Trivedi (1976). *Bull. Bot. Soc. Univ. Sougar* **24** : 77.

Khan, N.H., Rahman M. and M.S.A. Nur-e Kamal (1988). *Indian. J. Med. Res.* **87**: 395.

Lanhers M.C., Fleurentin J., Dorfman P., Mortier F. and J.M. Pelt (1991). *Planta. Med.* **57** (3): 225.

Lee, T. T, Gen-Ichirononaka and I. Nishioka (1990). *Phytochemistry* **29** (11): 3621.

Lin, C.C., Cheng H.Y, Yang C.M, and T. C. Lin (2002). *J. Biomed. Sci.* **9** : 656.

Mabry, T.J., Markham, K.R. and M.B. Thomas (1970). *The systematic identification of Flavonoids.* Spirnger Verlag, Berlin.

Satyanarayana ,T. and S.R. Kawdikar (1978). *Indian Drugs* **15** : 167.

Sharma, G.P. and P.V. Sharma (1971). *J. Res. Indian. Med.* **6** : 118.

Sharma, G.P. and P.V. Sharma (1972). *J. Res. Indian. Med.* **7** : 24.

Sharma, G.D., Upadhyay, B.N. and S.N. Tripati (1982). *J. Res. Ayur. Siddha.* **3**:109.

Sivarajan, V.V. and Indira Balachandran (1994). *Ayurvedic Drugs and their Plant Sources.* Oxford & IBH Publishing Co. Pvt. Ltd. New Delhi.

15

SILICA BODIES AS BIOMARKERS OF A FEW MEDICINAL PLANTS OF THE POACEAE

Priya, V. Nair and M. Daniel

Department of Botany, The Maharaja Sayajirao University of Baroda, Vadodara, 390 002, Gujarat, India

ABSTRACT

Silica bodies were found to be a characteristic feature of the Poaceae members. In the present study three medicinal herbs, *Cymbopogon citratus, Cynodon dactylon* and *Triticum aestivum* have been screened for their silica bodies. Silica bodies were found to be saddle shaped in *Cynodon,* cross-shaped in *Cymbopogon* and horizontally elongated in *Triticum*. These structures would help the identification of the plant in vegetative state also.

INTRODUCTION

Biomarkers are of paramount importance in the quality control procedures of medicinal plants. The absence of these important parameters leads to adulteration, often purposely and in a few cases unknowingly. Both phytochemical and pharmacognostic biomarkers are equally important, the former more useful in assessing the quality of finished product and the latter, for analysing the raw material. It is heartening to see that data on phytochemical and pharmacognostic analysis of more and more plants are being researched and published of late.

Silica bodies are one group of biomarkers found in a number of monocotyledons. Though silica is the second most abundant element in earth's crest, only a few families such as Poaceae, Arecaceae, Orchidaceae, Commelinaceae, Bromeliaceae, Haemodoraceae etc. are able to accumulate them. They are generally seen in the epidermal cells as in grasses or in bundle sheath cells as in palms and orchids. They possess varied shapes like dumb-bells, saddle, cross, oval or rectangular-shaped. They are assigned a number of roles within the plant they contain such as maintaining the rigidity in stems and linear leaves, protecting the plants from herbivores, offering resistance to pests like stem borer, leaf roller and certain fungal diseases like brown spots or blast diseases.

In the present paper silica bodies and other epidermal features such as trichomes of three medicinally important grasses have been studied. The plants selected are *Cymbopogon citratus, Cynodon dactylon* and *Triticum aestivum*. The medicinal properties of these plants are as follows.

1. *Cymbopogon citratus* Stapf.

This is lemon grass or locally known as "Lilichai" because this is added to tea for flavour and taste. Lemongrass is known for its medicinal volatile oils which are expectorants and mosquito repellents. In addition, this plant is used as a stimulant, diaphoretic and anti-spasmodic. The leaves are chewed for sore gums and for elephantiasis. The infusion of leaves is said to be sudorific, stimulant and anti-periodic.

2. *Cynodon dactylon* Pers.

This is "Durva", a sacred herb used in religious rites. It is a reputed remedy for epistaxis, haematuria and scabies. It is demulcent, diuretic, ophthalmic haemostatic and is useful in fever, diarrhoea, dropsy, bleeding piles, epilepsy etc.

3. *Triticum aestivum* Linn.

Wheat is the basis of famous "Wheat Grass Therapy" in which the infusion of seedlings is taken internally to cure many ailments. This is said to be a "panacea" and considered particularly useful in anal fissure, eczema, psoriasis, Thelesemia and asthma.

MATERIALS AND METHODS

All the three plants are collected from the Botanical Garden of the M. S. University and around. The voucher specimens are deposited in the herbarium (BARO) of the Maharaja Sayajirao University of Baroda, Vadodara.

For locating silica bodies epidermal peel of fresh plant materials were fixed in Jeffrey's fluid for 30 min., boiled for 1-2 min. The peels were washed with water and mounted in phenol which stains silica bodies as glistening pink structures. In another method the peels were boiled with 10% KOH for 1-2 min, washed with water and mounted in phenol. The size of silica bodies was measured with an ocular micrometer. Nature of trichomes also was studied from the epidermal peels.

RESULTS AND DISCUSSION

All the three plants are found to contain variously shaped silica bodies in epidermal peels (Fig. 1). The details are described below.

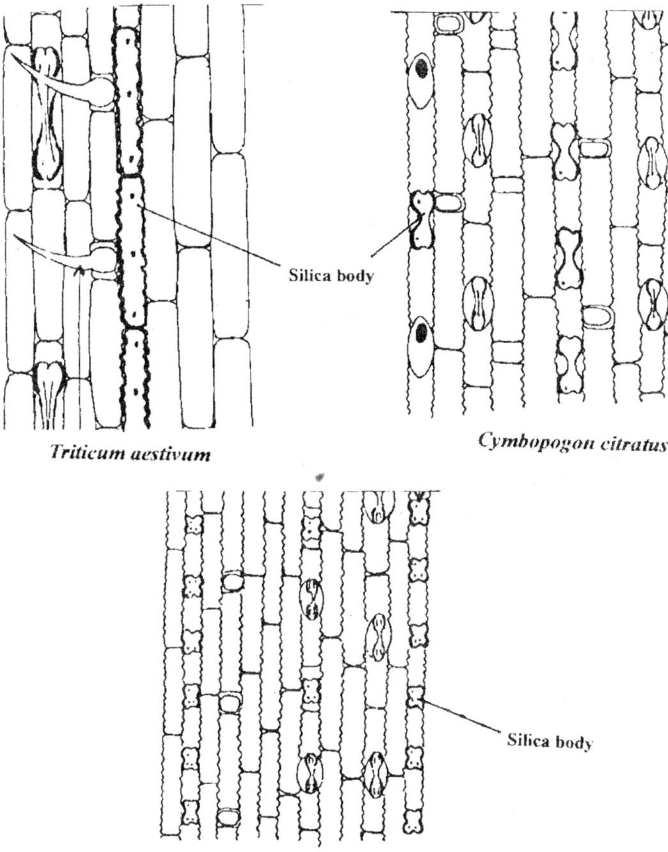

Fig. 1. Silica bodies of some medicinal members of Poaceae.

Cymbopogon

Silica bodies were found to be in rows alternate with those of stomata. They were mostly intermediate between cross-shaped and dumb-bell shaped. The length of the silica body is 17.28 μm and breadth is 11.07 μm. The trichomes were with elongated bases frequent over the veins.

Cynodon

Here the silica bodies were saddle shaped, found in rows alternating with those of stomata or at times in between stomata in a single row. The length of the silica body is 15.1 μm and breadth is 13.12 μm. Trichomes are absent here.

Triticum

Rectangular, horizontally elongated silica bodies were seen here. The outlines of these structures were sinuous. The length of the silica body is 52.44 μm and breadth is 11.22 μm. Trichomes were unicellular with swollen bases.

The presence of different silica bodies in all the three medicinal plants will help its utility as a biomarker. But the report of silica bodies of *Setaria italica* as carcinogenic chemicals (Parry and Hodson, 1982) is to be treated with caution because almost all our cereals contain silica and this is the first report on such an effect.

REFERENCES

Parry D. W. and M.J. Hodson (1982). *Ann. Bot.* **49:** 531.

16

STUDIES ON THE BIOMARKERS OF *AMMANIA BACCIFERA* LINN. AND *LAWSONIA INERMIS* LINN.

Elizabeth M. Daniel, Susy Albert and M. Daniel

Phytochemistry Laboratory, Department of Botany,
The Maharaja Sayajirao University of Baroda, Vadodara, 390 002, Gujarat, India

ABSTRACT

Pharmacognostic and phytochemical studies on *Ammania baccifera* Linn. and *Lawsonia inermis* Linn. two medicinal plants of the family Lythraceae have been carried out. The leaves of *Ammania* are characterised by bicollateral vascular bundle, a characteristic pattern of ridges and network on the cuticle, gallic acid and 3',4'-diOMe quercetin while its stem contained outer ring of larger xylem tracheids and inner ring of smaller tracheids, 3'-OMe quercetin and 3'-OMe kaempferol. Leaves of *Lawsonia* contain stone cells, lawsone, gallic acid and 3'-OMe luteolin whereas the stem contains stone cells surrounded by cells containing brown deposits, lawsone, gallic acid and 3', 4'-diOMe luteolin.

INTRODUCTION

Ammania baccifera Linn. and *Lawsonia inermis* Linn. are two medicinal plants belonging to the Lythraceae. The former plant is included in the tribe Ammaniae and the latter in the tribe Lythreae. *A. baccifera* is a glabrous, erect branching herb, found as a weed in rice fields and marshy localities throughout India. Leaves of this plant are acrid and used in treatment of rheumatic pains and fever. It is prescribed as stomachic and laxative. Fresh bruised leaves are used as rubefacient and external remedy for ringworm and other skin affections. Leaves are mixed with oil and applied to cure herpetic eruptions. The herb is rich in vitamin C and is reported to possess anti-typhoid and antitubercular properties (Anon., 1960). Leaves are used to treat rheumatism and fever (Pandey, 2005). *Lawsonia inermis* (Henna, Mehandi) is a glabrous, much branched shrub or a small tree with grayish brown bark. Henna leaves are used as a prophylactic against skin diseases and have astringent properties. They are used externally in the form of a paste or decoction against boils, burns, bruises and skin inflammations. Decoction is used as gargle for relaxed sore throat. Alcoholic extracts of leaves show mild antibacterial activity against *Micrococcus pyogenes* var. *aureus* and *E. coli*. Mehandi oil is used in perfumery. Leaves are also used to treat chest and lung diseases while stem is used for jaundice, growth of hair, headache, jaundice, leprosy, leucoderma, ulcers, lumbago, ophthalmia, skin diseases, stomachic etc. (Pandey, 2005). Triterpenoids have been isolated from *L. inermis* (Siddiqui, 2001). A naphthaquinone from stem bark has been isolated and 24β-Ethylcholest – 4- en-3β-ol from, the roots (Gupta, 1992) is the other compound reported.

Only the leaves of *Lawsonia* had been studied for their pharmacognostic characters. The pharmacognostic data on stem of this plant as well as the leaves and stem of *Ammania* are not available. The chemistry of *Ammania* also is not at all studied. Therefore, the present paper contains a detailed phytochemical and pharmacognostic studies on the leaves and stem of both these medicinal plants. It is hoped that the data presented here will be of great help in locating the biomarkers needed for the quality control programmes of these plants.

MATERIALS AND METHODS

Both plants were collected from the various localities of Baroda. Voucher specimens of these plants are deposited in the Herbarium of Department of Botany, M.S. University, Baroda (BARO). Standard procedures are followed for the extraction, isolation and identification of the various plant products such as flavonoids, phenolic acids, quinones, steroids etc. (Harborne, 1984; Daniel, 1991). For anatomical studies, sections were taken, stained with saffranin and mounted in glycerin. For micromorphological studies, epidermal peels were taken by direct peel or by heating in 10% KOH. The peels were then stained with saffranin and mounted on a clean microscopic slide. The quantitative and qualitative data comprising of stomatal index, trichome index, type of stomata, epidermal cells etc. were based on the average of 20 readings (Wallis, 1953).

RESULTS

(a) Pharmacognosy

(1) *Ammania baccifera*

Leaf of this plant contains mesophyll differentiated into single layer of palisade and closely packed spongy cells. In the midrib region, 3-4 layers of chlorenchyma were there on upper side of vascular bundle. There were two patches of phloem on both upper and lower sides of xylem. 6-7 Layers of parenchyma were seen with sphaeraphides on the lower side followed by a layer of collenchyma. Cuticle exhibited a pattern of net-like ridges and meshes. Leaf was found to be amphistomatic with anomocytic stomata. Stomatal index of upper epidermis and lower epidermis was 13-14 and 14-15 respectively (Fig. 1).

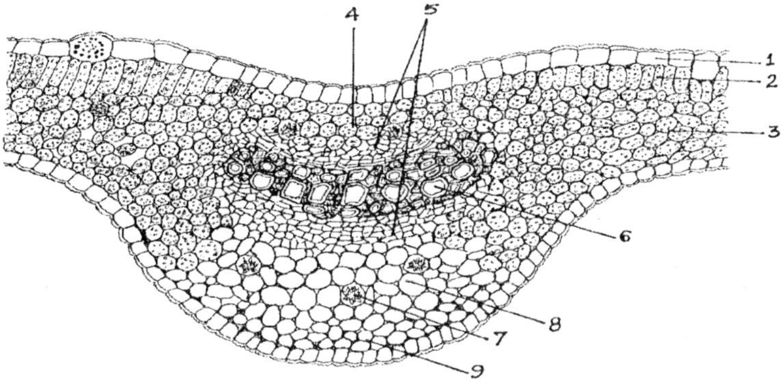

Fig. 1a. *Ammania baccifera* (leaf): 1. Epidermis, 2. Palisade, 3. Spongy, 4. Chlorenchyma, 5. Phloem, 6. Xylem, 7. Sphaeraphides, 8. Parenchyma, 9. Collenchyma

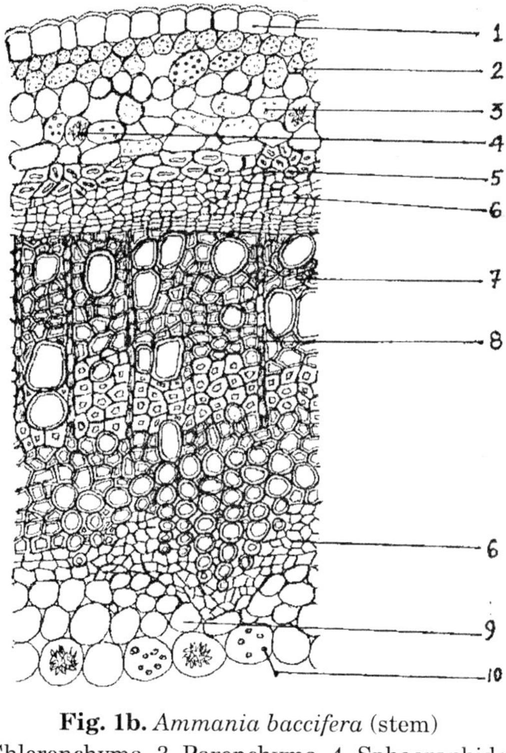

Fig. 1b. *Ammania baccifera* (stem)
1. Epidermis, 2. Chlorenchyma, 3. Parenchyma, 4. Sphaeraphides, 5. Sclereids, 6. Phloem, 7. Xylem, 8. Medullary rays, 9. Pith, 10. Starch grains

Stem of this plant has epidermis made of square shaped cells covered with cuticle having net like ridges and furrows. Cortex consisted of 5-6 layers of chlorenchyma where first 2-3 layers were full of chloroplasts as compared to the rest of the layers. Patches of stone cells encircled the upper phloem. Here also, inner phloem was seen. Xylem tracheids of outer and inner regions were with a large lumen while those of middle region were having narrow lumen giving an appearance of annual rings. Pith consisted of parenchymatous cells with sphaeraphides and very few starch grains (Fig. 2).

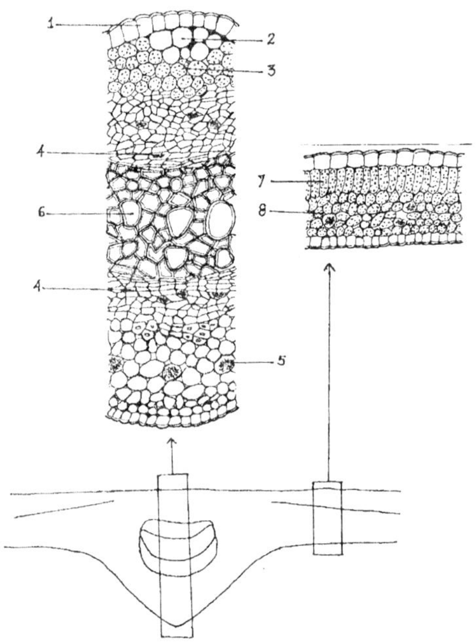

Fig. 2(a) *Lawsonia inermis* (leaf)

1. Epidermis, 2. Collenchyma, 3. Chlorenchyma, 4. Phloem, 5. Sphaerahides, 6. Xylem, 7. Palisade, 8. Spongy

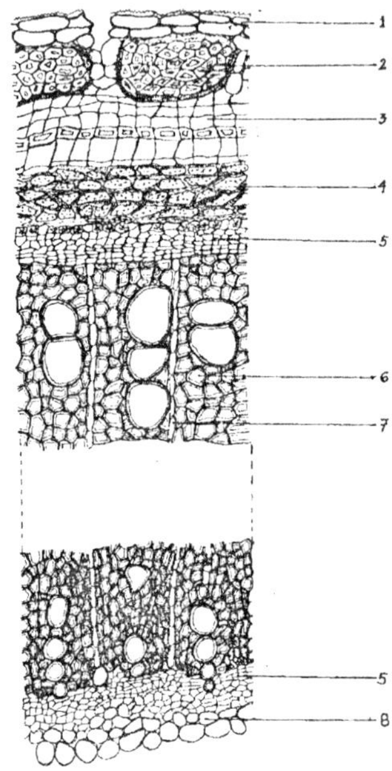

Fig. 2(b) *Lawsonia inermis* (stem)

1. Epidermis, 2. Sclereids, 3. Collenchyma with chloroplasts, 5. Phloem, 6. Xylem, 7. Medullary rays, 8. Pith

Powdered drug of the leaf consisted of fragments of anomocytic stomata, palisade, epidermal cells, sphaeraphides, vessel with double spiral thickening, cuticle with ridges and furrows, phloem fibres and crystal while that of stem consisted of sclereids, vessels with bordered pits alongwith medullary rays and tracheids, pieces of phloem, vessels with spiral and helical thickenings, cells containing chloroplasts and sphaeraphides and cork cells.

(2) *Lawsonia*

In **leaf,** the mesophyll was differentiated into one layered palisade and closely packed spongy cells containing sphaeraphides. Midrib on the upper side contained a patch of collenchyma surrounded by chlorenchymatous cells. Here also, there were two patches of phloem on both upper and lower sides of xylem. Lower phloem elements were less in number than upper phloem. Patches of stone cells were seen below the lower phloem. On the lower side of midrib, parenchymatous cells of 5-6 layers followed by collenchyma cells are seen. Upper epidermis was found larger than the lower epidermis. Leaves were amphistomatic containing ranunculaceous stomata. Stomatal index of upper epidermis was 14-16 while that of lower epidermis was 21-23 (Fig. 3).

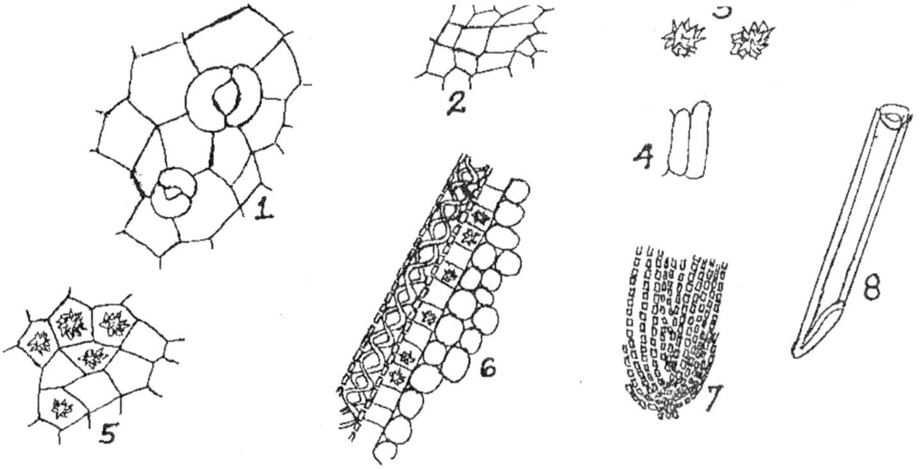

Fig. 3a. *Ammania baccifera* (leaf):
1. Anomocytic stomata, 2. Epidermal cells, 3. Sphaeraphides, 4. Palisade, 5. Cells containing sphaeraphides, 6. Vessel with double spiral thickening, 7. Cuticle with ridges and furrows, 8. Phloem fibre, 9. Crystal.

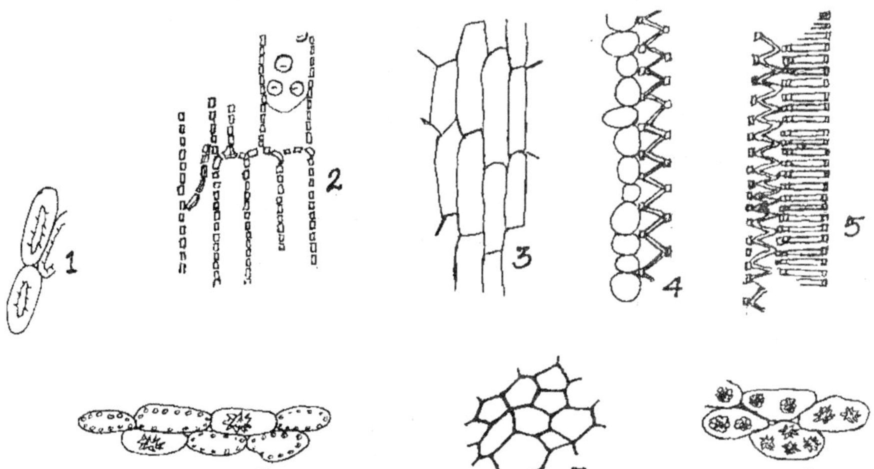

Fig. 3b. *Ammania baccifera* (stem)
1. Sclereids, 2. Vessel with bordered pits alongwith medullary rays and tracheids, 3. Phloem, 4. Vessel with spiral thickening, 5. Vessels with spiral and helical thickenings, 6. Cells containing chloroplasts and sphaeraphides, 7. Cork cells, 8. Cells with sphaeraphides

Stem – The outermost layer was cork followed by patches of stone cells surrounded by brown colored cells. Secondary cortex were found to be of 5-6 layers made up of colorless square to polygonal shaped cells. Primary cortex consisted of collenchyma cells with chloroplasts in it. Phloem was seen on the inner side of xylem also. The outermost layer of upper phloem showed presence of sphaeraphides. Medullary rays were also present. The inner phloem was followed by pith made of parenchymatous cells (Fig. 4).

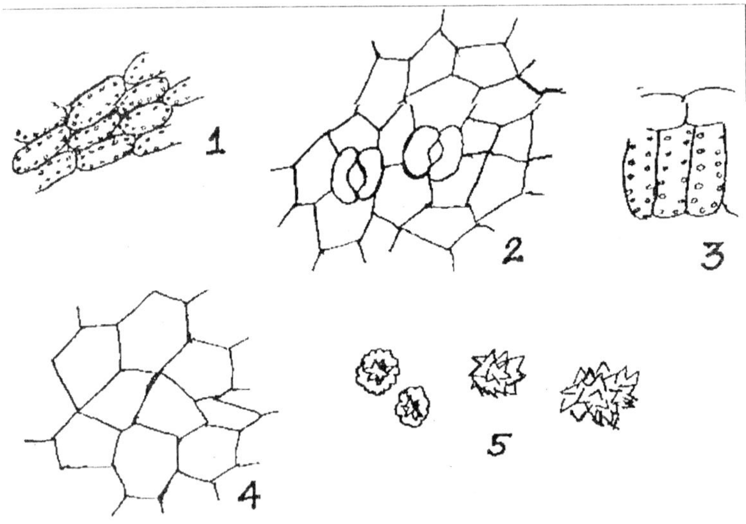

Fig. 4a. *Lawsonia inermis* (leaf)
1. Chlorenchyma, 2. Anomocytic stomata, 3. Palisade, 4. Epidermal cells, 5. Sphaeraphides

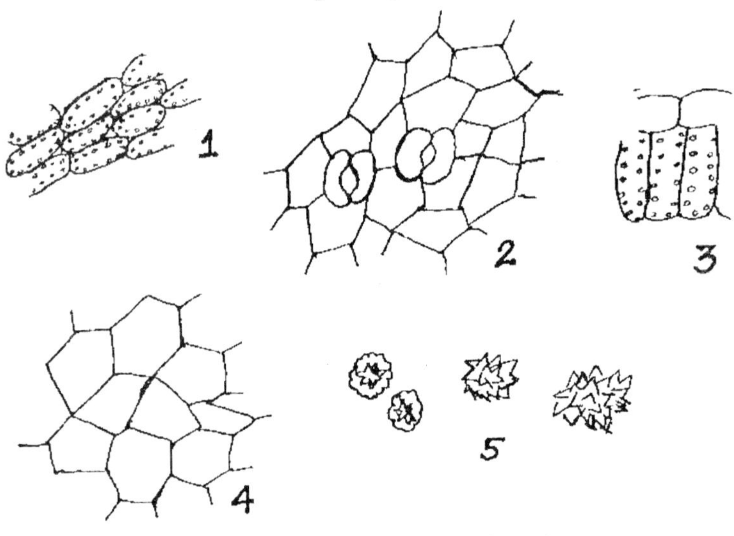

Fig. 4b. *Lawsonia inermis* (stem)
1. Cork cells, 2. Phloem cells with sphaeraphides, 3. Chlorenchyma cells, 4. Phloem fibre, 5. Sclereids, 6. Vessel with bordered pits alongwith medullary rays and tracheids

Powder of the leaf consisted of fragments of chlorenchyma, anomocytic stomata, palisade, epidermal cells and sphaeraphides and that of stem consisted of cork cells, phloem cells with sphaeraphides, chlorenchyma, phloem fibres, sclereids and vessel with bordered pits alongwith medullary rays and tracheids.

(b) Phytochemistry

The results of screening of leaves and stem of the two plants for their phytochemicals are presented in **Table 1**. Leaves of *A. baccifera* were found to contain 3', 4'-diOMe quercetin and phenolic acids like syringic, vanillic, gallic and melilotic acids. Quinones (other than lawsone, the compounds are yet to be identified) were present while steroids and glycoflavones were absent. Stem contained 3'-OMe quercetin and 3'-OMe kaempferol and phenolic acids like melilotic, gallic, syringic and vanillic acids.

Table 1. Distribution of phytochemicals in the leaves and stems of *Ammania baccifera* and *Lawsonia inermis*

S.N.	Name of Plant		1	2	3	4	5	6	7	8	9	10	11	12	13	14	15	16	17
1	*Ammania baccifera*	(leaf)	+	-	-	-	-	-	+	+	+	+	-	-	-	-	+	-	-
		(stem)	-	+	+	-	-	-	+	+	+	+	+	-	-	-	+	+	-
2	*Lawsonia inermis*	(leaf)	-	-	-	+	+	-	-	+	+	-	-	+	-	-	+	+	-
		(stem)	-	-	-	-	-	+	+	+	+	-	+	-	+	+	+	+	-

(1) 3', 4'-diOMe Quercetin, (2) 3'-OMe Quercetin, (3) 3'OMe Kaempferol, (4) Luteolin, (5) 3'-OMe Luteolin, (6) 3', 4'-diOMe Luteolin, (7) Syringic acid, (8) Vanillic acid, (9) Gallic acid, (10) Melilotic acid, (11) cis- & trans-ferulic acid, (12) *p*-OH benzoic acid, (13) *p*-coumaric acid, (14) Gentisic acid, (15) Quinones, (16) Steroids, (17) Glycoflavones

Leaves of *L. inermis* contained luteolin and 3'-OMe luteolin and vanillic, gallic and p-OH benzoic acids. Stem contained 3',4'-diOMe luteolin and phenolic acids such as syringic, vanillic, ferulic, gallic, *p*-coumaric and gentisic acids. Quinone (lawsone) and steroids were present and glycoflavones were absent in both leaves and stem.

DISCUSSION

The present study brings out the biomarkers of both these plants clearly. *Ammania* has peculiar mesh like thickenings on the cuticle, quinones, gallic acid and flavonols. *Lawsonia* possesses stone cells, lawsone, gallic acid and 3'-OMe luteolin in the leaves and stone cells surrounded by cells containing brown deposits, lawsone, gallic acid and 3', 4'-diOMe luteolin in the stem.

REFERENCES

Anonymous (1960). *The wealth of India, Raw Materials,* Vol : 6, CSIR, New Delhi.

Daniel, M. (1991). *Methods in Plant Chemistry and Economic Botany*, Kalyani Publishers, New Delhi.

Gupta, S., Ali, M. and S. Alam (1992). *Phytochemistry,* **31**(7) : 2558.

Gupta S., Ali, M. and S. Alam (1993). *Phytochemistry,* **33**(3) : 723.

Harborne, J.B. (1984). *Phytochemical methods*, (2nd Edn.), Chapman & Hall London.

Johansen, D.A. (1940). *Plant Microtechnique,* McGraw Hill Book Company, New York.

Metcalfe, C.P. and L. Chalk (1950). *Anatomy of the Dicotyledons*, 1: Clarendon Press. Oxford.

Pandey, C.N., B.R. Raval, Seema, M. and H. Salvi (2005). '*Medicinal plants of Gujarat*', GEER Foundation, Gandhinagar.

Panigrahi S.G. (1980). *Phytomorphology,* **30**(4) : 320.

Raghunathan, K. and R. Mitra (1982). In : *Pharmocognosy of indigenous drugs,* Central Council of Research in Ayurveda and Siddha, New Delhi **1**: 614.

Siddiqui, B.S. and M.N. Kardar (2001). *Phytochemistry,* **58** (8):1195.

Wallis, T.E. (1953). *Practical Pharmocognosy*, J. & A. Churchill Ltd., Gloucester Place, W.I. p. 187.

17

PHARAMACOGNOSTIC STUDIES AND HYPOGLYCEMIC ACTIVITY OF *ARTOCARPUS HETEROPHYLLUS* LAM.

Remya R. and M. Daniel

Phytochemistry Laboratory, Department of Botany, Faculty of Science The Maharaja Sayajirao University of Baroda, Vadodara-390002.

ABSTRACT

The present study was carried out to evaluate the antidiabetic activity of *A. heterophyllus* on alloxan induced diabetic rats. Oral administration of fresh leaf extract (400 and 600 mg/kg body weight) for 60 days treatment resulted in significant decrease in blood glucose level, there was also significant changes observed in carbohydrate metabolizing enzyme and antioxidants. The study clearly shows that the effect of the drug was less than the standard drug glibenclamide. Pharmacognostic study was carried out to find out the biomarkers. Characteristic latex cells, two layered palisade, collenchymatous patches among palisade and sclereids are the distinguishing features of the leaf. Micromorphological characters of the leaf also are given.

INTRODUCTION

Plants have been the major source of drugs for the treatment of diabetes mellitus in Indian system of medicine. Out of many plants used thus, only a few have been evaluated as per modern standardization of medicine for this disease. Because of the widespread belief that "Green medicine" is safe and more dependable than the synthetic drugs, many of which exert adverse side effects such as Hematological, Cardiovascular and Gastrointestinal reactions. Diabetic coma, damage to kidney and liver are the side effects which can endanger the life of diabetic patients. Recently attention has been focused on the relationship between production of activated oxygen and diabetes. Activated oxygen species such as hydrogen peroxide, superoxide anions, and singlet oxygen and hydroxyl radicals can be formed in cells not only during ionizing radiation, but also during aerobic metabolism of either endogenous or exogenous substance. Cells have enzymatic and non-enzymatic scavenger system against these free radicals. Nevertheless if free radical production and scavenger systems some how become unbalanced cells are exposed to oxidative damage resulting in cell injury (Parinandi *et al.*, 1990; Griesmacher *et al.*, 1995). Hyperglycemia results in the generation of free radicals (Giardino *et al.*, 1996). The concentrations of the reactive oxygen species are modulated by antioxidant enzymes. Marked alterations in antioxidant enzyme activities have been reported in diabetes (Wohaieb and Godin 1987; Oberley, 1988). As new drugs are being developed, it is appropriate to consider ingredients of indigenous herbs (Cherian and Augusti, 1993).

Because of the high reliability and effectiveness of the herbal medicines the herbal industries are mushrooming every where. Since the ayurvedic texts which are the main basis of plant sources give the name of the plants only in *Sanskrit,* there is a great confusion on the botanical identity of these plants. Therefore the market is flooded with spurious drugs derived from many different sources. These drugs are available as raw drugs, powders or extracts, the later two are often easily adulterated

The identity of the medicinal plants in powder or in extract form can be established by pharamacognostic methods. But the data on pharmacognosy and for individual plants are lacking for many medicinal plants. Therefore the urgent need today is to provide such data which would help the manufacturer or the consumer to assess the purity of the medicinal preparations or raw materials. There are many antidiabetic herbs which had been reported pharmacologically active and still the search continues to find out an efficient herb which retards diabetic complications. The matured leaf of *Artocarpus heterophyllus* Lam. (Family : Moraceae, Syn. *Artocarpus integrifolius* Auct.) exhibited hypoglycemic activity in experimental animals (Fernado and Thabrew, 1989). The hot water extract of the leaves enhances the glucose tolerance capacity in diabetic patients (Fernado *et al.*, 1991).

To our knowledge, no systematic studies have been carried out to find out the effect of *A. heterophyllus* on antioxidant enzymes in diabetic induced rats. Hence, the present investigation was undertaken to assess the effect of *A. heterophyllus* leaf on enzymatic and non-enzymatic antioxidants and carbohydrate metabolizing enzyme in alloxan induced diabetic rats. And also pharamacognostic study was done to provide biomarkers to check the purity of drug.

MATERIALS AND METHODS

The Leaf of *Artocarpus heterophyllus* (Moraceae) was collected from Vadodara City, Gujarat. A Voucher specimen was deposited in the herbarium of the Botany Department of M.S. University. For preparing the extract, 500g of Fresh matured leaves were boiled in water for 30 min. The extract was then filtered and the process of boiling was repeated three times with the residue, each time collecting the extract. The collected extract was pooled and passed through a fine muslin cloth. The filtrate upon evaporation at 40°C yielded 14.5% semi solid extract. Male Sprague Dawley rats (Weighing 150-200g) were used in the experiment under standard environmental conditions, and the animals were fed on standard pellet diet and allowed water *ad libitum*. Adult male Sprague Dawley rats were made diabetic with an intra peritoneal injection of alloxan (150 mg/kg body weight) dissolved in phosphate buffer saline (0.1M, pH 7.4). Alloxan injected animals exhibited massive glycosuria and hyperglycemia with in few days. Diabetes was confirmed in Alloxan rats by measuring the fasting blood glucose concentration 96 h after injection with alloxan. Sprague Dawley rats with blood glucose level above 250 mg/dl were considered to be diabetic and were used in the experiment. The rats injected with 2% gum acacia alone then served as control.

After the induction of diabetes the rats were divided into five groups of six animals each.

Group 1 Control rats received vehicle solution containing 2 % gum acacia

Group 2 Diabetic control

Group 3 Normal rats received an aqueous extract of *A. hetrophyllus* (400mg/Kg body weight in 2% gum acacia) using an intragastric tube for 60 days.

Group 4 Normal rats received an aqueous extract of *A. hetrophyllus* (600mg/Kg body weight in 2% gum acacia) using an intragastric tube for 60 days.

Group 5 Diabetic rats received glibenclamide orally (600μg/Kg body weight) as aqueous solution using an intragastric tube for 60 days.

After 60th day of treatment with leaf extract rats were fasted for 12 h then sacrificed by cervical dislocation and the blood samples were collected in tubes containing anticoagulant. Plasma samples were obtained by centrifugation and stored at -20°C till the experimental analysis.

Weighed kidney, liver, brain samples were homogenized on ice in the appropriate buffer (1/10 parts (w/v)). The homogenate were centrifuged at 10,000 x g for 15 min at 4°C to discard any cell debris. The supernatant was immediately used for enzyme assay.

Biochemical assay

Glucose levels were estimated by colorimetrically available glucose kit based on glucose oxidase method. The level of peroxidation was estimated by the methods of Nichans and Samuelson (1968). GSH was measured according to the method of Beutler and Kelly (1963). Ascorbic acid was measured according to the method of Omaye *et al.* (1979). α-Tocopherol was estimated by the method of Baker *et al.* (1980). Superoxide dismutase (SOD) was assayed by the method of Marklund and Marklund (1974). Catalase (CAT) was assayed colorimetrically by the method of Sinha (1972). Glutathione

Peroxidase (GPx) was estimated by the method of Paglia and Valentine (1967). Hexokinase was assayed according to Branstrup *et al.* (1957) by estimation of residual glucose in the reaction supernatant. Glucose -6-Phosphatase was assayed according to Koide and Odo (1959) by estimation of inorganic phosphate liberated from glucose -6- phosphate. Fructose -1, 6-bisphosphatase was assayed by the method of Gancedo and Gancedo (1971). Values were represented as mean ± SD. Data were analyzed using analysis of variance (ANOVA). And group mean was compared with Duncan's multiple range test (DMRT).

Pharmacognosy

Micromorphological studies were carried out on fresh materials. Epidermal peels were taken and stained with toluidine blue (0.5%) and mounted in glycerin. Leaf constants such as stomatal index, stomatal frequency, stomatal size, vein islet number, vein termination number and palisade ratio were calculated. Transections of petiole, lamina and midrib were taken by free hand and were stained in toluidine blue and mounted in glycerin. The slides were examined under the microscope and Camera Lucida sketches were drawn and the cells measured using ocular micrometer. The quantitative data are based on the average of 20 readings. The finely powdered drug was scanned under 40 x10 x magnifications for recording the cell elements.

RESULTS

Table 1 illustrates the hypoglycemic activity of *A. heterophyllus* in alloxan induced diabetic rats. There was a significant decrease in the blood glucose level and increase in the body weight in treated rats when compared to the untreated diabetic rats. The effect of the aqueous extract was less than the standard drug glibenclamide. Tables 2, 3, 4, and 5 show the levels of TBARS, GSH, Vitamin C and Vitamin E in plasma, liver, kidney and brain. In plasma Alloxan induced diabetic rats have increased TBARS, vitamin E and there is a decreased level of Vitamin C and GSH when compared to the control rats. Treatment with the leaf extract significantly decreased the level of vitamin E but there was no significant change in the level of TBARS.

Table 1. Effect of *A. heterophyllus* on blood glucose level and body weight

Gr	Treatment (Dose/Kg body weight)	Blood glucose (initial)	Blood glucose (final)	Body weight (initial)	Body wt. (final)
1	Control (received 2% gum acacia)	81.8 ± 6.17	97.5 ± 4.2^{a}	185.8 ± 26.15	189 ± 12.8^{a}
2	Diabetic + control	283 ± 14.35	337.5 ± 20.9^{b}	190 ± 17.8	129 ±19^{b}
3	Diabetic + *A. heterophyllus* (400mg)	287 ± 19.93	181.8 ± 16.9^{c}	205.8 ± 23.3	203 ±18^{c}
4	Diabetic + *A. heterophyllus* (600mg)	284 ± 19	180.8 ± 35^{c}	184 ± 17.4	192 ± 19.4^{c}
5	Diabetic + Glibenclamide (600 μg/kg body weight)	305 ± 27	135 ± 17.3^{a}	182 ± 20.4	185 ± 28^{c}

Values are means ± S.D for six animals in each group.
Values not sharing a common superscript differ significantly at $p<0.05$.
Duncan's Multiple Range Test (DMRT)

Table 2. Effect of *A. heterophyllus* on TBARS, GSH, Vitamin C, and Vitamin E in plasma of alloxan induced diabetic rats

Gr	Treatment (Dose/Kg body weight)	TBARS	GSH	Vitamin C	Vitamin E
1	Control (received 2% gum acacia)	1.58 ± 0.56^{a}	23 ± 3.2^{a}	1.95 ± 0.18^{a}	0.71 ± 0.04^{a}
2	Diabetic + control	3.6 ± 0.39^{b}	14.98 ± 2.3^{b}	$.81 \pm 0.098^{b}$	1.07 ± 0.038^{b}
3	Diabetic + *A. heterophyllus* (400mg)	3 ± 0.44^{b}	18.08 ± 2.3^{d}	1.0 ± 0.22^{c}	0.83 ± 0.045^{c}
4	Diabetic + *A. heterophyllus* (600mg)	2.9 ± 0.70^{b}	18.08 ± 2.4^{d}	1.12 ± 0.13^{c}	0.83 ± 0.040^{c}
5	Diabetic + Glibenclamide (600μg/kg body weight)	1.7 ± 0.51^{a}	21.1 ± 2.3^{a}	1.7 ± 0.18^{a}	0.86 ± 0.016^{d}

Values are means ± S.D for six animals in each group.

Values not sharing a common superscript differ significantly at $p<0.05$.

Duncan's Multiple Range Test (DMRT)

Table 3. Effect of *A. heterophyllus* on TBARS, GSH, Vitamin C, and Vitamin E in liver of alloxan induced diabetic rats

Gr	Treatment (Dose/Kg body weight)	TBARS	GSH	Vitamin C	Vitamin E
1	Control (received 2% gum acacia)	1.52 ± 0.52^{a}	13.9 ± 3.25^{a}	0.75 ± 0.01^{a}	5.4 ± 0.28a
2	Diabetic + control	3.6 ± 0.39^{b}	7.24 ± 2.5^{b}	0.45 ± 0.04^{b}	3.4 ± 0.19^{b}
3	Diabetic + *A. heterophyllus* (400mg)	3.1 ± 0.68^{b}	10.8 ± 1.6^{c}	0.47 ± 0.02^{c}	4.5 ± 0.37^{c}
4	Diabetic + *A. heterophyllus* (600mg)	3.2 ± 0.66^{b}	12.4 ± 1.9^{c}	0.47 ± 0.02^{c}	4.6 ± 0.43^{c}
5	Diabetic + Glibenclamide (600μg/kg body weight)	1.8 ± 0.40^{a}	13.9 ± 3.24^{a}	0.69 ± 0.03^{d}	5.2 ± 0.04^{d}

Values are means ± S.D for six animals in each group.

Values not sharing a common superscript differ significantly at $p<0.05$.

Duncan's Multiple Range Test (DMRT)

Table 4. Effect of *A. heterophyllus* on TBARS, GSH, Vitamin C, and Vitamin E in Kidney of alloxan induced diabetic rats

Gr	Treatment (Dose/Kg body weight)	TBARS	GSH	Vitamin C	Vitamin E
1	Control (received 2% gum acacia)	1.6 ± 0.47^{a}	11.8 ± 3.6^{a}	$0.60 \pm .024^{a}$	3.8 ± 0.46^{a}
2	Diabetic + control	3.7 ± 0.33^{b}	3.1 ± 1.67^{b}	$0.30 \pm .054^{b}$	1.40 ± 0.14^{b}
3	Diabetic + *A. heterophyllus* (400 mg)	3.2 ± 0.25^{b}	7.7 ± 1.6^{d}	$0.34 \pm .060^{b,c}$	2.6 ± 0.29^{c}
4	Diabetic + *A. heterophyllus* (600 mg)	3.4 ± 0.69^{b}	7.8 ± 1.6^{d}	$0.37 \pm .016^{c}$	2.5 ± 0.46^{c}
5	Diabetic + Glibenclamide (600 μg/kg body weight)	1.8 ± 0.66^{a}	9.83 ± 2.3^{a}	0.56 ± 0.015^{a}	3.0 ± 0.17^{d}

Values are means ± S.D for six animals in each group.

Values not sharing a common superscript differ significantly at $p<0.05$.

Duncan's Multiple Range Test (DMRT).

Table 5. Effect of *A. heterophyllus* on TBARS, GSH, Vitamin C, and Vitamin E in Brain of alloxan induced diabetic rats

Gr	Treatment (Dose/Kg body weight)	TBARS	GSH	Vitamin C	Vitamin E
1	Control (received 2% gum acacia)	1.6 ± 0.70^{a}	14.9 ± 4.5^{a}	0.84 ± 0.05^{a}	5.5 ± 1.14^{a}
2	Diabetic + control	3.79 ± 0.65^{b}	5.6 ± 2.3^{b}	0.44 ± 0.045^{b}	3.3 ± 0.11^{b}
3	Diabetic + *A. heterophyllus* (400mg)	3.8 ± 0.91^{b}	10.7 ± 2.5^{b}	0.54 ± 0.01^{c}	4.47 ± 0.34^{c}
4	Diabetic + *A. heterophyllus* (600mg)	2.9 ± 0.72^{b}	10.9 ± 4.2^{b}	0.56 ± 0.02^{c}	4.8 ± 0.28^{c}
5	Diabetic + Glibenclamide (600μg/kg body weight)	1.7 ± 0.65^{a}	12.9 ± 3.05^{a}	0.69 ± 0.02^{d}	5.08 ± 0.32^{d}

Values are means ± S.D for six animals in each group.
Values not sharing a common superscript differ significantly at $p<0.05$.
Duncan's Multiple Range Test (DMRT)

In liver, kidney, and brain the level of TBARS is increased in the diabetic rats where as the levels of GSH Vitamin C and Vitamin E is decreased when compared to the normal rats. Treatment with aqueous extract of *A. heterophyllus* resulted in increased levels of GSH, Vitamin C and Vitamin E while there was no significant changes in the level of TBARS

The activity of SOD, Catalase and GPX in Liver, Kidney and Brain are depicted in Table 6, 7, and 8. The activity of SOD, CAT and GPX were decreased in liver and brain of diabetic rats when compared to control rats. In kidney there was an increased activity of GPX and decreased activity of SOD and CAT in diabetic rats. Treatment with *A. heterophyllus* increased the activity of SOD in liver, kidney and brain but there was no significant changes seen in the activity of CAT and GPX when compared with untreated diabetic rats.

Table 6. Effect of *A. heterophyllus* on SOD, CAT and GPX in brain of alloxan induced diabetic rats

Gr	Treatment (Dose/Kg body weight)	SOD	CAT	GPX
1	Control (received 2% gum acacia)	11.77 ± 0.21^{a}	166 ± 34.85^{a}	8.7 ± 0.63^{a}
2	Diabetic + control	5.47 ± 1.7^{b}	71.5 ± 10^{b}	4.8 ± 0.79^{b}
3	Diabetic + *A. heterophyllus* (400 mg)	7.48 ± 1.29^{c}	80.5 ± 9.7^{b}	5.7 ± 0.57^{b}
4	Diabetic + *A. heterophyllus* (600 mg)	8.34 ± 1.08^{c}	79.8 ± 4.07^{b}	6.0 ± 0.62^{b}
5	Diabetic+Glibenclamide (600μg/kg body weight)	10.65 ± 0.37^{a}	130.5 ± 17.8^{c}	7.7 ± 0.97^{a}

Values are means ± S.D for six animals in each group.
Values not sharing a common superscript differ significantly at $p<0.05$.
Duncan's Multiple Range Test (DMRT)

Table 7. Effect of *A. heterophyllus* on SOD, CAT and GPX in liver of alloxan induced diabetic rats

Gr	Treatment (Dose/Kg body weight)	SOD	CAT	GPX
1	Control (received 2% gum acacia)	12.01 ± 0.32^{a}	163.2 ± 24.36^{a}	8.9 ± 1.8^{a}
2	Diabetic + control	4.33 ± 0.68^{b}	68.76 ± 8.24^{b}	4.1 ± 0.78^{b}
3	Diabetic + *A. heterophyllus* (400mg)	8.07 ± 0.96^{c}	75.01 ± 6.9^{b}	5.3 ± 0.73^{b}
4	Diabetic + *A. heterophyllus* (600mg)	9.1 ± 0.47^{c}	79.8 ± 20.6^{b}	5.7 ± 0.57^{b}
5	Diabetic + Glibenclamide (600μg/kg body weight)	10.7 ± 0.38^{d}	134.01 ± 13.25^{c}	8 ± 0.82^{a}

Values are means ± S.D for six animals in each group.
Values not sharing a common superscript differ significantly at p<0.05.
Duncan's Multiple Range Test (DMRT)

Table 8. Effect of *A. heterophyllus* on SOD, CAT and GPX in Kidney of alloxan induced diabetic rats

Gr	Treatment (Dose/Kg body weight)	SOD	CAT	GPX
1	Control (received 2% gum acacia)	11.72 ± 0.18^{a}	75.7 ± 7.1^{a}	3.88 ± 0.50^{a}
2	Diabetic + control	4.65 ± 1.17^{b}	32.6 ± 6.68^{b}	7.2 ± 0.88^{b}
3	Diabetic + *A. heterophyllus* (400mg)	$5.8 \pm 1.5^{b,c}$	41.6 ± 5.8^{b}	5.2 ± 0.65^{b}
4	Diabetic + *A. heterophyllus* (600mg)	7.07 ± 0.62^{c}	45.8 ± 5.26^{b}	5.4 ± 0.77^{b}
5	Diabetic + Glibenclamide (600μg/kg body weight)	10.64 ± 0.21^{a}	69.45 ± 17.8^{c}	4 ± 0.59^{c}

Values are means ± S.D for six animals in each group.
Values not sharing a common superscript differ significantly at p<0.05.
Duncan's Multiple Range Test (DMRT)

Table 9 shows the activity of carbohydrate metabolizing enzyme Hexokinase, Glucose- 6-phosphatase, and Fructose-1, 6-bisphosphatase. There was a decreased activity of Hexokinase and increased activity of Glucose-6-Phosphatase and Fructose-1, 6-bis Phosphatase in diabetic rats compared to control rats. Treatment shows no significant activity in Hexokinase whereas there was deceased activity of Glucose-6-Phosphatase and Fructose-1, 6-bis Phosphatase when compared with untreated diabetic rats.

Table 9. Effect of *A. heterophyllus* on Hexokinase, Glucose-6-phosphatase and Fructose-1, 6-bisphosphatase of alloxan induced diabetic rats

Gr	Treatment (Dose/Kg body weight)	Hexokinase	Glucose-6-phosphatase	Fructose-1,6-bisphosphatase
1	Control (received 2% gum acacia)	0.22 ± 0.01^{a}	0.16 ± 0.01^{a}	0.55 ± 0.02^{a}
2	Diabetic + control	0.11 ± 0.01^{b}	0.31 ± 0.06^{b}	1.08 ± 0.04^{b}
3	Diabetic + *A. heterophyllus* (400mg)	0.12 ± 0.008^{b}	0.22 ± 0.02^{c}	0.73 ± 0.06^{c}
4	Diabetic + *A. heterophyllus* (600mg)	0.12 ± 0.01^{b}	0.22 ± 0.01^{c}	0.68 ± 0.025^{c}
5	Diabetic + Glibenclamide (600 μg/kg body weight)	0.15 ± 0.05^{a}	0.17 ± 0.01^{a}	0.60 ± 0.05^{a}

Values are means ± S.D for six animals in each group.
Values not sharing a common superscript differ significantly at p<0.05.
Duncan's Multiple Range Test (DMRT)

For all the parameters studied, aqueous extract *A. heterophyllus* at a dose 400 and 600 mg/kg body weight had the same effect and the effect was less than the standard drug glibenclamide.

Pharmacognostic characters

Micromorphology

The leaves are hypostomatic with anomocytic stomata. Latex ducts were found on the lower surface. Spharaphides were found on the epidermal cell. The average size of the epidermal cell was from 23.10 x 15.18 μm on lower epidermis and 25.08 x 21.12 μm on upper epidermis. The average size of the stomata was 21.12 x 11.88 μm and the size of the crystal was 9.9 x 8.58 μm on lower epidermal layer and the average cell size on the upper epidermal layer was 10.9 x 9.22 μm.

Stomatal complex

Stomatal Index/mm^2 was 16.2, Stomatal Frequency/mm^2 was 31.7, Palisade ratio was 10.5, Vein Islet number/mm^2 was 15.9 and Vein Termination number/mm^2 was 14.

Anatomy

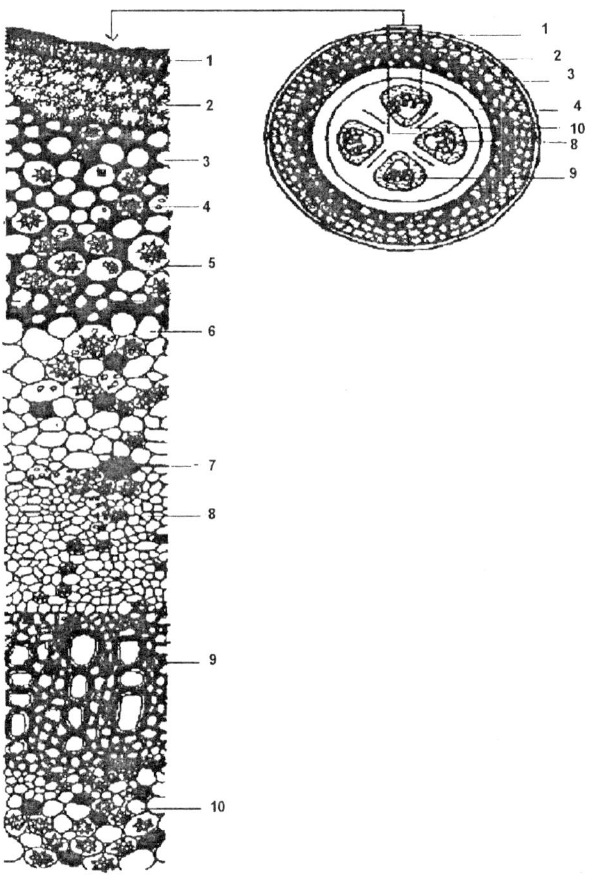

Fig. 1. (1) Epidermal layer, (2) Chlorenchyma, (3) Collenchyma, (4) Spharaphides, (5) Starch grains (6) Parenchyma (7) Latex cell, (8) Phloem, (9) Xylem vessel, (10) Pith

Midrib

In the midrib the epidermis consists of square cells containing starch grains. The cortex contains hypodermal collenchymatous and chlorenchyma tous cells with starch grains and sphaeraphides; below this were parenchymatic cells containing latex cells and sphaeraphides. These layers were also

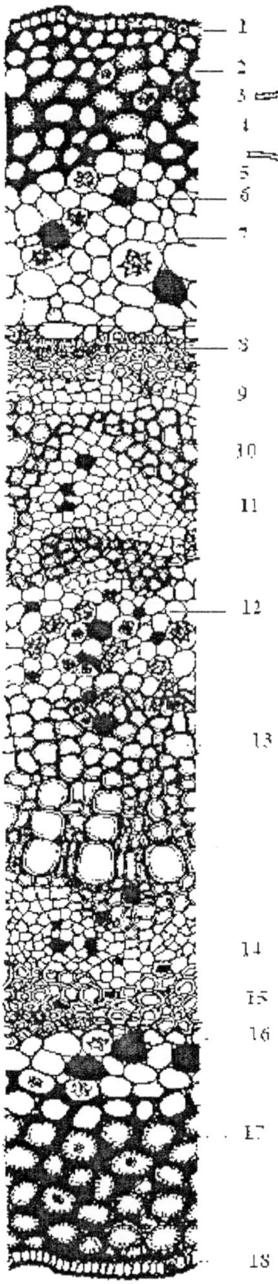

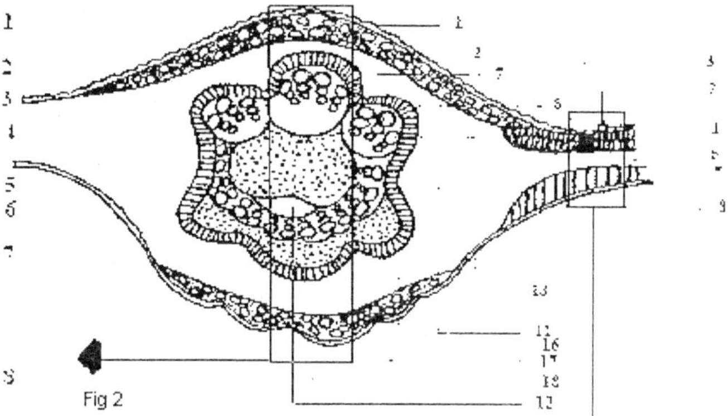

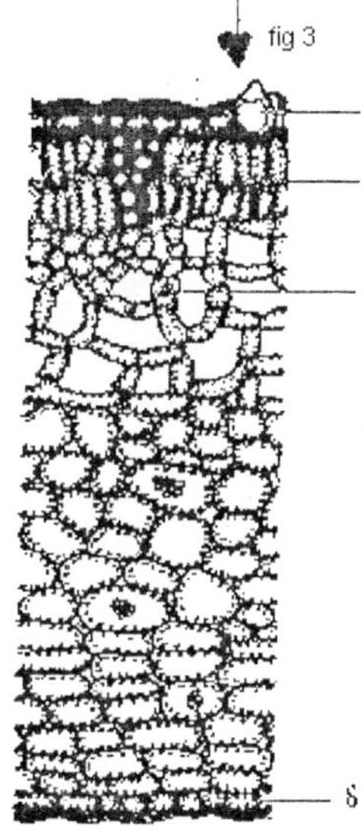

Fig 3: (1) Upper epidermal layer, (2) Latex Sac, (3) Patches of Collenchyma, (4) Double layered Palisade, (5) Loosely arranged Spongy cells, (6) Starch grains, (7) Closely arranged Spongy cells, (8) Lower epidermal layer.

Fig 2: (1) Epidermal cell, (2) Collenchyma, (3) Spharaphides, (4) Chlorenchyma, (5) Crystals, (6) Latex cell, (7) parenchyma cells, (8) Sclerides, (9) Phloem, (10) Xylem Rays, (11) Latex Cell, (12) Parenchyma layer, (13) Xylem Vessel, (14) Starch grains, (15) Sclerides, (16) Parenchyma layer, (17) Chlorenchyma, (18) Elongated lower epidermal layer.

found on the lower side of the midrib. The sclereids are seen surrounding the whole vascular bundle. In the vascular bundle the phloem was seen on both sides of xylem which contains latex cells and starch grains. The latex cell, starch grains and sphaeraphides were also seen in the xylem medullary rays. The lower epidermis contains elongated cells with starch grains. The size of cells were: upper epidermal cells 14.52 x 8.58 μm; lower epidermal cells 15.56 x 5.08 μm; sclereids 11.22 x 14.52 μm; collenchyma 13.20 x 14.52 μm; parenchyma 20.46 x 23.10 μm; xylem rays 19.14 x 19.80 μm; crystals 17.80 x 22.40 μm; xylem vessels 23.10 x 22.40 μm (Fig. 2).

Lamina

The mesophyll was differentiated into palisade and spongy cells containing starch grains. Palisade tissue was two layered. Patches of collenchyma were seen between the palisade cells. The epidermal layer contains latex sacs. The spongy cells were of two types arranged in two regions, the cells of upper region was loosely packed while the lower layer contains closely packed cells (Fig. 3).

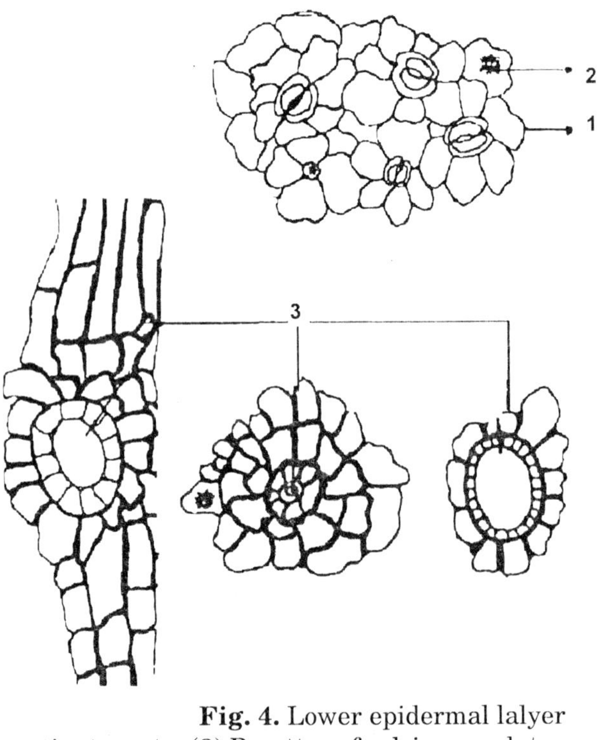

Fig. 4. Lower epidermal lalyer
(1) Anomocytic stornata, (2) Rosettes of calcium oxalate crystals, (3) Latex sac

Petiole

The epidermal layers contain chloroplasts. The cortex consists of hypodermal chlorenchyma and inner parenchyma containing latex cells, sphaeraphides and starch grains. The middle layer collenchyma also contains sphaeraphides and starch grains. In the vascular bundle, phloem and xylem rays contain sphaeraphides, starch grains and latex cells. The pith region is abundantly filled with sphaeraphides; it also contains starch grains and latex cells. The length and breath of the cells were: upper epidermal cell 17.82 x 8.58 μm, lower epidermal cell 17.82 x 9.10 μm, collenchyma 30.36 x 31.68 μm, parenchyma 28.38 x 35.64 μm, xylem rays 25.08 x 15.84 μm, crystals 17.80 x 21.10 μm, xylem vessel 19.80 x 19.80 μm, and pith cells 40.90 x 4.90 μm (Fig. 1).

Powder study

Powdered leaf was analyzed under the microscope. Fragments of palisade tissues, stomata, veins, fibres, latex sacs, starch grains, rosettes of calcium oxalate crystals, spirally coiled tracheids and fragment of sclereids are depicted in (Fig. 5).

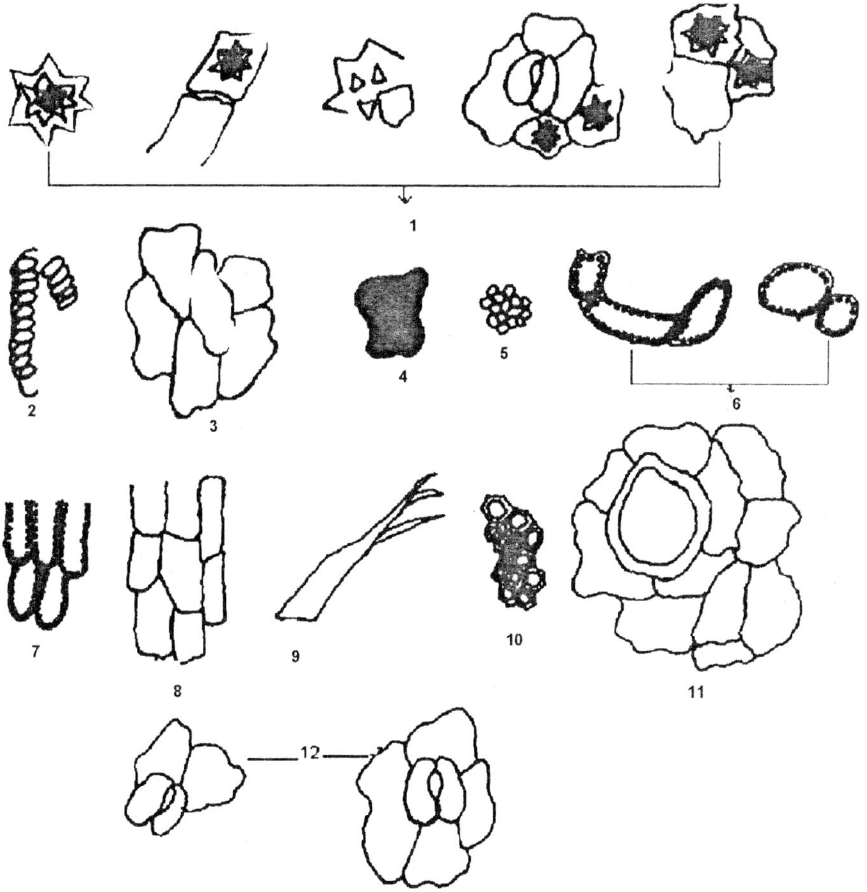

Fig. 5. Powder characters: (1) Rosettes of calcium oxalate crystals, (2) Tracheids, (3) Epidermal cells (4) Latex cells, (5) Starch grains, (6) Spongy cells, (7) Palisade, (8) Vein, (9) Sclerides, (10) Latex sec, (11) Stomata

DISCUSSION

In the present study there is an increase in the level of TBARS and decrease in the level of enzymatic and non enzymatic antioxidants in plasma, liver, kidney and brain of alloxan induced diabetic rats. Increased level of TBARS suggests increasing OFRs. Lipid peroxide mediated tissue damage has been observed in the development of Type I and Type II diabetes mellitus. (Sundaram *et al.,* 1996) have reported that the concentration of lipid peroxides increases in the kidney of diabetic rats. An increased level of TBARS is an index of lipid peroxidation. Our study shows that administration of *A. heterophyllus* did not show any significant changes in the level of TBARS in plasma, liver, kidney and brain.

Glutathione is known to protect the cellular system against the toxic effects of lipid peroxidation. The total glutathione concentration declines in the plasma, liver, kidney and brain of alloxan induced diabetic rats, which represent increase utilization due to oxidative stress. Treatment with *A. heterophyllus* glibenclamide treated rats show significant reversal of glutathione in plasma, liver kidney and brain. Other workers have also reported decreased level of plasma GSH in STZ diabetic rats (Garg *et al.*, 1996) Vitamin C is known to act as an antioxidant in both *in vivo* and *in vitro*. It functions as a free radical scavenger and successfully prevents detectable oxidative damage under all types of oxidative stress. Vitamin C plays an important role in detoxification of reactive intermediates produced by cytochrome P_{450} which detoxifies xenobiotics. Reduction in tissue ascorbic acid was observed in alloxan induced diabetic rats the reduction may be due to increased utilization of vitamin C as an antioxidant defense against increased reactive oxygen species or due to decrease in GSH level, since GSH is required for the recycling of vitamin C (Hunt, 1996).

α-Tocopherol is the most important lipid soluble, radical scavenging antioxidant. It reduces lipid hydroperoxide generated during the process of peroxidation and protects the cell structures against damage (Kinalski *et al.*, 2000). The elevated levels of α- tocopherol in the plasma of diabetic rats play a protective role against increased peroxidation in diabetes (Stanely and Menon, 1998). Similar results have been observed by other workers in STZ-diabetic rats. In tissues there was decrease in the level of α-Tocopherol in alloxan induced diabetic rats this may be due to the increased utilization of α- Tocopherol due to increased level of TBARS.

The antioxidant defense system is significantly altered in diabetes. A significant decrease in SOD and CAT activities is observed in the liver, kidney and brain of diabetic rats when compared with control animals, which may result in a number of deleterious effects due to the accumulation of superoxide radicals ($O2^-$) and hydrogen peroxides (H_2O_2) (Kono and Fridovich, 1982). Catalase causes reduction of H_2O_2 where as GPX reduces H_2O_2 and lipid peroxides. The GPX increased in the kidney of diabetic rats as compared to control. The decreased catalase activity and increased GPX activity in kidney suggests that there may be a compensatory mechanism among the antioxidant enzymes in response to increased oxidative stress so that tissues lacking significant catalase activity may be critically dependent on the activity of GPX. Administration of the leaf extract showed significant changes in the activity of SOD but there was no effect in the level of CAT and GPX.

Hexokinase is universally present in the cells of all types. Hepatocytes also contain a form of hexokinase called hexokinase D or Glucokinase which is more specific for glucose and differs from other forms of hexokinase in kinetic and regulatory properties (Lehninger *et al.*, 1993). Glucokinase (also called hexokinase IV) catalyzes the conversion of glucose to glucose -6-phosphate and plays an important role in the maintenance of glucose homeostasis. In liver, the enzyme is an important regulator of glucose storage and disposal (O'Doherty *et al.*, 1999). In our study the hexokinase activity was decreased in alloxan induced diabetic rats which may be due to insulin deficiency (insulin stimulates and activates hexokinase D). Treatment with the fresh leaf extract and glibenclamide elevated the activity of hexokinase D in the liver. *A. heterophyllus,* like glibenclamide, may stimulate insulin secretion which may activate hexokinase D, thereby increasing utilization of glucose leading to decrease blood sugar level. As there is no accumulation of glucose-6-phosphate, hexokinase activity is not inhibited by auto-regulation, as in the case of diabetic rats. Hence the activity of hexokinase D is decreased in *A. heterophyllus* and glibenclamide treated diabetic rats.

Insulin decrease gluconeogenesis by decreasing the activity of key enzyme, such as glucose-6-phosphatase, fructose-1, 6-bisphosphatase and phosphoenolpyruvate carboxykinase (Murray *et al.*, 2000). In our study, the increase of glucose-6-phosphatse and fructose-1,6-bisphosphate in the liver of alloxan induced diabetic rats may be due to insulin insufficiency. In *A. heterophyllus* and glibenclamide treated groups, the activities of these two enzymes were significantly reduced; this may be due to increased insulin secretion which is responsible for the repression of gluconeogenic key enzyme.

In conclusion, the fresh leaves extract showed hypoglycemic activity but the effect was less than the standard drug both the doses (400 and 600 mg/kg body weight) had the same effect. Though it did not show any significant changes in CAT and GPX there was significant changes observed in antioxidants such as GSH, Vitamin C, Vitamin E and SOD but there was no decrease in the level of TBARS.

There were significant changes in the activity of carbohydrate metabolizing enzyme. The hypoglycemic activity may be due to the changes in these enzymes. Traditionally ayurvedic formulation is mostly a combination of drugs. The fresh matured leaves of A. *heterophyllus* can be used in combination since there is no much antioxidant effects observed in the treatment of the extract the effect can be compensated with a combined formulation.

REFERENCES

Baker, H.O., Fuanko, B., De Angelis, S. and F. Feingold (1980). *Nutr. Rep Int.* **21**: 531.

Beutler, E. and D.M. Ditrono Kelly (1963). *J. Lab clin med.* **61**:882.

Branstrup, N., Kirk, J.K. and C. Bruni (1957). *J. Gerontol.* **12**: 166.

Briviba, K. and H. Sies (1994). In. *Antioxidants in human health and diseases* (Bfrie. Ed)., Academic press Inc. San. Diego.CA: USA: p.107

Cherian, S. and K.T. Augusti (1993). *Indian J Exp Biol.* **31**:26.

Fernando, M.R. and M.I. Thbrew (1989). *Ceylon J med Sci.* **32**:1

Fernando, M.R., Nalinie wickyamasingle, S.M.D., Thbrew, M.I., Ariya Nanda, P.L., and E.H. Karunanayake (1991). *J. Ethanopharmacol.* **31**: 277.

Gancedo, J.M. and C. Gancedo (1971). *Arch. Microbiol.* **76**:132.

Garg, M.C., Ojha, S. and D.D. Banasal (1996). Indian *J. Exp. Biol.* **34(3)**: 264.

Giardino, I., Edelstein, D. and M. Brownlee (1996). *J. Clin. Invest.* **97(6)**:1422.

Griesmacher, A., Kindhauser, M. and E.S. Andert *et al.* (1995). *Am. J. Med.* **98**:469.

Hunt, J.V. (1996) *Subcell. Biochem.* **25**: 369.

Kinalski, M., Sledziewski, M., Telejko, B., Zarzycki, W. and I. Kinalska (2000). *Acta Diabetologia.* **37**: 179.

Koide, H. and T. Odo (1959). *Clin. Chim. Acta.* **4**: 554.

Kono, Y. and I. Fridovich (1982). *J. Biol. Chem.* **257**: 5751.

Lehninger, L., David, L. and M.M. Nelson (1993). *Principles of biochemistry*: Second edition, CBS Publishers & Distributors, New Delhi. P. 406.

Marklund, S. and G. Marklund (1974). *European journal of Biochemistry.* **47**: 469.

Murray, R.K., Granner, D.K., Mayes, P.A. and V.W. Rodewell (2000). In. *Harper's Biochemistry,* 25th edition. Appleton and Lange Stanford, Connecticut.

Nichans, K.G.J. and B. Samuelson (1968). *Brit. J. Biochem.* **6**: 126.

Oberley, L.W. (1988). *Free Radic Biol Med.* **5**:113.

O' Doherty, R.M., Lehman, D.L., Telemaque Potts, S. and C.B. Newgard (1999). *Diabetes.* **48**: 2022.

Omaye, S.T., Turnabull, J.C. and H.E. Sanberlick (1979). *Methods in Enzymology.* **62**: 3.

Parinandi, N.L., Thompson, E.D.W. and H.O.H. Schmid (1990). *Biochem. Biophys. Acta.* **1047**:63.

Paglia, D.E. and W.N. Valentine (1967). *J. Lab. Clin. Med.* **70**: 158.

Sinha, A.K. (1972). *Anal. Biochem.* **47**: 389.

Stanly M.P. and V.P. Menon (1998). *J.Clinical Biochemistry and Nutrition.* **25**: 81.

Sundaram, R.K., Bhasakar, A., Vijilingam, S., Viswanathan, M., Mohan, R. and K.R. Shanmugasundram (1996). *Clin. Sci.* **90**: 255.

Wohaieb, S.A and D.V. Godin (1987). *Diabetes.* **36**: 1014.

18

EFFECT OF *CASSIA ALATA* L. LEAF EXTRACT ON ENZYMES INVOLVED IN CARBOHYDRATE METABOLISM IN NORMAL AND STREPTO-ZOTOCIN INDUCED DIABETIC RATS

R. Remya, S. Sathya and A. Prakasam

Department Of Biochemistry, Vivekananda College of Arts and Science, Periyar University, Salem, Tamilnadu, India

ABSTRACT

Hypoglycemic effect of *Cassia alata* leaf extract and the activity of hexokinase, glucose-6-phosphatase, fructose-1,6-bisphosphatase on liver of normal and Streptozotocin induced diabetic male wister rats were investigated. Rats were divided in to seven groups with six animals each. The rats were made diabetic by single intraperitonial injection of Streptozotocin. The rats were treated by oral administration of *Cassia alata* leaf extracts (400 and 600 mg/kg body weight) for 45 days. The results indicated that there were significant changes in hexokinase, glucose-6-phosphatase and fructose-1, 6-bisphosphatase in the liver of *Cassia alata* treated rats. The study clearly shows that aqueous leaf extract of *Cassia alata* show potent enzyme activity but weaker than that of standard drug glibenclamide and more activity was observed in 600mg/kg body weight of leaf extract than 400 mg/kg body weight.

INTRODUCTION

Diabetes mellitus is a metabolic disorder affecting carbohydrate, fat and protein metabolism; attributed to the diminished production of insulin or mounting resistance to its action. Chronic hyperglycemia during diabetes causes glycation of body proteins that in turn leads to secondary complications affecting eyes, kidney, nerves and arteries (Sharma, 1993). The oral hypoglycaemic agents currently used in clinical practice have characteristic profiles of serious side effects (Prout *et al.*, 1974; Holman and Turner, 1991; Williams and Pickup, 1991). Hence there is a need to search for newer anti-diabetic agents that retain therapeutic efficacy and are devoid of side effects that could be important sources of such agents.

Several studies have been conducted regarding the oral hypoglycaemic activity of different medicinal herbs. Antidiabetic herbs are an important therapy for diabetic patients in India. It is found that use of juices of bitter gourd, decoction of chiraita, neem leaves, betel leaves etc cause 10-20% lowering of blood glucose. In the present study *Cassia alata* Linn (Caesalpinaceae) has been chosen, for the leaves of this plant, by far the most extensively investigated and most widely acclaimed remedy for the treatment of diabetes mellitus since ancient times. Palanichamy *et al.* (1988) reported the hypoglycaemic activity of *C. alata* in Streptozotocin diabetic rats in different time intervals.

No systematic studies have been carried out to find the effect of this plant on carbohydrate metabolic enzymes, in control and experimental animals. Therefore in the present investigation, optimum dose determination and the related biochemical parameters in normal and Streptozotocin diabetic rats are studied. The effects of the extract were compared with glibenclamide, a well known hypoglycaemic drug.

MATERIALS AND METHODS

Leaves of *Cassia alata* were collected from Namakkal district, Tamilnadu and the voucher specimen [VICAS-03-1674] was deposited in survey medicinal plant unit, Regional research laboratory, Bangalore. The air dried leaves were dried at 60°C and powdered. 100g of this powder was suspended in 250 ml water for 2h and then warmed to 60-65°C for 30 minutes. The extract was filtered out and the residual powder was subsequently extracted for three times with water. The pooled extracts were evaporated to get a semi solid fraction (14.5%). Streptozotocin (STZ) was obtained from sigma chemical company.

Male Wister albino rats (weighing 150-160) were procured from the central animal house, Department of experimental medicine, Government medical college, Perundurai, Erode, Tamilnadu, India. The animals were fed on standard diet and water *ad libitum*. Diabetes was induced by a single intraperitoneal injection of STZ (50 mg/ kg body weight) dissolved in citrate buffer (0.1 M, pH 4.5). After 96h fasting blood glucose were considered diabetic and used for the experiment.

Study design

After the induction of diabetes, the rats were divided in to seven groups with six animals each. The duration of the treatment was for 45 days, the extract was given in 2% gum acacia using an intragastric tube.

Group [1] - Control rats (received vehicle solution- 2% gum acacia).

Group [2] - Diabetic control.

Group [3] - Normal rats + 400 mg *C. alata* leaf extract.

Group [4] - Normal rats + 600 mg extract.

Group [5] - Diabetic rats + 400 mg extract.

Group [6] - Diabetic rats + 600 mg extract.

Group [7] - Diabetic rats + Glibenclamide (600 μg / kg body weight) in 2% gum acacia.

At the end of the experiment, the animals were killed by cervical dislocation. The liver was excised immediately and homogenized in ice cold, Tris- HCL buffer (0.1 N). The homogenate was used for the enzyme assay.

Biochemical analysis

Spectrophotometric assays were used to measure the activities of Hexokinase, Glucose -6- Phosphatase, and Fructose 1, 6- bisphosphatase

Enzyme assays

Hexokinase [EC 2.7.1.1] was assayed according to Branstrup *et al.,* (1957) by estimation of residual glucose in the reaction supernatant. Glucose -6- Phosphatase [EC 3.1.3.9] was assayed according to Koide and Odo (1959) by estimation of inorganic phosphate liberated from glucose -6- phosphate. Fructose-1, 6-Phosphatase [EC 3.1.3.9] was assayed by the method of Gancedo and Gancedo (1971). Values were represented as means ± SD, and group mean were compared with Duncan's Multiple Range Test (DMRT).

RESULTS AND DISCUSSION

Table 1.

Gr	Treatment (Dose/Kg body wt)	Hexokinase (U^a / mg protein)	Glucose-6-phosphatase (U^b / mg protein)	Fructose-1, 6-bisphosphatase (U^c / mg protein)
1.	Control (received 2% gum acacia)	0.22 ± 0.06^a	0.16 ± 0.01^a	0.56 ± 0.18^a
2.	Diabetic Control	0.12 ± 0.01^b	0.28 ± 0.02^b	1.08 ± 0.64^b
3.	Normal + *C. alata* I	0.23 ± 0.04^a	0.15 ± 0.03^a	0.55 ± 0.06^a
4.	Normal + *C. alata* II	0.24 ± 0.06^a	0.14 ± 0.02^a	0.53 ± 0.02^a
5.	Diabetic + *C. alata* I	0.16 ± 0.03^c	0.24 ± 0.03^c	0.78 ± 0.06^c
6.	Diabetic + *C. alata* II	0.18 ± 0.02^c	0.22 ± 0.04^c	0.66 ± 0.04^d
7.	Diabetic + Glibenclamide (600 μg/ Kg body wt)	0.18 ± 0.01^c	0.19 ± 0.03^d	0.62 ± 0.05^d

Values are means ± S.D for six animals in each group

Values not sharing a common superscript after significantly at $P < 0.05$

Duncan's Multiple Range Test (DMRT)

The activity of Hexokinase, Glucose -6- Phosphatase, and Fructose 1, 6-bisphosphatase was represented in the Table 1. A decreased activity of hexokinase is observed. On the other hand the activities of Glucose-6-Phosphatase and Fructose -1, 6- Phosphatase are significantly increased in diabetic rats. Aqueous extracts of (400 and 600 mg / kg body weight) caused a significantly increased Hexokinase activity and decreased Glucose-6-Phosphatase and Fructose -1, 6- bisphosphatase activity.

The decreased Hexokinase activity in STZ diabetic rats may be attributed to insulin deficiency, treatment with *Cassia alata* and glibenclamide elevated the activity of Hexokinase, there by increasing utilization of glucose which leads to decreased blood glucose level. Whereas the increased activity of Glucose-6-Phosphatase and Fructose-1,6- bisphos-phatase may be due to insufficiency of insulin in the diabetic rats. The treatment significantly reduces the activity of these enzymes. This may be due to increased insulin secretion that is responsible for the repression of the gluconeogenic key enzymes. The extract of 600 mg/kg. body weight showed better results but less than that of glibenclamide.

Therefore it can be concluded that the aqueous extract of *Cassia alata* leaf exhibits hypoglycaemic effect in Streptozotocin induced diabetic rats. The effect on carbohydrate metabolic enzymes changes significantly. Further studies are necessary to determine the extract nature of the active principles and the mechanism of action.

REFERENCES

Atta-Ur-Rahman and K. Zaman (1989). *J. Ethonopharmacol.* **20**: 553.

Branstrup, N., Kirk, J.K. and C. Bruni (1957). *J. Gerontol.* **12**: 166.

Gancedo, J. M. and C. Gancedo (1971). *Arch. Microbiol.* **76**:132.

Holman, R.R. and R.C. Turner (1991). In: *Text book of diabetes.* (Pickup, J. and G. Williams, Eds.), *Black* well, Oxford, p. 467.

Koide, H. and T. Odo (1959). *Clin. Chim. Acta.* **4**: 554.

Palanichamy, S., Nagarajan, S. and M. Devasagayam (1988). *J. Ethonopharmacol.* **22**: 81.

Prout, T.E., Malaisse, W.J and J. Pirart (1974). *Proceedings VIII Congress of International diabetic federation*, Excerpta medica, Amsterdam, p. 162.

Sharma, A.K. (1993). In: *Diabetes mellitus and its complications: An update*, Galadari, E.O. and M.K. Mehra. [Eds.]. 1st Ed, Macmillan, New Delhi.

Williams, G and J. Pickup (1991). In. *Text book of diabetes* II. (Pickup, J.C and Williams, G. Eds.). Black well, Oxford, p. 977.

19

STUDIES ON PHARMACOGNOSY AND PHYTOCHEMISTRY OF *JATROPHA CURCAS* LINN AND *JATROPHA GOSSYPIFOLIA* LINN.

Sallykutty Thomas and M. Daniel

Phytochemistry Laboratory, Department of Botany, Faculty of Science, The Maharaja Sayajirao University of Baroda, Vadodara-390 002

ABSTRACT

Pharmacognostic and Phytochemical studies on leaves, stem and root of *J. curcas* and *J. gossypifolia* were carried out. Leaf constants such as stomatal index/mm^2, stomatal frequency/mm^2, size of guard cell, palisade ratio, vein islet number/mm^2, vein termination number/mm^2 and measurements of various cell elements in transections were enumerated. Leaves of *J. curcas* were characterized by glabrous nature, contiguous stomata, rosettes of calcium oxalate crystals in spongy tissue, abundance of starch grains uniformly distributed in all parts of the stem along with rosettes and prisms of calcium oxalate crystals. The root showed patches of cortical fibres and diarch primary xylem. Proteins, resin, starch, saponins, tannins and alkaloids were histochemically located. Vessel elements with simple perforation plates, bordered pits, reticulate and scalariform thickenings were seen on the lateral walls. Both the stem and leaves were found to contain flavones apigenin and its methoxy derivatives and luteolin, phenolic acids vanillic, syringic, *p*-OH Benzoic acid, gentisic, *cis* and *trans* ferulic, *o*-Coumaric, melilotic acids, steroids, glycoflavones and proanthocyanidins. *J. gossypifolia* was different in having simple uniseriate covering, multicellular and gland tipped sticky brown hairs on the leaf margins, petiole and stipules, brachyparacytic and adjoining stomata, smaller sized starch grains and rosettes of calcium oxalate crystals. The leaves were found to possess apigenin and it methoxy derivative, luteolin and its methoxy derivatives and *p*-Coumaric acid in addition to the flavones and phenolic acids of the former plant.

INTRODUCTION

Jatropha curcas L. (Dravanti), commonly known as *Physic nut* is a medicinal plant, used traditionally, for the treatment of sciatica, dropsy, paralysis, rheumatism, dysentery, jaundice, malarial fever and certain skin diseases. The latex of this plant is reported to have anticancerous properties. The leaves are reported to be galactagogue, rubifacient, insecticidal and are used in foul ulcers, tumours and scabies. The seeds are powerful purgative, acrid and useful in piles, wounds and enlargement in spleen (Anon., 1959; Kirthikar and Basu, 2003; Pandey *et al.*, 2005; Sivarajan and Balachandran, 1994). The roots of this plant are reported to have anti-inflammatory activity (Mujumdar and Misar, 2003). The leaves are reported to contain apigenin, vitexin, isovitexin, stigmasterol and β-sitosterol (Khafagy *et al.*, 1977). Pharmacognostic characters of stem bark of *J. curcas* have been described (Gupta, 1985) and these were compared with those of *J. gossypifolia*, which is sometimes used as Dravanti (Datta and Datta, 1983). Reghunathan and Mitra (1982) have carried out the pharmacognostic studies on Dravanti. Detailed micromorphological, anatomical, histochemical, powder analysis and phytochemical data have been found wanting for both these plants. Therefore the present investigation is undertaken to study the detailed pharmacognostic and phytochemical characters of *J. curcas* and its related species *J. gossypifolia* that are useful in identifying the plant sources and adulterants.

MATERIALS AND METHODS

The plant materials for the present study were collected from different localities of the campus of the M.S. University of Baroda, Gujarat. The voucher specimens of both these plants are submitted in BARO, the herbarium of Department of Botany, The Maharaja Sayajirao University of Baroda, Vadodara. Plant materials were washed, shade dried and completely dried by keeping in an oven at 60°C. The dried materials were powdered and stored in airtight plastic bags. This powder was used for the analysis of all the chemical constituents. Micromorphological studies were carried out on fresh materials. Epidermal peels were taken and stained with Toluidine blue (0.5%) and mounted in glycerin. Leaf constants such as stomatal index, stomatal frequency, size of guard cellum, vein islet number, vein termination number and palisade ratio were calculated. Transections of leaf, stem and root were taken by free hand sections, and were stained in Toluidine blue and mounted in glycerin (Johansen, 1940).

For the study of vessel elements fragments of stem and root were macerated using a mixture of nitric acid (10%) and potassium dichromate (10%) solution in equal proportions. The vessel elements were stained with Toluidine blue and mounted in glycerin and the margins of the cover slips were sealed with DPX. The slides were examined under the microscope and Camera Lucida sketches were drawn and the cells measured using an ocular micrometer. The quantitative data were based on the average of 20 readings. Histochemical tests were conducted on fresh materials and were located in sections according to Johansen (1940); Gibbs (1974) and Jensen (1962). The finely powdered drug was scanned under 400 x magnifications for recording the cell elements. For Phytochemical analysis dried powdered materials were analysed for flavonoids, phenolic acids, steroids, alkaloids, proanthocyanidins and glycoflavones using standard procedures (Mabry *et al.*, 1970; Harborne, 1984; Daniel, 1991). The identification of compounds were done by colour reactions and spectral data and confirmed by co-chromatography with authentic samples.

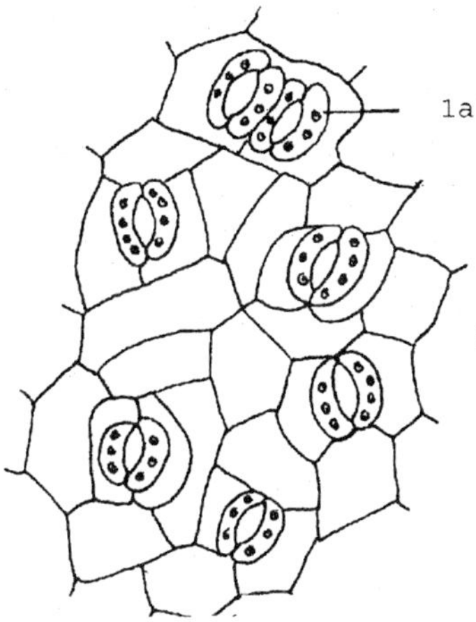

Fig. 1a. Surface view of J. curcas leaf lamina, contiguous stomata

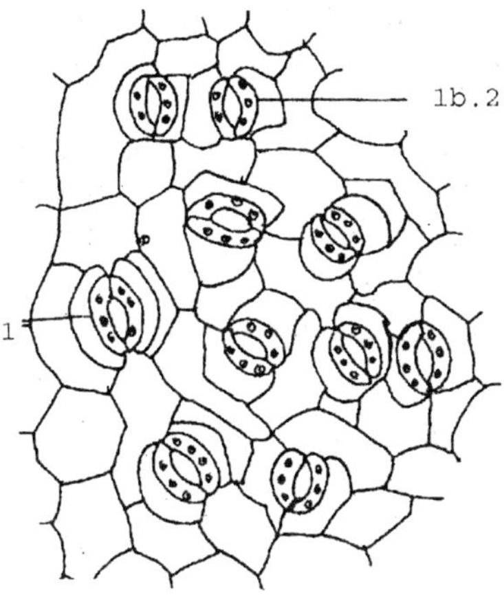

Fig. 1b. Surface view of J. gossypifolia leaf lamina stomata
1b1. brachyparacytic 1b2. Adjoining

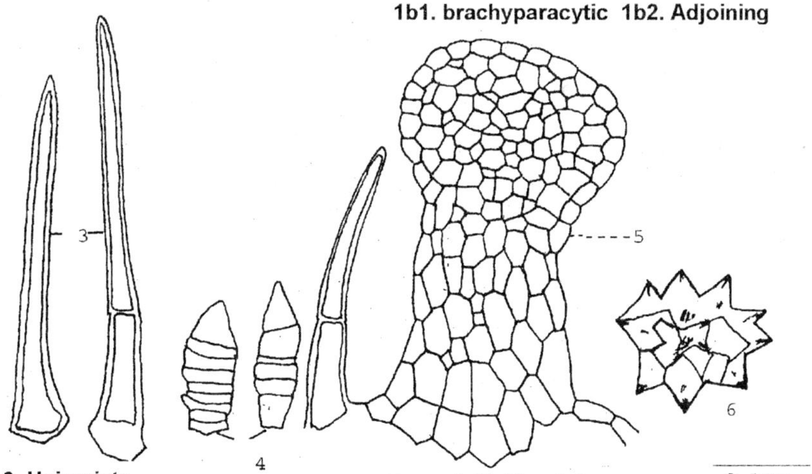

3. Uniseriate covering hair **4. Multi cellular hair** **5. Multiseriate glandular hair**

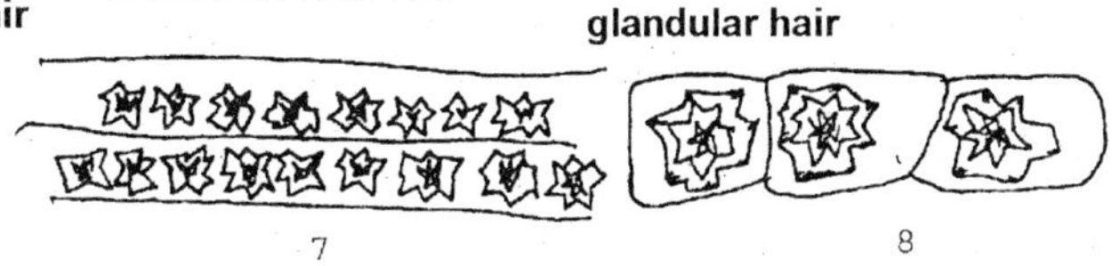

6. Rosettes of calcium oxalate crystal
7. Crystalliferous parenchyma strand
8. Crystalliferous parenchyma cell

RESULTS

Pharmacognostic characters

Micromorphology: Epidermal cell complex: Adaxial and abaxial epidermal cells were polyhedral with straight walls. The average size of the cells was from 13.3 x 16.1 μm, in *J. curcas* and 14.35 x 18.55 μm in *J. gossypifolia*.

Stomatal complex : Leaves were glabrous, multicostate, amphistomic bearing paracytic stomata. Paracytic type was prominent in both the species where as contiguous type (Fig. 1a) was found in *J. curcas*. Rosettes of calcium oxalate crystals were scattered in the leaves of both the taxa (Fig. 6). Leaf constants such as stomatal index/mm^2, stomatal frequency/mm^2, size of guard cellum, vein islet number/mm^2, vein termination number /mm^2 and palisade ratio were enumerated and are represented in Table 1. Vein termination number (30.7/mm^2) and size of guard cell (36.8 x 28 μm) were maximum in *J. curcas*.

Table 1. Leaf Constants

Sr. No.	Name of Plant	Adaxial epidermis			Abaxial epidermis			
		V. I. N/ mm^2	V. T. N / mm^2	P. R.	S. I/ mm^2	S. F/ mm^2	S. G. C (μ m)	
							L	B
1.	*J. curcas*	6.8	30.7	16.4	21.6	5.4	36.8	28
2.	*J. gossypifolia*	7.9	29.2	17.8	26.6	11.4	26.5	19.3

S. I. - Stomatal Index; S. F. - Stomatal Frequency; S.G.C. - Size of Guard Cell ; V.I.N - Vein Islet Number; V. T. N. - Vein Termination Number; P. R. - Palisade Ratio; L - Length; B - Breadth

Trichome Complex : Uniseriate covering (Fig. 1b.3), multicellular (Fig. 1b.4) trichomes were confined to the marginal surfaces of the leaves. Mutiseriate glandular hairs were found only on the margins of the leaves (Fig. 1b.5).

Anatomy : (Fig. 2a-2b) In T.S. epidermis consisted of a single layer of cubical to tangentially elongated parenchyma cells with cuticle. Trichomes were absent on both the surfaces and margins of the leaves. The mesophyll was differentiated into single layered cylindrical palisade cells and a narrow zone of spongy tissues. Spongy tissues consisted of loosely arranged parenchyma cells with intercellular spaces containing chloroplast. Non-articulate branched laticifers were ramified into the mesophyll tissues. Latex cells occurred in costa, mesophyll, and spongy tissues. Crystalliferous parenchyma cells (Fig. 8) were also found in the mesophyll tissues of both the taxa. T.S. through each costa consisted of single layered epidermis followed by a wide zone of cortex, the outer layers as collenchyma followed by parenchyma cell of various sizes. A number of these cells contained rosettes of calcium oxalate crystals. There was a single crescent shaped median vascular bundle with xylem facing upwards. Xylem consisted of radially arranged vessels, tracheids, xylem parenchyma and xylem rays. Phloem cells contained sieve tubes, companion cells, phloem parenchyma a number of which contained rosettes of calcium oxalate crystals, the size of which varied from 23.45 x 25.2 μm and starch grains size varied from 10.3 x 9.8 μm.

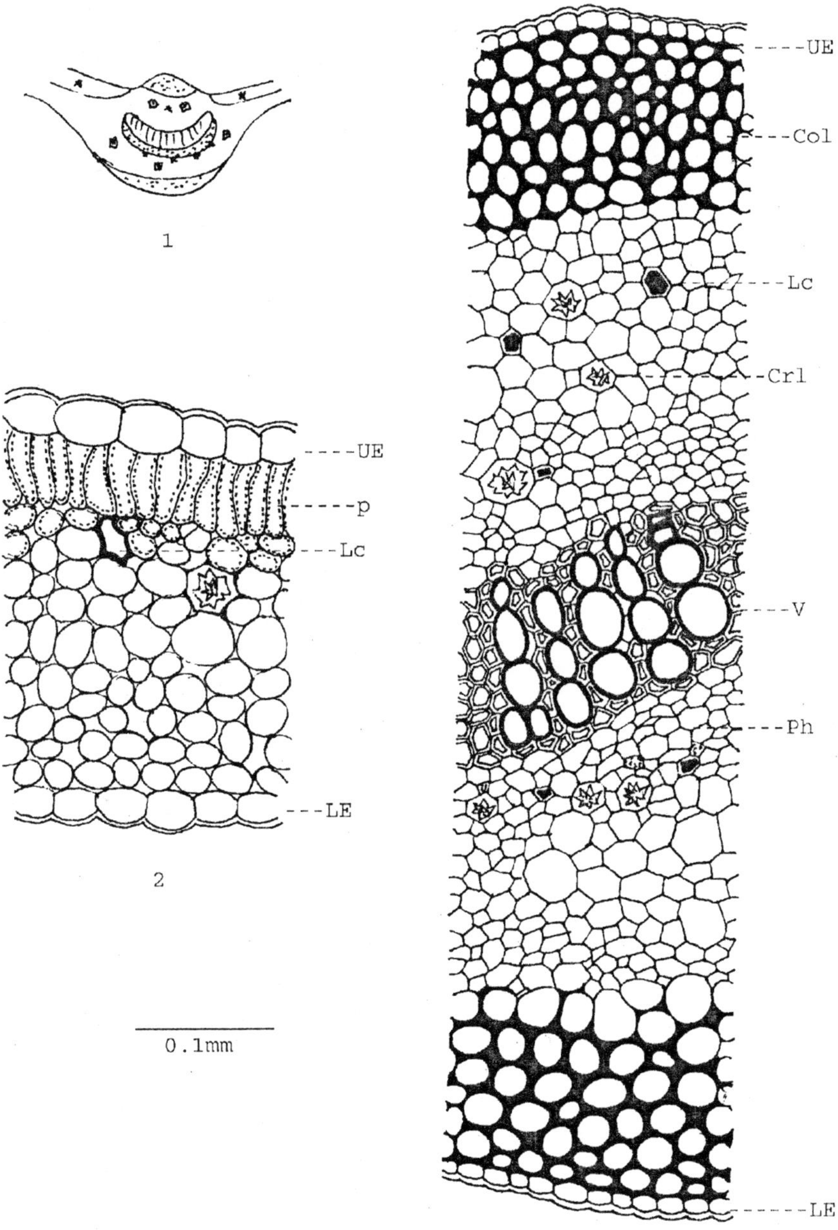

Fig. 2a. Histology of leaf of *J. curcas*

1. Diagramatic sketch of the T.S. of the leaf, 2. A portion of the T.S. of leaf lamina, 3. A portion of the T.S. costa UE, upper epidermis; Col, collenchyma, LC, latex canal; Crl, crystal, V, vessel; Ph, phloem, LE, lower epidermis, P, palisade

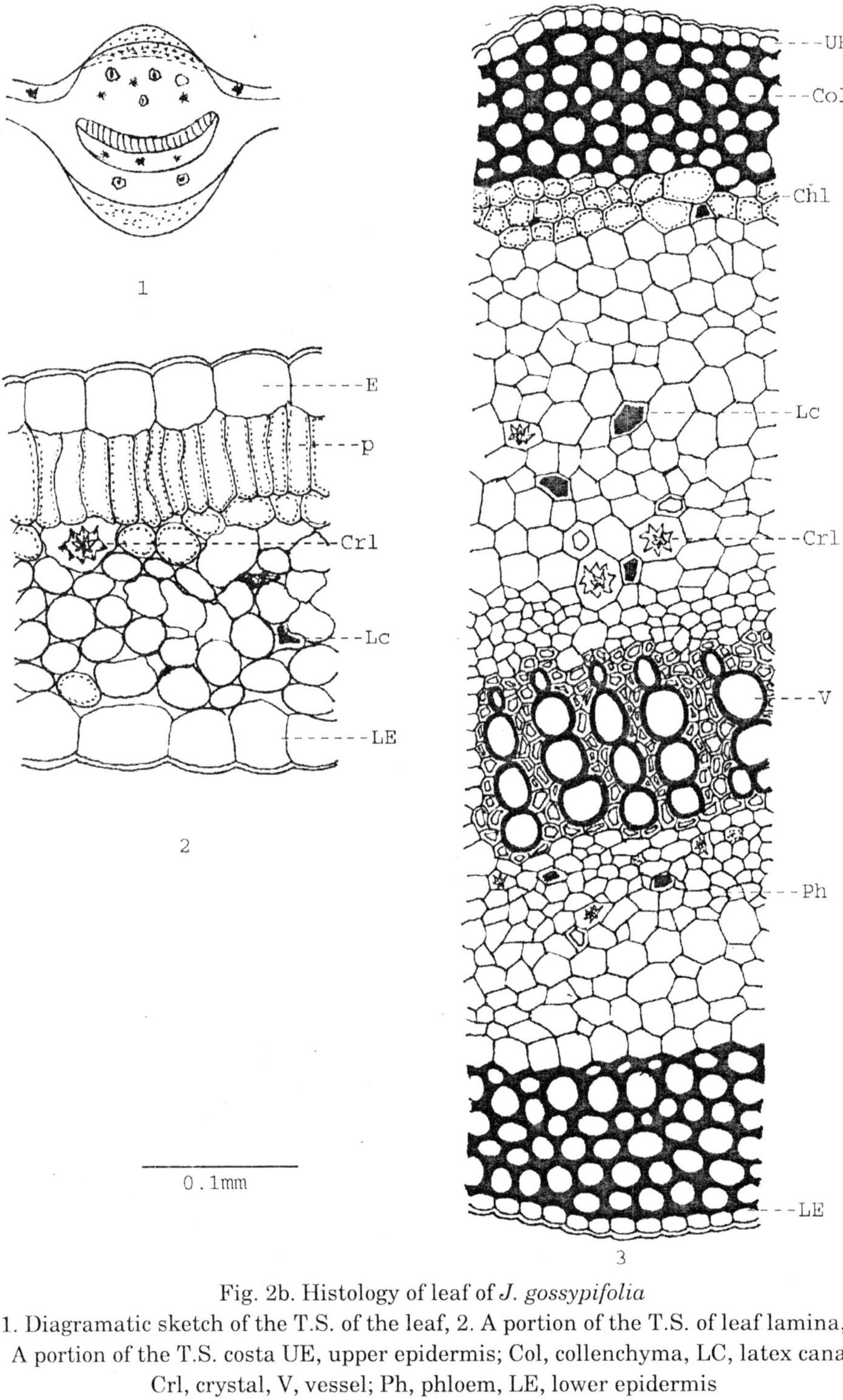

Fig. 2b. Histology of leaf of *J. gossypifolia*

1. Diagramatic sketch of the T.S. of the leaf, 2. A portion of the T.S. of leaf lamina, 3. A portion of the T.S. costa UE, upper epidermis; Col, collenchyma, LC, latex canal; Crl, crystal, V, vessel; Ph, phloem, LE, lower epidermis

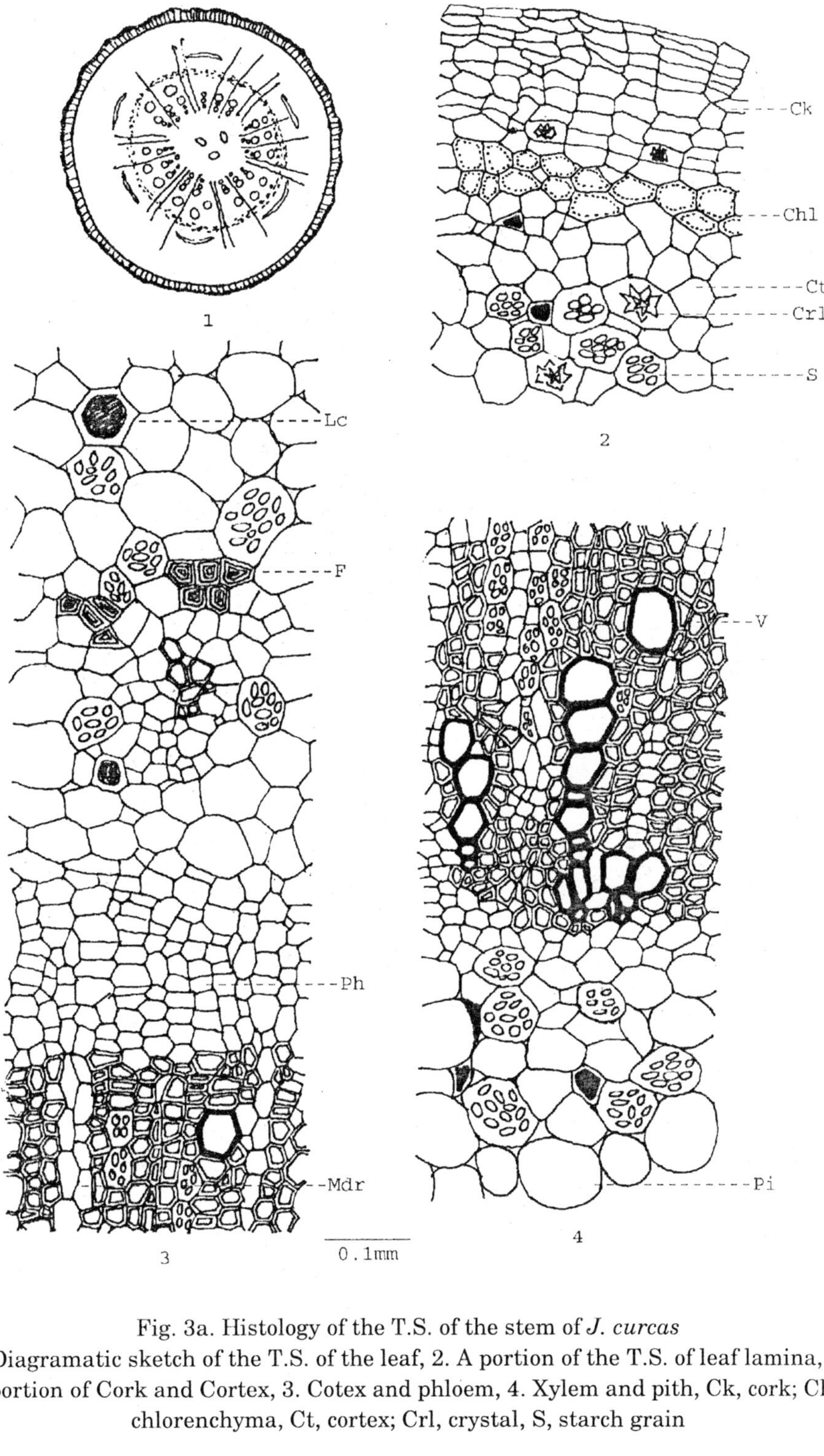

Fig. 3a. Histology of the T.S. of the stem of *J. curcas*
1. Diagramatic sketch of the T.S. of the leaf, 2. A portion of the T.S. of leaf lamina, 3. A portion of Cork and Cortex, 3. Cotex and phloem, 4. Xylem and pith, Ck, cork; Chl, chlorenchyma, Ct, cortex; Crl, crystal, S, starch grain

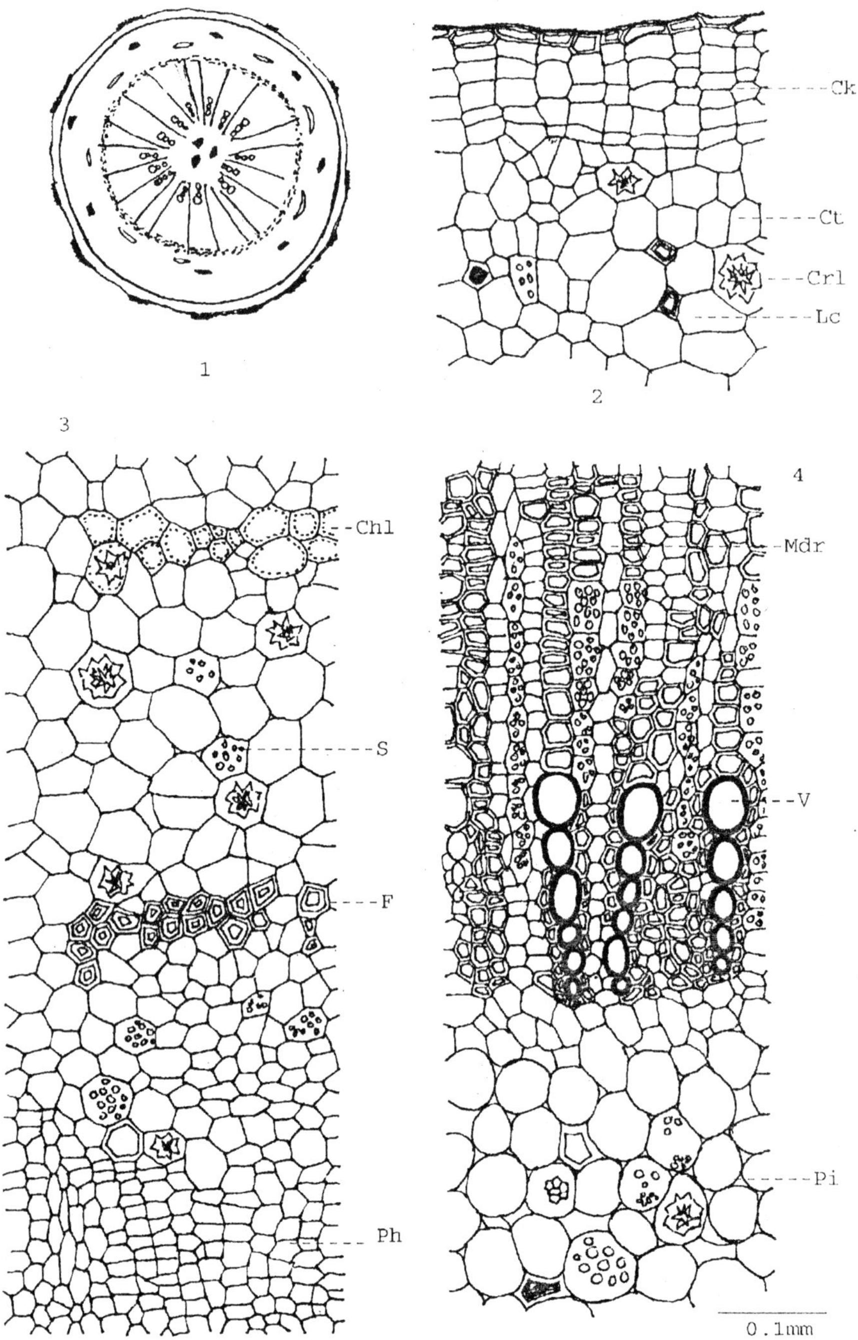

Fig. 3b. Histology of the T.S. of the stem of *J. gossypifolia* 1. Diagramatic sketch of T.S. of the stem, 2. A portion of cork and cortex, 3. cortex and phloem, 4. Xylem and pith, Ck, cork, Ct, cortex, Crl, crystal, Chl, chlorenchyma, S, strach grain, F, fibres; Ph, phloem; Mdr, medullary ray; V, vessel; Pi, pith

In T.S. of stem (Fig. 3a-3b) cork cells exhibited varied structures, mainly tangentially elongated thin walled cells arranged in radial rows. Cortex was differentiated into collenchyma containing chloroplasts and parenchyma. Rosettes of calcium oxalate crystals and starch grains were uniformly distributed in all parts of the stem. Latex tubes filled with brown contents and crystalliferous parenchyma strands (Fig. 7) were also seen in both the taxa. Secondary phloem consisted of sieve tubes, companion cells, phloem parenchyma and phloem fibres. Medullary rays were uni or biseriate. Solitary and small groups of phloem fibres were regularly arranged on the radial rows of phloem element. Phloem fibres were long thick walled having rosettes of calcium oxalate crystals in some of them. Xylem consisted of xylem vessels, tracheids, fibres, xylem parenchyma and xylem rays. Pith was composed of polyhedral oval to almost circular parenchymatous cells containing starch grains and calcium oxalate crystals.

In root (Fig. 4a-4b) cork consisted of tangentially elongated thinwalled parenchyma cells which was followed by phelloderm composed of thin walled parenchyma cells of various sizes and shapes contained rosettes of calcium oxalate crystals in some of them. Secondary phloem consisted of sieve tubes, companion cells, phloem parenchyma, phloem fibres and phloem rays. Solitary and small groups of phloem fibres were irregularly arranged on radial rows of phloem elements. Phloem parenchyma were thin walled and of various shapes and sizes contained rosettes of calcium oxalate crystals in some of them. Latex tubes filled with dark brown contents were distributed in the cork, cortical and vascular tissues. Xylem contained trachieds, fibres, xylem parenchyma and xylem rays. Xylem vessels were distributed singly and in groups and were arranged radially. Primary xylem was diarch.

In macerations, the vessels were found to be of different shapes and sizes. Some were cylindrical having peg like outgrowths at the ends while others were short wide and drum shaped having scalariform, reticulate and bordered pits arranged alternatively. Trachieds were longer than vessels and were having tapering ends.

Vessel elements of the stem (Fig. 5a) were cylindrical drum shaped, their length ranging from 450-1050 μm and 45 -75 μm in diameter. They were moderately small to extremely large. Perforation plate was simple, oval to round placed at the terminal or sub terminal ends. Scalariform thickenings, reticulate, crowded angular pit pairs and bordered pits were seen on the lateral walls. Tail was short with pointed end.

In root, (Fig. 5b) vessel elements were short cylindrical, their length ranging from 150-270 μm and 60 -75 μm in diameter. They were moderately small to extremely large. Perforation plate was simple and oval placed at terminal or sub terminal ends. Bordered pits, scalariform, reticulate and crowded angular pit pairs were arranged on the lateral walls. Tail was short with pointed end.

The six tests were conducted in leaf, stem and root are presented in Table 3.

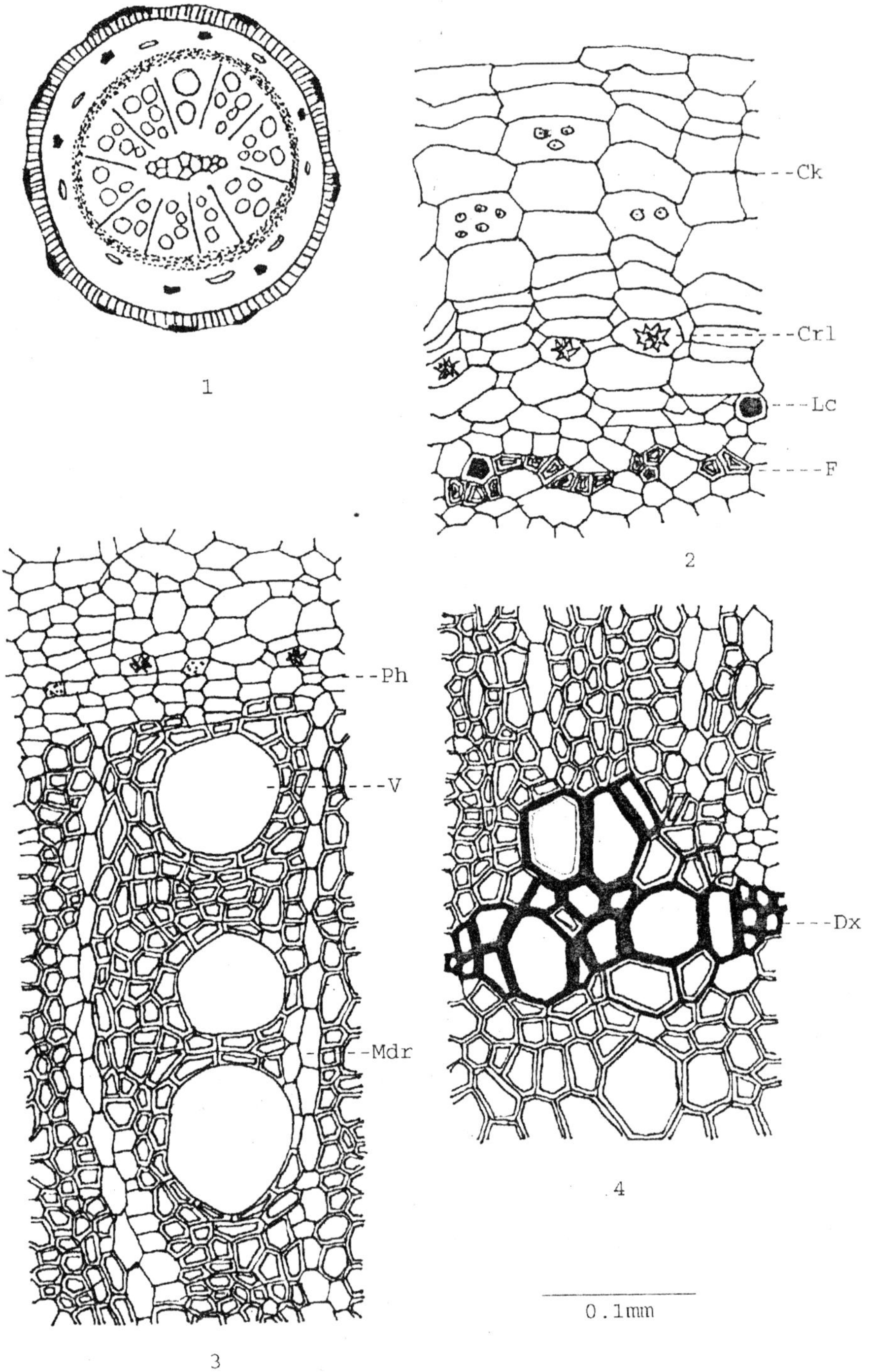

Fig. 4a. Histology of root of *J. curcas.* 1. Diagramatic sketch of the T.S. of root, 2. A portion of cork and cortex, 3. Phloem and xylem, 4. Xylem Ck, cork; Crl, crystal, Lc, latex cell; F, fibres; Ph, phloem; V, vessel; Mdr, medullary ray; Dx, diarch xylem

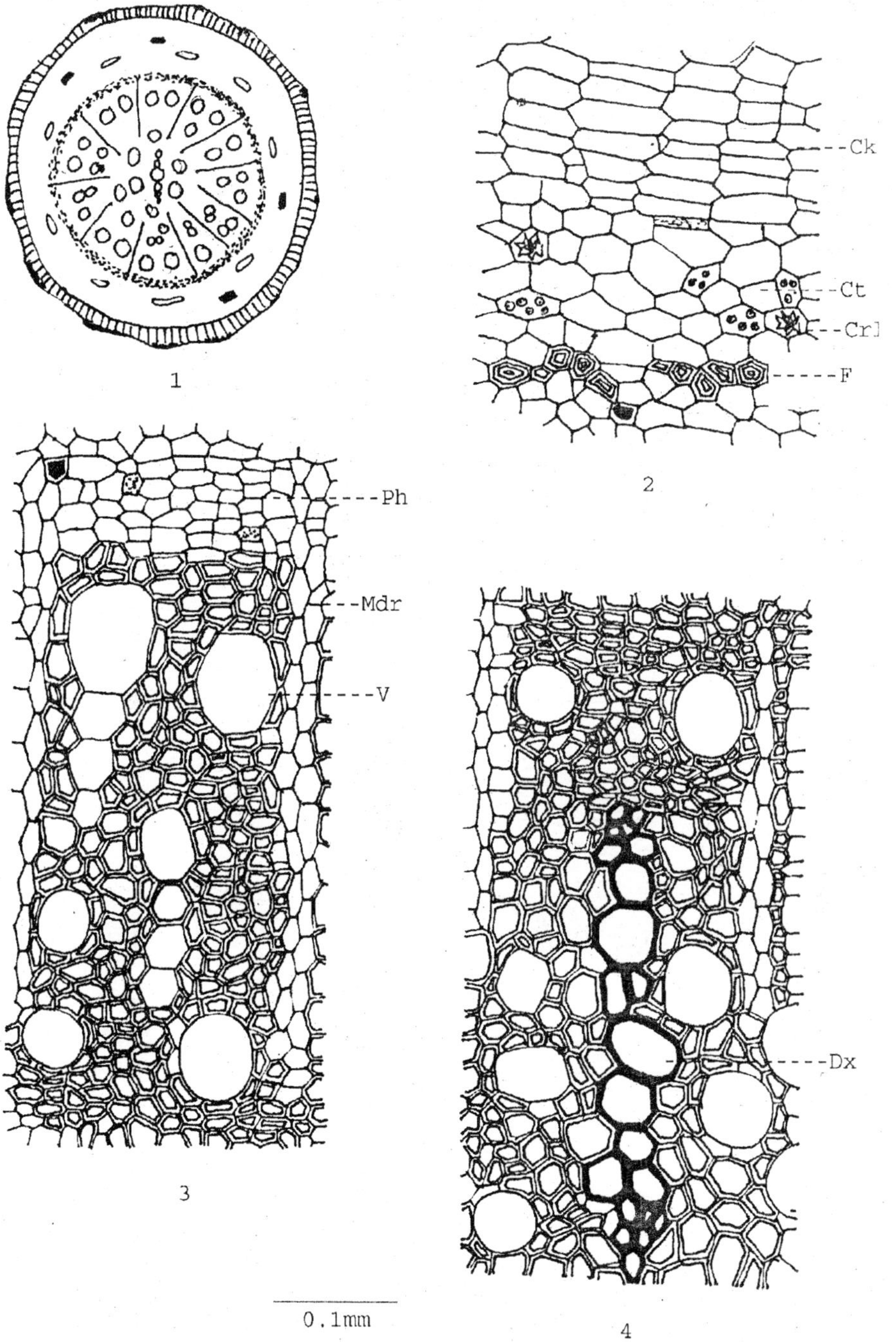

Fig. 4b. Histology of root of *J. gossypifolia*, 1. Diagramatic sketch of the T.S. of the root, 2. A portion of cork and cortex, 3. Phloem and xylem, 4. xylem Ck, cork, Ct, cortex; Crl, crystal F, fibres, Ph, phloem; Mdr, medullary ray; V, vessel; Dx, diarch xylem.

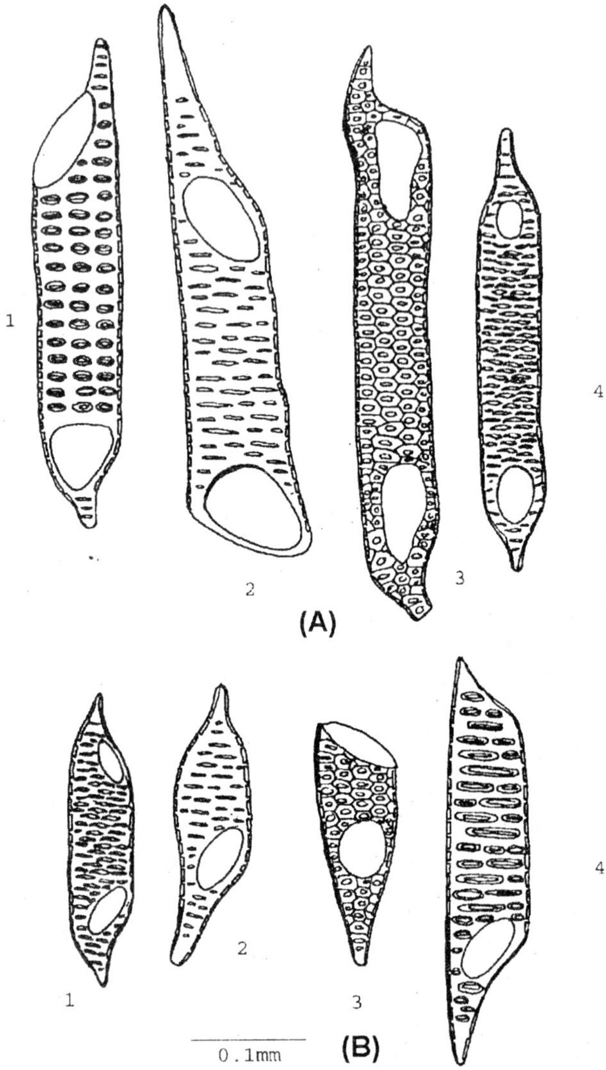

Fig. 5a. Vessel elements of *J. curcas* stem, 1. bordered pits, 2. scalariform, 3. crowded angular pit pairs, 4. reticulate thickening,

Fig. 5b. Vessel elements of *J. curcas* root, 1. reticulate, 2. scalariform, 3. crowded angular pit pairs, 4. bordered pit.

Powdered drug of leaf, stem and root showed fragments of palisade tissues, stomata, broken parts of tracheids, fibres, latex tubes, starch grains, rosettes of calcium oxalate crystals, vessels with single and double spirals, helical, scalariform, crowded angular pit pairs and bordered pits (Fig. 7a).

J. gossypifolia was different in having simple uniseriate covering and multiseriate gland tipped sticky brown hairs on the leaf margins, petiole and stipules, multicellular hair and small sized simple, semi compound starch grains and rosettes of calcium oxalate crystals (Fig. 1b.8), brachyparacytic (Fig. 1b.1) and adjoining stomata (Fig. 1b.2). Leaf constants such as stomatal index (26.6/mm^2), stomatal frequency (11.4 /mm^2), vein islet number (7.9/ mm^2) and palisade ratio (17.8) were higher in this plant. Vessel elements with oblique end walls. Perforation plates were oval to elongate and scalariform in stem. Crowded angular pit pairs, bordered pits and scalariform thickenings were seen on the lateral walls of the element (Fig. 6a) their length ranging from 450-1200 μm and 45-65 μm in diameter. They were moderately small to extremely large. In root their length ranging from 300-600 μm and 45-75 μm in diameter. They were moderately small to extremely large. Scalariform and bordered pits arranged along the lateral walls (Fig. 6b). Perforation plates was oval placed at the terminal or sub terminal ends.

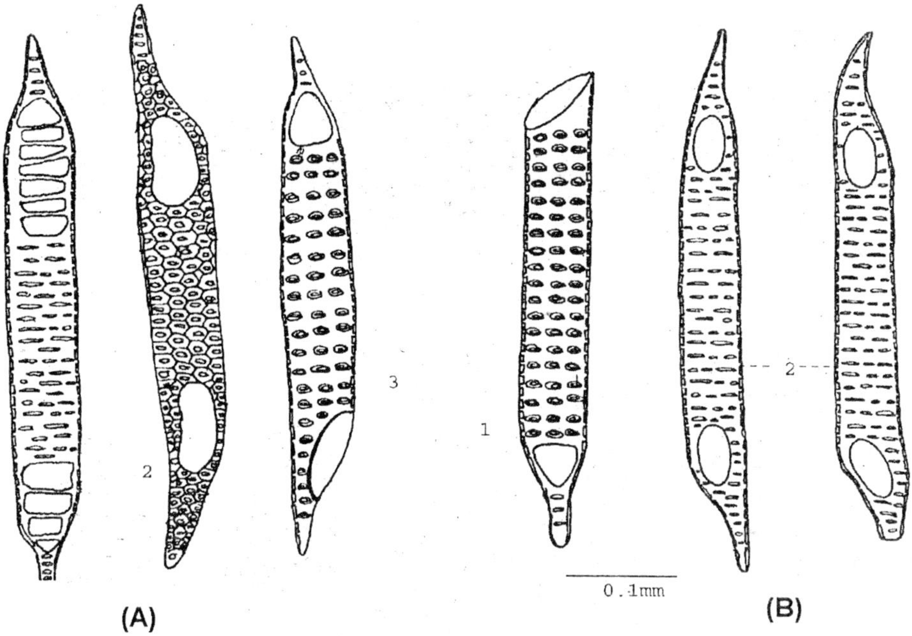

Fig. 6a. *J. gossypifolia* stem 1. scalariform pits, 2. crowded angular pit pairs, 3. round bordered pits,

Fig. 6b. Vessel elements of *J. gossypifolia* root 1. root bordered pits, 2. scalariform pits.

Powdered drug of leaf, stem and root of this plant showed simple, semi, compound starch grains, prisms and rosettes of calcium oxalate crystals, fragments of trichomes, vessel elements with crowded angular pit pairs, elongate pit pairs, scalariform pits (Fig. 7b).

Phytochemical analysis

Leaves and stem were taken for phytochemical analysis and were found to possess flavonoids, phenolic acids and steroids. Proanthocyanidins and glycoflavones were identified in the leaves only. In leaves the flavonoids located as flavones apigenin and 4'-OMe apigenin, 7'-OMe apigenin and luteolin; phenolic acids were vanillic, syringic, *p*-OH Benzoic acid, *o*-Coumaric, melilotic, *cis* & *trans* ferulic and gentisic acids. In stem only 3'4'- di

OMe luteolin and phenolic acids vanillic, syringic, *cis* and *trans* ferulic and *o*-coumaric were located.

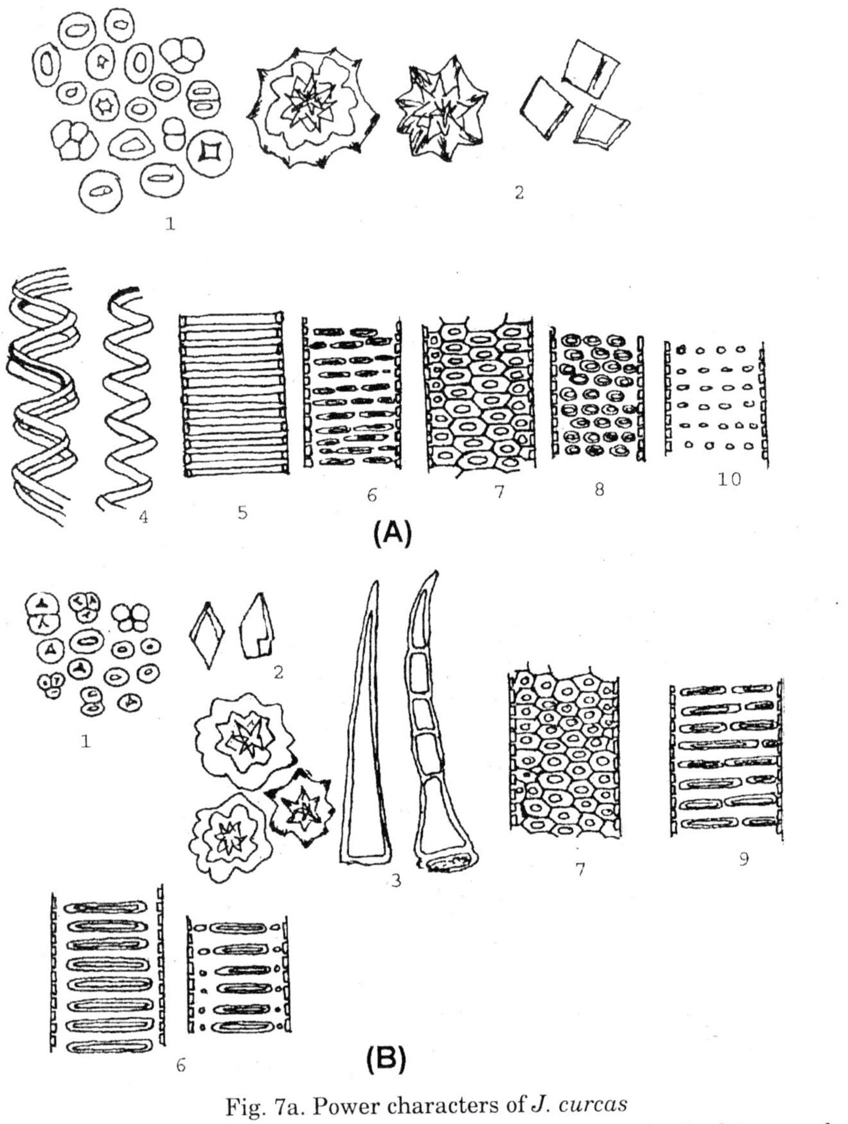

Fig. 7a. Power characters of *J. curcas*

Fig. 7b. Simple and compound starch grains, 2. rosettes and prisms of calcium oxalate crystals, 3. uniseriate covering trichome, 4. single and double spiral vessels, 5. helical, 6. scalariform, 7. crowded angular pit pairs, 8. bordered pits 9. elongate pit pairs, 10. simple pits.

In *J. gossypifolia* the leaves were found to possess luteolin, 3'-OMe luteolin, 3'4'-diOMe luteolin, 7' OMe luteolin and *p*-Coumaric acid in addition to the flavones and phenolic acids present in the former plant. In stem the flavonoids located as 3'4'-di OMe luteolin and phenolic acids vanillic, syringic and *p*-OH Benzoic acids (Table 2).

Table 2. The Distribution of various phytochemicals in leaves / stem of *J. curcas* and *J. gossypifolia*

Sr. No.	Name of Plant	1	2	3	4	5	6	7	8	9	10	11	12	13	14	15	16	17	18	19
1.	*J. curcas* (leaves)	+	+	+	+	-	-	-	+	+	-	+	-	+	+	+	-	+	+	+
	Stem	-	-	-	+	-	-	-	+	+	+	-	+	+	-	+	-	+	-	-
2.	*J. gossypifolia* (leaves)	+	+	+	+	+	+	+	+	+	+	+	+	+	+	+	+	+	+	+
	Stem	-	-	-	-	-	+	-	+	+	+	-	-	-	-	-	+	+	-	-

(1) Apigenin (2) 4'-OMe apigenin (3) 7' -OMe apigenin (4) Luteolin (5) 3'-OMe luteolin (6) 3'4'-di OMe luteolin (7) 7'-OMe luteolin (8) Vanllic acid (9) Syringic acid (10) *p*-OH Benzoic acid (11) Melilotic (12) *cis* ferulic acid (13) *trans* ferulic acid (14) Gentisic acid (15) *o*-Coumaric acid (16) *p*-Coumaric Acid (17) Steroid (18) Proanthocyanidins (19) Glycoflavone

Table 3. Histochemical Tests

Sr. No.	Tests	*J. curcas*			*J. gossypifolia*		
		Leaf	Stem	Root	Leaf	Stem	Root
1.	Alkaloids	-	-	-	-	+	-
2.	Starch	+	+	+	+	+	+
3.	Tannin	-	+	-	+	+	-
4.	Saponin	-	+	-	+	+	-
5.	Resin	-	+	+	-	+	+
6.	Protein	-	+	-	-	+	-

DISCUSSION

The drug dravanti (*J. curcas*) has been used in the treatment of various diseases including rheumatism, abdominal complaints and certain skin diseases. Detailed pharmacognostic and phytochemical studies were conducted on various parts of two species of *Jatropha* and it was found that each part possesed specific characteristic features useful in its identification. The shape and size of different tissues, starch grains, rosettes of calcium oxalate crytals, stomata and vessel elements varied among the species. The present investigation has shown that glabrous leaf, both paracytic and contiguous stomata, highest vein termination number and size of guard cell, bigger sized starch grains and calcium oxalate crystals, luteolin, were the pharmacognostic and phytohemical markers of *J.curcas* whereas simple uniseriate, multicellular, multiseriate glandtipped brown hairs, brachy-paracytic paracytic and adjoining stomata, highest stomatal index, stomatal frequency, vein islet number, palisade ratio smaller sized starch grains and calcium oxalate crystals, methoxy derivatives of luteolin and *p*-Coumaric acid were the pharmacognostic and phytochemical markers of *J. gossypifolia*.

REFERENCES

Anonymous, (1959). *Wealth of India Raw Materials* **5**. C.S.I.R, New Delhi.

Datta, S.K. and P.C. Datta (1983). *Bull. Med. Ethnobot.* **2** (3): 336.

Daniel, M. (1991). *Methods in Plant Chemistry & Economic Botany*. Kalyani Publishers, Delhi.

Gibbs, R.D. (1974). *Chemotaxonomy of Flowering Plants*. Mc Gill Queens Univ. Press, Montreal, London.

Gupta, R.C. (1985). *Proc. Indian Acad. Sci.* (Pi. Sci). **94** (1): 65.

Harborne, J.B. (1984). *Phytochemical Methods* (2ed), Chapman and Hall, 104-Gloucester, London.

Jensen, W.A. (1962). *Botanical Histochemistry Principles and Practice*, Freeman and Company U.S.A.

Johansen, D.A. (1940). *Plant Microtechnique*, Tata McGraw-Hill publishing company Ltd. New Delhi.

Khafagy, S.M., Mohamed, Y.A. Abdel Salam and Z.F. Mohammed, (1977). *Planta. Med.* **31**: 274.

Kirthikar K.R. and B.D. Basu (2003). *Indian medicinal plants with illustrations*, Oriental enterprises, Dehra Dun.

Mabry, T.J., Markham, K.R. and M.B. Thomas (1970). *The Systematic identification of Flavonoids* Springer-Verlag, New York.

Mujumdar, A.M. and A.V. Misar (2003). *Indian J. Pharm. Sci.* **65** (2): 554.

Pandey, A.K., Patra A.K. and P.K. Shukla (2005). *The Indian. Forester.* **131:** 857.

Reghunathan, K. and R. Mitra (1982) Eds. In: *Pharmacognosy of indigenous drugs*, Central Council of Research in Ayurveda and Siddha, New Delhi, **1**: 262.

Sivarajan, V.V. and Indira Balachandran (1994). *Ayurvedic Drugs and their Plant Sources*. Oxford & IBH. Publishing, Co. Pvt. Ltd. New Delhi.

20

TLC SCREENING OF *IN VITRO* PRODUCED PHYTOSTEROLS OF *CLERODENDRUM PHLOMIDIS* L.

Zankhana Rathod, Monika Christian and O.P. Saxena

Tissue Culture Laboratory, Botany Department, School of Sciences, Gujarat University, Ahmedabad-380 009, Gujarat, India

ABSTRACT

Clerodendrum phlomidis L. (Arani) is a very important medicinal plant used in curing different diseases such as diabetes, gonorrhoea, measles etc. Tissue culture showed the maximum growth index in eight weeks old cultures as compared with two, four and six weeks old cultures. The total steroidal content was also higher in eight weeks old. Maximum content of isolated compounds was in callus compared to *in vivo* plant parts. *In vivo* material of *C. phlomidis* shows β-sitosterol, stigmasterol, tigogenin, diosgenin and 9 unidentified steroids, while *in vitro* shows β-sitosterol, stigmasterol, tigogenin, diosgenin and 11 unidentified steroids.

INTRODUCTION

Clerodendrum phlomidis L. (Arani) is a very important medicinal plant used in curing different diseases. It is used in diabetes, gonorrhoea, measles etc. This plant has aromatic, astringent, demulcent, anti-convulsant, anti-diarrhoeal activities. In India parts of the plant are used in post-natal conditions in women and in gastrointestinal disorders. The roots are employed as an appetite stimulant.

Plant tissue culture has opened up new vistas as one of the most promising technique to study biosynthesis of natural herbal products like primary metabolites and secondary metabolites. Plant tissue culture is being used very successfully to obtain mass production of callus and bioactive compounds. Murashige and Skoog's (MS) medium was used as a basal nutritive medium for *in vitro* culture. Different plant parts (leaf, stem) were used as an explant. Callus was obtained on MS + 6mg/l 2,4-D + 6mg/l KIN. Phytosterols are of alcoholic and hydrocarbon in nature and are triterpenes (C_{23} to C_{29}). It is used as starting materials for synthesis of pharmaceutically active steroids, corticosteroids, sex hormones, contraceptives and anti-tumour properties. Hence, the present investigation was taken up to estimate *in vivo* and *in vitro* produced phytosterols through Thin Layer Chromatography (TLC). *In vivo* material of *C. phlomidis* shows β-sitosterol, stigmasterol, tigogenin, diosgenin and 9 unidentified steroids, while *in vitro* shows β-sitosterol, stigmasterol, tigogenin, diosgenin and 11 unidentified steroids. Thus it supports the view that *in vitro* maintained tissue retains the potentialities of biosynthesis of steroids.

MATERIAL

Medicinal Plants Studied

1. *Clerodendrum phlomidis* L.

Family : Verbenaceae

Common name : Arani

Useful parts : Leaves, roots

METHODOLOGY

The work is divided into two parts. (A) Tissue culture work (B) Thin Layer Chromatography

(A) Tissue culture work

It involves different steps:

1. Preparation of MS (Murashige and Skoog's media 1962) as a basal medium.
2. Inoculation of explant (leaf, stem, bud).
3. Maintaining of culture vials in culture room.
4. Periodic observation must be noted.
5. Maintaining a culture through transferring the material from test tube to test tube (Subculture).
6. Maintaining of plantlets or multiple shoots and successfully transfer from test tube to pot and filed (Hardening or Acclimatization).

(B) Thin layer chromatography (TLC)

The operation performed in TLC is essentially the same as in paper chromatography. This technique involves several steps:

1. Preparing thin layer (e.g. silica gel G)
2. Choice of solid for support (e.g. glass plate)
3. Sample application
4. Choice of solvent (mobile phase)
5. Development of chromatograph
6. Detecting or spraying reagent
7. Identification and calculation of Rf value

$$Rf = \frac{\text{Distance travelled by sample}}{\text{Distance travelled by solvent}}$$

RESULTS

Table 1. Tissue culture results

Explant	Level of Auxin		Level of cytokinin		Callus initiation (in weeks)				Characters
	Name	Concentration (mg/l)	Name	Concentration (mg/l)	1	2	3	4	
Leaf, stem and Bud	2,4-D	1	KIN	1	-	-	-	-	White coloured, friable callus
		2		2	-	-	-	-	
		3		3	-	-	-	-	
		4		4	-	-	+	+	
		5		5	-	-	+	+	
		6		6	++	+++	+++	+++	
Callus	2,4-D	0		23	-	++	+++	+++	Compact red coloured nodular callus

Table 2. TLC results of phytosterols

Treatment	Steroids present					
	Tigogenin Rf 0.52	Diosgenin Rf 0.57	Cholesterol Rf 0.61	Stigmasterol Rf 0.91	β-Sitosterol Rf 0.95	Unidentified steroids
In vivo						
Spraying reagent	+	+	+	+	+	8
Iodine vapour	+	+	+	+	+	7
In vitro						
Spraying reagent	+	+	+	+	+	9
Iodine vapour	+	+	+	+	+	9

DISCUSSION

Callus was obtained on MS + 6mg / l,2,4-D + 6mg/l KIN. There are many reports on the presence of flavonone and their glycosides in *C. phlomidis* (Anam, 1999) also sterols by (Joshi *et al.,* 1999) where they have reported the presence β-sitosterol and other ceryl alcohols. Phytosterols are present in *in vivo* (leaf) and *in vitro* (2, 4, 6, 8 week old callus) plant materials and was analysed with help of chromatography. Thus it supports the view that *in vitro* maintained tissue retains the potentialities of biosynthesis of steroids.

REFERENCES

Anam, E. M. (1999): *Ind. J. of Chem.,* **38(B)**: 1307.

Joshi, K. C., P. Singh, and A. Mehra, (1979): *Medica plant research,* **37**: 4.

21

IN VITRO PRODUCED FLAVONOIDS OF *BOERHAVIA DIFFUSA* L.

Monika Christian, Zankhana Rathod and O.P. Saxena

Tissue Culture Laboratory, Botany Department, School of Sciences, Gujarat University, Ahmedabad – 380 009, Gujarat (India)

ABSTRACT

Boerhaavia diffusa L. (Satodi or Punarnava) is very important medicinal plant used in curing different diseases. Plant tissue culture (*in vitro*) is used for the comparison of the production of flavonoids along with natural condition (*in vivo*). For it, Murashige and Skoog's (MS) medium was used as a basal nutritive medium. Different Plant Growth Regulators (PGRs) were added singly or in combinations to the basal medium. Different types of growth promoters like 2,4-D, IAA, IBA, NAA, Kinetin, BAP were used to prepare hormonal medium. Even it can be prepared with alone auxin or cytokinin or in combination of both. Callus was obtained on MS + 1.5 mg/l NAA +1.5 mg/l Kinetin from leaf and stem as an explant. TLC analysis of *in vitro* (8 week old callus) material showed 9 unidentified free flavonoids and kaempferol in leaf as well as callus. 11 unidentified bound flavonoids and quercetin were present in leaf as well as callus. Quantitative analysis was done with HPTLC. HPTLC analysis of *in vitro* (8 week old callus) was scanned at different wavelength (200 nm, 254 nm, 366 nm). Calibration spectrum, Rf values and maximum absorbance at particular wavelength of each substance were recorded through scanning. Total 13 falvonoids at 200 nm, 9 at 254 nm, 10 at 366 nm were analysed. Some of these are confirmed while others need more confirmation.

INTRODUCTION

Boerhaavia diffusa L. (Satodi) is very important and promosing drug to regenerate new cells in body so it is known as punarnava also. It is also useful in jaundice, scanty urine, internal inflammation and as anti-dote to snake venom (Nadkarni, 1954). These properties are mainly because of its primary and secondary metabolites of drugs. This drug is naturally present in the parts like leaf, stem, root, seed or some times in the whole plant.

These parts can be artificially cultured on the media and can get maximum amount of drugs. For it, Murashige and Skoog's (MS) medium is used as a basal nutrient medium. Media can be prepared by two ways (1) hormonal and (2) non-hormonal or plain. Different types of growth promoters like 2,4-D, IAA, IBA, NAA, Kinetin, BAP are used to prepare hormonal medium. Even it can be prepared with auxin alone or cytokinin or in combination of both (Vasil, 1984). Amount of drugs, which are artificially produced in 8 week old callus can be compared with naturally occur in plant (whole plant). Plant tissue culture approach has been found to be advantageous as it provides a continuous and reliable source of artificial product year around without the destruction of entire plant. With the help of tissue culture, high quantities of desired compounds can be obtained.

Flavonoids are important herbal drugs category. Flavonoids have 15-carbon skeleton. They are widely spread, soluble phenolic derivatives that are variously coloured. They are restricted to the vacuole though some are found in chloroplast and chromoplast. Flavonoids are glycosides and their glycones vary depending upon the flavan structure, which comprises two aromatic rings joined in a chroman structure by a 3-carbon unit (C_6-C_3-C_6). Free aglycones are found in dead woody tissue and formed from free flavonoids through hydrolysis and are variously classified depending upon the state of oxidation of the C_3 unit in the molecule. The flavonoids, chalcones etc. are chief classes of flavnoids, which contribute to the color scheme of the plants. Biflavonoids have as many as C_{30} atoms in their basic skeleton and are obtained from dimerisation of flavones by C-C linkage. They are found in the gymnosperms. Flavonoids directly form the C_6-C_3 portion of phenylpropanoids. Flavonoids are distinguished by various increasing state of the C_3 chain. Hence, the present investigation was taken up to estimate flavonoids through Spot test, TLC and HPTLC.

Experimental Plant Material

Scientific Name : *Boerhaavia diffusa* L.

Family : Nyctaginaceae

Common Names : Satodi, Punarnava

Chemical constituents reported

Alkaloids : Punarnavine-1, Punarnavine-2

Steroids : β-Sitosterol, β-Sitosterol-D-glucoside, Sitosteryl oleate, Sitosteryl palmitate

Flavonoids : C-methyl flavone, Kaempferol, Quercetin

Rotenoids : Boerhaaqvinone A, Boerhaavinone B, Boerhaavinon C, Boerhaavinone D, Boerhaavinone E, Boerhaavinone F

Lignans (Iridoids) : Liriodendrin, Stringarsionol, Mono-D-glucoside, β-D-glucoside

METHODOLOGY

(A) Spot test analysis (Daniel, 1991)

1. Mg ribbon test
2. Fluorescence test

(B) Thin layer chromatography (TLC)

The operation performed in TLC is essentially the same as in paper chromatography. This technique involves several steps:

1. Preparing thin layer (e.g. silica gel G)
2. Choice of solid for support (e.g. glass plate)
3. Sample application
4. Choice of solvent (mobile phase)
5. Development of chromatograph
6. Detecting or spraying reagent
7. Identification and calculation of Rf value

$$Rf = \frac{\text{Distance travelled by sample}}{\text{Distance travelled by solvent}}$$

(C) High performance thin layer chromatography (HPTLC)

- Formulation : Herbal drugs
- Classification : Flavonoids
- Dosage form : Liquid
- Sample preparation : (Subramanian and Nagarajan, 1969)
- Conditions of chromatography:
    1. Test plate: HPTLC precoated plate, silica gel 60 F_{254} – aluminium (Merk)
    2. Format: 10 x 10 cm^2
    3. Thickness : 250 μm
    4. Spotting volume: 2,4,6,8 μl (Linomat IV)
    5. Separation technique : Ascending
    6. Development chamber : Twin-through glass chamber (10 X 10 cm^2) (Camag) (saturate for 10 minutes prior to development)
    7. Mobile phase: Benzene : Acetic Acid (Flavonoids)

8. Spraying reagent : Alcoholic aluminium chloride for flavonoids
9. Relative Humidity : 52%
10. Temperature : 24°C
11. Migration distance : 80 mm
12. Migration time: 30 minutes for Flavonoids
13. Detection : UV

Densitometric Scanning (Camag Scanner III)

1. Wavelength and Mode:
   a. 200 / 254 nm – Absorbance / Reflectance
   b. 366 nm – Florescence / Reflectance
2. Slit Dimension : 6 X 0.45 mm

RESULTS

Table 1. Results of Tissue culture

Sr. No.	Media	Results
1	MS + 1.5mg/l NAA + 1.5mg/l KIN	Callus formation
2	MS + 1.5mg/l NAA + 1.5mg/l KIN	Sub culture
3	Plain MS	*In vitro* seed germination
4	MS + 1.5mg NAA + 10mg/l KIN	Micropropagation – Shooting
5	MS + 10mg IAA + 5 mg/l KIN	Micropropagation – Rooting
6	MS + 5mg NAA + 5 mg/l KIN	Multiple shoots and roots

Table 2. Spot Test analysis of flavonoids

Test	Plant materials			
	Leaf	Stem	Root	8 week old callus
Mg ribbon test	+++	++	-	+
Florescence test	++	+	-	+

Table 3. TLC analysis of flavonoids

Plant material	Kaempferol Rf 0.86	Quercetin Rf 0.78	Unidentified free flavonoids	Unidentified bound flavonoids
Leaf	+	+	9	11
Callus	+	+	9	11

Table 4. HPTLC results of 8-week-old callus flavonoids

Secondary Metabolite	Spot Numbers	Scanning at different wavelengths					
		200 nm		254 nm		366 nm	
		Rf	Area %	Rf	Area %	Rf	Area %
Flavonoids	Spot 1	0.02	52.82	0.02	64.64	-	-
	Spot 2	-	-	-	-	0.04	35.11
	Spot 3	-	-	-	-	0.11	14.78
	Spot 4	0.14	4.57	-	-	0.13	14.78
	Spot 5	0.22	5.35	0.22	8.27	0.22	11.91
	Spot 6	0.28	21.03	0.28	2.97	-	-
	Spot 7	0.35	1.23	0.36	1.99	-	-
	Spot 8	0.38	1.34	-	-	-	-
	Spot 9	0.42	0.86	-	-	0.43	8.60
	Spot 10	0.54	1.30	-	-	-	-
	Spot 11	-	-	0.56	1.73	0.57	5.18
	Spot 12	0.59	2.28	-	-	-	-
	Spot 13	0.71	4.01	0.71	7.52	-	-
	Spot 14	-	-	-	-	0.74	3.81
	Spot 15	0.77	2.40	0.77	5.14	-	-
	Spot 16	-	-	-	-	0.80	2.71
	Spot 17	-	-	-	-	0.85	2.33
	Spot 18	0.87	1.35	0.87	3.63	-	-
	Spot 19	0.90	1.45	0.90	1.45	-	-
	Spot 20	-	-	-	-	0.92	0.61
Total Numbers		13		9		10	

DISCUSSION

Plant tissue culture or the aseptic culture of cells, tissues and organs is an important tool in both basic and applied studies. Callus was obtained on MS medium supplemented with combinations of NAA + KIN in *B. diffusa*. The satisfactory result was obtained on MS + 1.5 mg/l NAA + 1.5 mg/l KIN, seed germination in plain MS, multiple shoot and root as well as micro-propagation on MS + 5 mg/l NAA + 5 mg/l KIN. Flavanoids are present in *B. diffusa* callus as well as leaves. Spot test gave positive, quick and qualitative results from *in vivo* and *in vitro* materials.

REFERENCES

Daniel, M. (1991). *Methods in plant chemistry and economic botany,* Kalyani Publishers, New Delhi.

Vasil, I.K. (1984). Cell culture and somatic cell genetics of plants, Vol. 1, Laboratory procedures and their applications, Academic Press, Inc. Harcourt brace Jovanovich Publishers.

22

CURING OF PLASMID MEDIATED VANCOMYCIN RESISTANCE IN *STAPHYLOCOCCUS AUREUS* USING HERBAL NAPHTHOQUINONES

Sheetal Jahagirdar and P.K. Dhakephalkar

Microbial Sciences Division, Agharkar Research Institute,
G.G. Agharkar Road, Pune - 411 004

ABSTRACT

Aim of this investigation was to eliminate plasmid mediated vancomycin resistance in *Staphylococcus aureus* using herbal naphthoquinones. Clinical isolates of *S. aureus* were screened for the resistance to multiple antibiotics and for the presence of plasmid. Antibacterial activity of herbal naphthoquinones such as plumbagin, juglone and lawsone was tested against a plasmid harboring vancomycin resistant *S. aureus* strain KEM94. Plumbagin and lawsone (but not juglone) cured vancomycin resistance and eliminated plasmid in KEM94. Physical loss of plasmid accompanied by subsequent loss of antibiotic resistance indicated elimination of vancomycin resistance due to plasmid curing by plumbagin or lawsone. Use of herbal naphthoquinones (employed routinely in alternative medicines) as plasmid curing agents is a novel approach towards containing the spread of vancomycin resistance especially in hospital environment.

INTRODUCTION

Since the 1940s, antibiotics have been the mainstay of treatment of infections caused by bacteria. However, despite the diverse types of antibiotics and the varied methods by which they work, many bacteria have developed means of resistance to the antibiotics (Shlaes *et al.,* 1991). Until recently, Vancomycin was one of the most powerful antibiotics that no bacterial cell had resistance to (Perichon and Courvalin, 2000). However, in recent years there have been increasing reports of vancomycin resistant staphylococci raising serious concerns within the scientific and health care communities (Chang *et al.*, 2003; Tenover, 1998). *S. aureus* is a major cause of hospital (nosocomial) infections where it survives in an environment rich in antibacterial agents. The Van A gene encoding vancomycin resistance in bacteria is located on conjugative plasmid that has the ability to transfer within and between species e.g. between enterococci and staphylococci (Flannagan *et al.*, 2003; Weigel *et al.*, 2003; Noble *et al.*, 1992). Vancomycin resistance in *Staphylococcus aureus* is very difficult to treat. Physicians have to resort to administering a cocktail of antibiotics at high dosages.

Theoretically an already ineffective antibiotic will become effective if R-plasmid encoding antibiotic resistance is removed from the bacterial population. As a result of earlier studies it is known that acridine orange, ethidium bromide and sodium dodecyl sulphate affect plasmid replication (Trevors, 1986; Molnar, 1988). However, acridine dyes and ethidium bromide cannot be tested *in vivo* because of their mutagenicity while SDS, because of its detergent action. There are many other compounds that cannot be considered as potential medicines. Thus, there is a constant need of developing a novel curing agent, which is more effective and nontoxic. In this respect the naturally occurring quinones constitute a major class. Besides, their vital role in biological systems, the quinones have shown potential application in chemotherapy (Lown, 1982). Plumbagin, a naturally occurring naphthoquinone, has been shown to have antibacterial, antifungal and antiviral properties (De Paiva *et al.*, 2003; Krishnaswami and Purushothaman, 1980).

In the present investigation, we have screened naturally occurring naphthoquinones such as plumbagin, juglone, lawsone etc. for their ability to eliminate plasmid mediated vancomycin resistance in *S. aureus*.

MATERIALS AND METHODS

Bacterial strains and culture conditions

Staphylococcus aureus strains used in this study were clinical isolates from King Edward Memorial Hospital, Pune (India). Strains were identified on the basis of their morphological, cultural and biochemical characteristics according to the Bergey's Manual of Systematic Bacteriology (Kloss and Schleifer, 1986). The identity of the isolates was further confirmed by sequencing the 16S rRNA genes.

Determination of resistance to antibiotics

Antibiotics resistance profile was determined by the disc diffusion method (Bhadbhade *et al.*, 2002). About 10^4 cells from overnight grown culture were spread on Luria agar plates. Multidiscs containing antibiotics

(Don Whitley Scientific Equipments, Mumbai, India) were placed on the plates. The zones of inhibition around the antibiotic discs were measured after incubation at 37°C for 24 h. The cultures were then assigned as resistance or sensitive by referring to the manufacturer's interpretation table.

Minimal inhibitory concentration (MIC) of antibiotics

MICs of antibiotics were determined by an agar dilution method as described previously (Dhakephalkar and Chopade, 1994). Luria agar containing antibiotics were used to determine the MIC of antibiotics. The agar plates were dried at 37°C for one hour and spot inoculated with ca. 10^5 cells per spot. The plates were incubated at 37°C for 24 h. The development of colonies on the plates indicated resistance to the antibiotic added. The lowest concentration of antibiotic that inhibited the growth was termed the minimal inhibitory concentration (MIC).

Curing of antibiotic resistance

The plasmid curing was performed as elucidated before (Deshpande *et al.*, 2001). In brief, the culture was grown in the presence of a curing agent at the specified concentration for 24 h at 37°C and then plated on Luria agar plates to obtain isolated colonies. The isolated colonies were then replica plated on to Luria agar and Luria agar containing antibiotics. The colonies that failed to grow in presence of antibiotics were considered as putative cured derivatives. The physical loss of plasmid in the cured derivative was confirmed by agarose gel electrophoresis of the plasmid DNA preparation of respective cultures. The percentage curing efficiency was expressed as number of colonies with cured phenotype per 100 colonies tested. Vancomycin sensitive cured colonies were also tested for loss of resistance to other antibiotics by disc diffusion assay described above.

Plasmid isolation

Plasmid DNA isolation was carried out by alkali lysis method (Sambrook *et al.* 1989). DNA was detected by horizontal agarose gel (0.7%) electrophoresis using Tris-boric acid-EDTA buffer (pH 8.0).

RESULTS

Several clinical isolates of *Staphylococci* were tested for their resistance to vancomycin. A coagulase positive *S. aureus* strain that was resistant to vancomycin (100 $\mu g/ml^{-1}$) was selected for further studies. The strain was designated as *S. aureus* KEM94. Table 1 shows the level of antibiotic resistance of KEM94, which was resistant to multiple antibiotics.

Antibacterial activity of plant-derived naphthoquinones was tested against *S. aureus* strain. It was observed that all three naphthoquinones i.e. plumbagin (5-hydroxy-2-methyl-1, 4-naphthoquinone), juglone (5-hydroxy-

1,4-naphthoquinone) and lawsone (2-hydroxy-1,4-naphthoquinone) inhibited growth of KEM94. Plumbagin was more potent than juglone and lawsone against KEM94. This was evident from the lower MIC of Plumbagin as compared to juglone and lawsone (Table 2). The minimal inhibitory concentration of naphthoquinones was found to be same as the minimum bactericidal concentration (MBC) thus indicating the bactericidal nature of these compounds.

Sub inhibitory concentration of naphthoquinones was used for the curing of plasmids in *S. aureus*. A culture control without naphthoquinone but with solvent (DMSO) was maintained to determine the rate of spontaneous loss of plasmids, which was found to be less than one in 10^8. Of the three naphthoquinones tested, only plumbagin and lawsone eliminated vancomycin resistance from *S. aureus* strain tested (Table 2). Even though plumbagin was a more potent bactericidal compound than lawsone, the later was found to be more effective plasmid-curing agent. Interestingly, both plumbagin and lawsone cured resistance to tetracycline, ampicillin, gentamycin and streptomycin along with resistance to vancomycin.

Table 1. Resistance of *S. aureus* and its cured derivatives to antibiotics

Antibiotics	MIC of antibiotics (μg.ml^{-1})		
	KEM94	KEM94C1*	KEM94C2*
Vancomycin	>100	<10	<10
Tetracycline	>100	<10	<10
Ampicillin	>100	<10	<10
Gentamycin	>100	<10	<10
Streptomycin	>100	<10	10
Amoxicillin	>100	>100	>100
Clindamycin	>100	>100	>100
Chloramphenicol	>100	>100	>100
Doxicyclin	>100	>100	>100
Erythromycin	>100	100	>100
Norfloxacin	>100	>100	>100
Pefloxacin	75	75	>100

*KEM94C1 and KEM94C2 were the cured derivative of KEM94 obtained after treatment with plumbagin and lawsone respectively.

Table 2. Elimination of vancomycin resistance in *S. aureus* KEM94 by herbal naphthoquinones

Curing agent	MIC (μg.ml^{-1})	SIC (μg.ml^{-1})	% Efficiency of curing
Plumbagin	50	25	4
Lawsone	400	200	20
Juglone	400	200	N.D*

*Below detection limit (none of the 100 colonies tested showed elimination of vancomycin resistance. Frequency of spontaneous mutation was found to be less than 0.000001%.

Agarose gel electrophoresis of plasmid DNA preparation revealed the presence of a single plasmid in KEM94 (Fig. 1). This plasmid was designated as pSAV94. The plasmid was found to be 16kb in length. The size of the plasmid was determined by comparing its mobility on agarose gel with that of CCC DNA ladder from Sigma (St Louis, USA). Plasmid DNA preparation of the cured strains did not show any bands on agarose gel electrophoresis (Fig. 1). The loss of vancomycin resistance in cured strains accompanied by loss of plasmid indicated that lawsone and plumbagin could cure plasmid encoded vancomycin resistance in *S. aureus*.

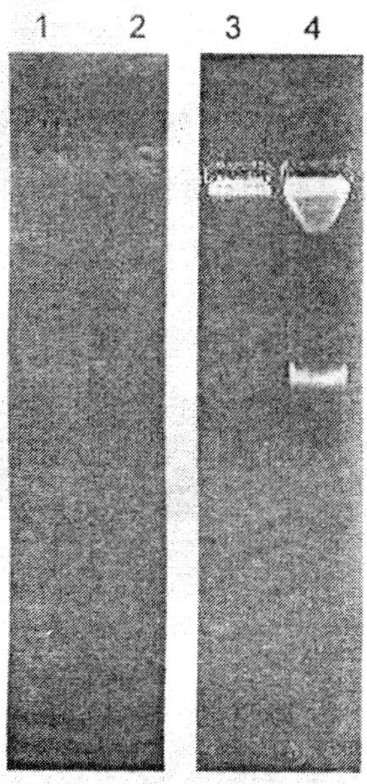

Agarose gel electrophoresis of plasmid DNA preparation of *S. aureus* KEM94 (Lanes 1 and 4) and the cured derivatives KEM94C1 (Lane 2) and KEM94C2 (Lane 3).

Fig. 1. Plasmid profiles of *S. aureus* KEM94 and cured strains

DISCUSSION

Vancomycin is widely used for the treatment of infection caused by staphylococci and enterococci (Tenover *et al.*, 1998). Development of vancomycin resistance along with resistance to beta lactams and amino-glycosides leaves medical community with very few therapeutic options. One of the ways to overcome this problem is to eliminate genes encoding antibiotic resistance in bacteria. In this study, the ability of the herbal naphthoq-uinones to eliminate antibiotic resistance in pathogenic staphylococci has been evaluated. It was observed that vancomycin resistance in *S. aureus* could be cured by plumbagin and lawsone with curing efficiency of 4% and 20% respectively. It must be noted here that frequency of spontaneous mutation was found to be less than one in 10^8 cells. In comparison the antibiotic resistance curing efficiencies obtained with plumbagin or lawsone were extremely high.

The antibiotic resistance elimination may occur due to mutations also. Mutagenic activity of any compound can be extremely dangerous, especially to human health. It was, therefore, necessary to confirm that loss of antibiotic resistance was due to curing of plasmids carrying such genes and not due to mutations. Hence, the cured derivatives were tested for the loss of plasmids. The physical loss of plasmid as evidenced by agarose gel electrophoresis of plasmid DNA preparation of cured strains indicated that plasmid curing resulted in subsequent loss of vancomycin resistance.

Curing of reference plasmids in *Escherichia coli* by plumbagin was reported earlier (Laxmi *et al.*, 1987). However, to the best of our knowledge, this is the first report of curing of vancomycin resistance in *S. aureus* by plumbagin or lawsone. It was observed that both plumbagin and lawsone cured resistance to tetracycline, ampicillin, gentamycin and streptomycin along with resistance to vancomycin. This suggested that plasmid carrying vancomycin resistance gene also carried genes encoding resistance to tetracycline, ampicillin, gentamycin and streptomycin.

As discussed earlier, bacteria have a tendency to develop resistance against any antimicrobial agent used against them. The concentration of so-called curing agents (plumbagin, lawsone) used for plasmid curing in this study were sub inhibitory. Thus bacteria were already resistant to these concentrations of compounds. Therefore, the question of bacteria developing resistance against them does not arise. Subsequently it is proposed that bacteria will not develop any mechanism to counter the plasmid curing property of these compounds. It must be considered here that plasmids are after all the dispensable element in any living cell.

Thus in conclusion we may say that this investigation has provided us novel curing agents that have the ability to cure plasmid mediated resistance to vancomycin and other antibiotics. Hence, it will have a tremendous potential in containing spread of antibiotic resistance especially in nosocomial environment.

REFERENCES

Bhadbhade, B. J., P.K. Dhakephalkar, S.S. Sarnaik, and P.P. Kanekar (2002) *Biotech. Lett.* **24**: 647

Chang, S. *et al.* (2003) *The New Eng. J. Med.* **348:** 1342

Paiva, S. R. *et al.* (2003). *Mem. Do Inst. Osw. Cruz,* **98**: 959

Deshpande, N.M., Dhakephalkar, P.K. and P.P. Kanekar (2001). *Lett. Appl. Microbiol.* **33**: 275.

Dhakephalkar, P.K. and B.A. Chopade (1994) *Biometals* **7**: 67.

Flannagan, S.E. *et al.* (2003). *Antimicr. Agents and Chemother.* **47:** 3954.

Kloss, W.E. and K.H. Schleifer (1986) *Bergey's Manual of Systematic Bacteriology.* (Sneath, P.H.A. *et al.*) p. 1013, Williams and Wilkins, Baltimore.

Krishnaswami, M. and K. Purushothaman (1980). *Ind. J. Exp. Biol.* **18**: 871.

Laxmi, V.V., P. Sridhar and H. Polasa (1987). *Curr. Microbiol.* **16**: 159.

Lown, T.W. (1982). *Acc. Chem. Res.* **15**: 381.

Molnar, J. (1988). *Exp. Clin. Pharmacol.* **10**: 467.

Noble, W.C., Z. Virani and R. Cree (1992). *FEMS Mcirobiology Letters* **93**: 195

Perichon, B. and P. Courvalin (2000). *Int. J. Clin. Pract.* **115**: 88.

Shlaes, D., S. Levy and G. Archer (1991). *ASM News* **57** : 455.

Tenover, F.C. *et al.* (1998). *J. Clin. Microbio.* **36**: 1020.

Trevors, J.T. (1986). *FEMS Microbiol. Rev.* **32**: 149

Weigel, L.M. *et al.* (2003). *Science* **302**: 1569.

B

NATURAL DYES

23

ROHIRA (*TECOMELLA UNDULATA*) LEAVES - A NEW SOURCE OF DYE FOR COTTON AND SILK FABRIC

Sabra Qureshi and Inderpal Rai

Department of Home Science, J.N.V. University, Jodhpur

INTRODUCTION

The modern society is exposing their daily increasing population to a variety of chemical substance in response to the pressure for high and better living standards in both the developed and developing countries. The invention of synthetic dyes almost limited the use of vegetable dyes to specific textiles only, such as Kalamkari, Ikats, tie and dyed chunaris etc. The unbridled use of synthetic dyes in the textile industry, many of which are known to be toxic and non-treatment of effluents contained in the wastewaters of the dyeing process has led to horrendous results.

Dyeing of textiles with natural dyes is an old practices. Vegetable dyes have several advantages over synthetic dyes from the point of view of health safety and ecology. Since last decade, application of natural colourants is gaining popularity all over the world.

Manufacturing of eco-textile through cleaner production concept is receiving new recognition today. It has been realized that the industrialization in all the countries has been at some cost to public health and environment. The fundamental point being, it is better for the society to prevent pollution, than to cure it, after its creation.

Designers have very effectively utilized natural dyes as a designer tool. The non-reproducibility and non-uniformity of shades makes each creation of unique piece.

Tecomella undulata

Local Name: Rohira, Rohiro, Roheda

Trees: With drooping branches and simple leaves. Flowers yellow-orange, in cymes terminating branches. Capsules linear, curved : seeds winged at the apex. Common in the west of Aravali in sandy habitats. *Fl. & Frts :* January – April.

Timber : The wood has high timber value in the desert and locally called as "desert teak or Marwar teak". It is chiefly used for furniture, toll-handles, boxes, well curbs, cartwheels, building items, huts (pillar, beam, rafter) and agricultural implements in the desert.

Medicine : The Bhils and Girasia tribals give orally the powder of root-bark with fresh honey or sugar and milk before sleeping at night to the women to cure leucorrhoea. The paste of old stem-bark with vegetable oil is applied over eczema. In case of old eczema, the inner bark of stem is placed in an earthen pot, the mouth of pot is sealed keeping a hole in the centre. This pot is placed upside down over the fire; the fluid that comes out through the hole of lid is collected and applied on the eczema. The local vaids prescribe seed-powder with pure "Ghee" to apply on abscesses. The tribals of Udaipur district inhale the vapors of crushed leaves to cure cough. Children use the twigs as toothbrush.

The present study has the following objectives —

1. To extract the dye from Rohira (*Tecomella undulata*) leaves.
2. To apply the extracted dye on premordanted cotton and silk fabrics.
3. To study the change in effects of colors with the change of mordant.
4. To test the colourfastness of the dyed fabrics.

MATERIALS AND METHODS

Material

For the present study plain weave cotton fabric and tussar silk were used for experimentation.

Mordants used

1. Natural Mordant : Myrobolan
2. Chemical Mordants : Alum, Tartaric Acid, Stannous Chloride and ferrous sulphate.

Selection of Dye

Rohira (*Tecomella undulata*) leaves were selected for the study, because after going through review, it was found that no much work has been done on them – as a source of environment, friendly natural dyes were selected.

Pre-Treatment of the fabrics

All textile material before dyeing has to be scoured, degummed and bleached.

Preliminary information of the fabric

S. No.	Test	Details
1.	Fiber content	Cotton
2.	Weave	Plain
3.	Ends and Picks per Inch	64/56
4.	Weight	280 gms/sq.mtr.
(i)	Fiber content	Tussar silk
(ii)	Weave	Plain
(iii)	Ends and picks per inch	120/110
(iv)	Weight	190 gms/sq.mtr.

Mordanting

Before dyeing pre-mordanting of the samples were done. The fabrics was mordanted by immersing it in a mordant bath prepared as per the required 1% and 3% concentration and the temperature was maintained at 60°-70°C for minutes. The fabric was rinsed in cold water and dried in the shade.

Extraction of the dye

For the preparation of dye solution, required amount of Rohira (*Tacomella undulata*) leaves were dried; through drying took around 8-10 days and the drying was carried out in shade.

Then 250 gms of dyes source was soaked in 1 liter of distill water overnight and then boiled for 60 to 90 minutes. Maintaining the level of solution in the container throughout.

After the heat treatment, the dye solution was allowed to cool in the dye bath. The contents were filtered to remove the unwanted residue. The solution was then further heated to achieve a concentration.

Dyeing Process

The dye bath was prepared with 100% concentration of the dye solution. The pre-mordanted samples were entered in the dye bath with M:L ratio. The temperature was raised 95-100°C for 45 minutes. After dyeing the sample was rinsed in water and dried in shade.

Colour fastness test

The samples were subjected for colourfastness tests using the following ISO standards:

1. Colourfastness to rubbing
2. Colourfastness to washing
3. Colourfastness to perspiration
4. Colourfastness to light

Table 1. Fastness Grades of the *Tecomella undulata* (Rohira) dye materials on cotton and silk fabrics

Sr. No.	Method of mordanting	Mordant	Conw. of mordrant	Washing fastness			Rubbing fastness		Perspiration fastness						Light fastness
				Changing in colour	Colour staining		Dry	Wet	Acidic			Alkaline			
					Cotton	Wool	C.S.	C.S.	Change in colour	Colour staining Cotton	Colour staining Wool	Change in colour	Colour staining Cotton	Colour staining Wool	
1	Premordanting	ALUM	1%	4	5	5	3/4	4	4	5	4	5	4/5	3	4
2			3%	4	5	5	3/4	2	5	5	4	4	4/5	4	4

Table 2.

Sr. No.	Method of mordanting	Mordant	Conw. of mordrant	Washing fastness			Rubbing fastness		Perspiration fastness						Light fastness
				Changing in colour	Colour staining		Dry	Wet	Acidic			Alkaline			
					Cotton	Silk	C.S.	C.S.	Change in colour	Colour staining Cotton	Colour staining Silk	Change in colour	Colour staining Cotton	Colour staining Silk	
1	Premordanting	ALUM	1%	4/5	5	5	3	4/5	4/5	4/5	5	5	4	4/5	Change in Tone
2			3%	5	5	5	3/4	3	5	4	5	5	4/5	5	5

RESULTS AND DISCUSSION

The results of the present investigation indicate that Rohira (*Tecomella undulata*) can be successfully used for dyeing to obtain a wide range of soft, lustrous shades.

Table 3.

Sr. No.	Method of mordanting	Mordant	Conw. of mordrant	Washing fastness: Changing in colour	Washing fastness: Colour staining Cotton	Washing fastness: Colour staining Wool	Rubbing fastness: Dry C.S.	Rubbing fastness: Wet C.S.	Perspiration fastness Acidic: Change in colour	Acidic: Colour staining Cotton	Acidic: Colour staining Wool	Alkaline: Change in colour	Alkaline: Colour staining Cotton	Alkaline: Colour staining Wool	Light fastness
1	Premordantingg	$SnCl_2$	1%	4	5	5	3	4	4	4/5	3	4/5	4/5	4	Change in Tone
2			3%	4/5	5	5	3	2	4	4/5	3/4	5	4/5	4	Change in Tone

Table 4

Sr. No.	Method of mordanting	Mordant	Conw. of mordrant	Washing fastness: Changing in colour	Washing fastness: Colour staining Cotton	Washing fastness: Colour staining Silk	Rubbing fastness: Dry C.S.	Rubbing fastness: Wet C.S.	Perspiration fastness Acidic: Change in colour	Acidic: Colour staining Cotton	Acidic: Colour staining Silk	Alkaline: Change in colour	Alkaline: Colour staining Cotton	Alkaline: Colour staining Silk	Light fastness
1	Premordantingg	$SnCl_2$	1%	5	5	5	3/4	4	5	5	4/5	5	4/5	4/5	Change in Tone
2			3%	4/5	5	5	2	2/3	5	4/5	4	4/5	4	4/5	Change in Tone

The leaves of Rohira are a rich source of natural colour ranging from golden yellow to dark brown.

In the table, that excellent colour fastness is shown by the samples dyes with premordanting with alum, stannous chloride, ferrous chloride and tartaric acid.

The mordant alum gave the best result on both cotton and silk fabric. The light fastness both silk and cotton fabrics very good to excellent colour fastness to sunlight. Washing fastness of the cotton and silk samples dyed with Rohira was found to be excellent to good fastness on the Gray scale. Rubbing fastness of the silk was very good but slight staining were their in cotton fabric. No colour changed was observed for acidic and alkaline perspiration condition.

Both cotton and silk samples pre-mordanted with 1% and 3% of Ferrous Sulphate and dyed in rohira showed very good washing fastness both silk and cotton fabrics. In case of dry running, samples showed slight staining. Where as in case of wet rubbing, there was no change in colour. Excellent fastness was observed for both acidic and alkaline perspiration of silk fabric. Very slight staining was noticed for cotton fabric. Both cotton and silk fabrics showed very good to excellent light fastness.

Table 5.

Sr. No.	Method of mordanting	Mordant	Conw. of mordrant	Washing fastness			Rubbing fastness		Perspiration fastness						Light fastness
				Changing in colour	Colour staining		Dry	Wet	Acidic			Alkaline			
					Cotton	Wool	C.S.	C.S.	Change in colour	Colour staining		Change in colour	Colour staining		
										Cotton	Wool		Cotton	Wool	
1	Premordanting	$FeSO_4$	1%	4	5	5	3	2	5	4	3	4/5	3/4	4	5
2			3%	Increased Hue	5	5	3/4	3	4/5	4/5	3/4	4	3	3	5

Table 6.

Sr. No.	Method of mordanting	Mordant	Conw. of mordrant	Washing fastness			Rubbing fastness		Perspiration fastness						Light fastness
				Changing in colour	Colour staining		Dry	Wet	Acidic			Alkaline			
					Cotton	Silk	C.S.	C.S.	Change in colour	Colour staining		Change in colour	Colour staining		
										Cotton	Silk		Cotton	Silk	
1	Premordanting	$FeSO_4$	1%	5	5	5	4	3	5	4/5	5	4/5	4	4/5	5
2			3%	4/5	5	5	3/4	2/3	5	4	4/5	5	4/5	5	5

Pre-mordanted cotton and silk samples with 1% and 3% of tartaric acid showed very good light fastness. In dry state both 1% cotton and silk samples showed negligible staining but 3% samples showed noticeable staining. In case of wet rubbing slight staining were showing both cotton and silk fabrics. All dyed silk and cotton samples indicated no change and excelled to good fastness to acidic and alkaline perspiration. Both cotton and silk samples have very good washing fastness.

Both cotton and silk fabrics premordanted with 1% and 3% of stannous chloride exhibited change in tone during exposure to sunlight may be due to photochemical changes in due molecules, which led to darkening of the shade.

Cotton and silk samples showed fair to good washing fastness. In dry rubbing test cotton and silk samples showed slight to noticeable staining. But in wet rubbing both samples showed considerable staining. Thus it can be concluded that the dyed samples both cotton and silk showed excellent colourfastness to acidic and alkaline perspiration.

Table 7.

Sr. No.	Method of mordanting	Mordant	Conw. of mordrant	Washing fastness			Rubbing fastness		Perspiration fastness						Light fastness
				Changing in colour	Colour staining		Dry	Wet	Acidic			Alkaline			
					Cotton	Wool	C.S.	C.S.	Change in colour	Colour staining		Change in colour	Colour staining		
										Cotton	Wool		Cotton	Wool	
1	Premordanting	Tartaric acid	1%	5	5	5	5	4/5	4	5	4	4/5	4/5	4	4
2			3%	4/5	5	5	3	2/3	3	5	4	4/5	5	4	4

Table 8.

Sr. No.	Method of mordanting	Mordant	Conw. of mordrant	Washing fastness			Rubbing fastness		Perspiration fastness						Light fastness
				Changing in colour	Colour staining		Dry	Wet	Acidic			Alkaline			
					Cotton	Silk	C.S.	C.S.	Change in colour	Colour staining		Change in colour	Colour staining		
										Cotton	Silk		Cotton	Silk	
1			1%	5	5	5	4	3	5	5	4	5	5	4/5	5
2	Premordanting	Tartaric acid	3%	5	5	5	4	3/4	5	4/5	4	5	4/5	4/5	5

CONCLUSION

Natural dyes are again getting importance due to harmful effects caused by synthetic dyes during their production and use. Vegetable dyeing industry can promote indirect employment in cultivation of plants and herbals matter in waste lands which are widely spread all over India. From the result of the present study, it may be concluded that Rohira (*Tecomella undulata*) gives golden yellowish shades.

1. Fabric dyed with Rohira and Premordanted with 1% and 3% alum show better results for washing, rubbing, sunlight and perspiration in alkaline medium.
2. Rohira dyed fabrics gives good fastness properties with stannous chloride but moderate colourfastness to acidic perspiration.
3. Tartaric acid and ferrous sulphate give excellent result for washing, light, rubbing and perspiration in both alkaline and acidic.

By considering the fastness grades of both silk and cotton, it is proved that *Tecomella undulata* leaves are a rich source of fast natural dye, suitable for cotton and silk dyeing. Its use on these fabrics will are a rich source of fast natural dye, suitable for cotton and silk dyeing. Its use on these fabrics will certainly satisfy the consumer in terms of serviceability.

REFERENCES

Ahmed S.S. and A. Gogan (1997). *A study of Indian silk 36/7* : 17.

Bansal A. and S. Anjali (2001). *Textile Magazine*. P. 81.

Batacharya N., B.A. Doshi and A.S. Sahasrabudhe (1998). *American Dye Staff reporter 87/4*. p. 26.

Geeta, M., R.K. Sunanda and S. Dakshi (2001). *Natural dyes IIT*, Delhi. P. 111.

Goel, A., S. Shinde and K.C. Gupta (1993). *Asi. Tex. J.* (8) : 34.

Gogoi, N. and B. Kalita (1999). *Colorage XLVI* (1): 23.

Gulrajani, M. L. (1999). *Colorage XLVI* (7): 19.

Katyayini, V.K. and J. Marry (1999). Colorage Vol. XLVI (7): 3.

Rai, I.P. and Deepti (2001). *Natural dyes* IIT Delhi : 111.

Samy, P.V. and B.S.M. Raja (1998). *Man Made Textiles in India 41/3*: 132.

Teli, M.D., R. Paul and P.D. Pardeshi (2001). *Colorage* XLVIII (4): 51.

24

DYEING OF COTTON FABRIC WITH KATHA AND RAVENCHI WOOD

Anjali Karolia and Falguni Patel

Dept. of Clothing and Textiles, Faculty of Home Science, M.S. University of Baroda, Vadodara.

INTRODUCTION

Natural dyes comprise of those colorants that are obtained form natural sources i.e. animal or vegetable matter with or without processing. For the past five decades or so the use of natural dyes for coloration of textiles has mainly been confined to craft's dyes that have used these colors for a restricted clientele. This trend has been followed by increased public awareness about environment issues over the past decade and more business houses have started exploring the use of natural colorants as possible means of producing ecologically sound products which would also appear to the "Green-Minded" consumer. Manufacturing of Eco-textile through cleaner production concept is receiving new recognition today. It has been realized that the industrialization in all the countries has been at some cost to the human health and environment.

This paper explores the dyeing behaviour and fastness properties of two dyes:

- Katha (*Acacia catechu*)
- Ravenchi wood (*Rheum emodi*)

Katha is a brown dye obtained chiefly from the wood of tropical trees of *Acacia*. Katha is used in dyeing and in tanning to make rich brown colour. Katha is one of the important natural dyes, which can be successfully used on cotton. Previous studies in dyeing cotton with Katha have shown that it produces similar shades to that of C.I. Brown, which is a banned chemical. Its fastness properties on cotton substrate with various agencies have been studied in the following paper.

Ravenchi (*Rheum emodi*), also popularly known as Dolu, Gandhini and Revatchini, the roots which are the colour containing parts. Dolu contains a number of anthraquinone derivatives based on emodin, chrysophenol and rhein.

MATERIALS AND METHODS

Cotton fabric with weft counts 68 picks/cm and warp count as 36 ends/cm and 130-g/sq. mt. was used for the study. Katha powder and ravenchi wood powder were used as the colorant material. The mordant selected for the work were alum, ferrous sulphate, copper sulphate and potassium dichromate. A mixed mordant system of alum and ferrous acetate was also selected.

Cotton fabric procured form the market was first scoured and then treated with harda (myrobalan) solution for 30 minutes at room temperature. The fabric was then taken out of the harda bath and squeezed to remove the excess solution from the fabric; it was then hung on a clothesline and allowed to dry completely. This fabric was then given a pre-mordanting treatment by boiling with the five mordents mentioned above for 30 minutes, with the following concentrations:

- Alum 40% on the weight of material
- Copper sulphate 20% on the weight of material
- Potassium dichromate 10% on the weight of material
- Ferrous Sulphate 5% on the weight of material
- Alum and Ferrous Acetate, 40% and 10% owm. Respectively

Extraction of the dye and dyeing of fabric

1:40 was the material liquor ratio of the dye bath and the colour yielding wood was added as 2% of the total liquor. Katha and Ravenchi wood were weighed according to the required 2% concentration and then powered separately. This powdered solution was then added to the respective dye bath and allowed to boil for 15 minutes, after which the pre mordanted cotton fabrics were immersed in the solution in which the dye has been extracted and dyeing was carried out for 45 minutes. After the dyeing process is over the fabric was rinsed thoroughly and air-dried.

Assessment of Fastness Properties of the Dye

Fastness to light

Colour fastness to light was evaluated by exposing the dyed samples to artificial light on a light fastness tester fabricated at ATIRA. The procedure followed was according to the standard test method IS 2454:1985 based on ISO 105/B-1984.

Fastness to washing

ISO test no. 3 assessed wash fastness of the dyed sample. IS 764:1979, based on ISO, 105/C-1982. The change in colour and staining on white was evaluated by standard grey scale and staining scale rating of 1-5.

Fastness to perspiration

Fastness of the dye to perspiration was tested as per IS 971:1983 based on ISO 105/E 1978, and evaluated for change in color and staining on white.

Fastness to rubbing

Dye fastness to rubbing was evaluated by IS 766 : 1988 based on ISO 105/X-1984.

Fastness to hot pressing

IS 689 1988 based on ISO 105/X – 1984, was the standard test method followed for assessing fastness of the dye to hot pressing.

Fastness to sublimation

IS 975-1988 standard test method was used to evaluate the fastness of coloured textile material to sublimation on storage.

RESULTS AND DISCUSSION

Cotton fabric was mordanted with various metallic salts, viz, alum, ferrous sulphate, copper sulphate, potassium dichromate and mixture of alum and ferrous acetate, according to the percent concentrations mentioned in the methodology.

Plate 1.

It was observed that cotton fabric mordanted with the above mordants and dyed with katha gave very good brown shades (Plate I), and ravenchi dye gave excellent shades of khaki (Plate II).

Plate 2.

The fastness properties of the dye was then evaluated according to standard procedures and the results were as follows Light fastness of katha and ravenchi dye ranged from average to good with fastness rating of 3-4. In case of katha when mordanted with copper sulphate and potassium dichromate mordants it ranked good with a fastness rating of 4 (Table 1).

Table 1. Fastness to light

Sr. No.	Sample Particulars	Result			
		Hours of Fading		Rating	
		Katha	Ravenchi	Katha	Ravenchi
1	Alum	33	33	3-4	3-4
2	Ferrous sulphate	33	33	3-4	3-4
3	Alum and ferrous acetate	33	33	3-4	3-4
4	Copper sulphate	48	33	4	3-4
5	Potassium dichromate	48	33	4	3-4

Wash fastness of katha dye was very good and rated between 4-5 in all the cases for staining on white and was 4 only for alum mordant for change in colour, compared to that, ravenchi dye exhibited average staining on white with fastness rating of 3-4 and performed poor for change in colour as the rating score was 2 for all the mordant dye combinations (Table 2).

Table 2. Fastness to washing

Sr. No.	Sample Particulars	Result			
		Hours of Fading		Rating	
		Katha	Ravenchi	Katha	Ravenchi
1	Alum	4-5	3-4	4	2
2	Ferrous sulphate	4-5	3-4	4-5	2
3	Alum and ferrous acetate	4-5	3-4	4-5	2
4	Copper sulphate	4-5	3-4	4-5	2
5	Potassium dichromate	4-5	3-4	4-5	2

Katha dye gave good to very good results for fastness to acid sweat (fastness rating 4-5) but did not perform well for fastness to alkaline sweat (fastness rating 3-4). Poor results for fastness to perspiration were also observed for ravenchi dye. The fastness rating for acid sweat for staining on white and change in colour was 3 for all the mordant dye combinations and was 2-3 for fastness to alkaline sweat (Table 3 and 4).

Table 3. Fastness to acid sweat

Sr. No.	Sample Particulars	Result			
		Hours of Fading		Rating	
		Katha	Ravenchi	Katha	Ravenchi
1	Alum	4-5	3	4-5	3
2	Ferrous sulphate	4-5	3	4-5	3
3	Alum and ferrous acetate	4-5	3	4-5	3
4	Copper sulphate	3-4	3	4	3
5	Potassium dichromate	4-5	3	4-5	3

Table 4. Fastness to alkaline sweat

Sr. No.	Sample Particulars	Result			
		Hours of Fading		Rating	
		Katha	Ravenchi	Katha	Ravenchi
1	Alum	3-4	2-3	3-4	2-3
2	Ferrous sulphate	4	2-3	3-4	2-3
3	Alum and ferrous acetate	4	2-3	3	2-3
4	Copper sulphate	3	2-3	3-4	2-3
5	Potassium dichromate	3-4	2-3	3	2-3

The result to rub fastness for katha dye was average to poor with different fastness ratings with all the mordants and but it could be said that the dry fastness of katha dye is poor as compared to its wet fastness. As for ravenchi dye alum and ferrous acetate mordanted sample rated 4 for fastness to dry rubbing and the rest had a rating of 3-4. The rating for all the other mordants the dry and rub fastness rating was fair that is 2-3 (Table 5).

Table 5. Fastness to rubbing

Sr. No.	Sample Particulars	Result			
		Staining on white			
		Hours of Fading		Rating	
		Katha	Ravenchi	Katha	Ravenchi
1	Alum	4-5	3-4	4-5	2-3
2	Ferrous sulphate	3-4	3-4	4-5	2-3
3	Alum and ferrous acetate	4-5	4	4	2-3
4	Copper sulphate	3-4	3-4	3-4	2-3
5	Potassium dichromate	4	3-4	3-4	2-3

Good results were obtained for fastness to hot pressing and sublimation for katha dye but not hot pressing and sublimation rating for ravenchi dye showed average results (Table 6 and 7).

Table 6. Fastness to hot pressing

S.N.	Sample Particulars	Result											
		Dry Press				Damp Press				Wet Press			
		s.w.		c.c.		s.w.		c.c.		s.w.		c.c.	
		Ka	Ra	Ka	Ra	Ka	Ra	Ka	Ra	Ka	Ra	Ka	Ra
1	Alum	4-5	3-4	4-5	4	4-5	3-4	4	4	4	3	4	4
2	Ferrous sulphate	4-5	4	4-5	4	4-5	3-4	4-5	4	4	3	4-5	4
3	Alum & ferrous acetate	4-5	4	4-5	4	4-5	3-4	4-5	4	4	3	4-5	4
4	Copper sulphate	4-5	4	4-5	4	4-5	3-4	4-5	4	4	3	4-5	4
5	Potassium dichromate	4-5	4	4-5	4	4-5	3-4	4-5	4	4	3	4-5	4

s.w. = Staining on white, c.c. = Change in colour, Ka = Katha & Ra = Ravenchi

Table 7. Fastness to Sublimation

Sr. No.	Sample Particulars	Result					
		Staining on white					
		150°C		180°C		210°C	
		Katha	Ravenchi	Katha	Ravenchi	Katha	Ravenchi
1	Alum	4-5	4-5	4-5	4-5	4-5	4
2	Ferrous sulphate	4-5	4-5	4-5	4	4-5	4
3	Alum and ferrous acetate	4-5	4-5	4-5	4-5	4-5	4
4	Copper sulphate	4-5	4-5	4-5	4-5	4-5	4
5	Potassium dichromate	4-5	4-5	4-5	4-5	4-5	4

Hence from the results it could be concluded that Ravenchi dye produces excellent shades of khaki on dyeing with cotton but the fastness properties are poor. An interesting fact noted while testing the dye for sublimation is that it sublimates easily leaving stains on the polyester fabric hence it behaves as a disperse dye. But the dyeing has been carried out as a normal exhaust dyeing procedure for direct dyes so it may be giving poor results due to insufficient dyeing.

Katha or Catechu dye gives average to good fastness to all the agencies of fading that it is possibly exposed during wear. Hence, it can be exploited widely for use in commercial application in the textile made-ups for value addition.

ACKNOWLEDGEMENTS

We thank the Director and Staff or Ahmedabad Textile Industry's Research Association for assisting in the testing and assessment of the dyed samples.

25

DYEING OF COTTON WITH SEEDS OF *BIXA ORELLANA* LINN.

E.M. Dedhia

Department of Textiles and Fashion Technology, College of Home Science, Nirmala Niketan, 49, New Marine Lines, Mumbai-20

ABSTRACT

Dye extraction from annatto seeds has been carried out in different solvents like water, methanol, acetone and petroleum ether. Suitable solvent has been selected for dyeing. The extract obtained was used for dyeing of unmordanted cotton fabric as well as for the fabric mordanted with basic alum and tannic acid by varying pH condition. Depth of the color was observed. Colour of the fabric has been evaluated on computer matching system in terms of K/S and L^* a^* b^* color coordinates. Fastness to washing, dry and wet rubbing and sunlight were tested using standard test methods. The optical density of the extract was found maximum with acetone solvent. All the samples dyed with Annatto seed extract were orange in colour. K/S values of dyed samples were found more or less equal for unmordanted-dyed and mordanted-dyed samples. Samples dyed in alkaline pH showed maximum K/S values in both the cases. Visually unmordanted-dyed and mordanted-dyed samples did not show any tonal variation. The values for each colour coordinates vary between the tolerances limits of ±5, which is acceptable, except very slightly higher value for the samples dyed in alkaline medium. Fastness to sunlight and washing was found excellent for both the cases however moderate to good fastness to rubbing was reported. Dyeing with annatto seed extract did not show the need for any pretreatment with mordant on the cotton fabric. The annatto dye itself has good quality of dye fixation on cotton fabric. This quality of annatto dye makes it less expensive and less time consuming in application, provides higher opportunity for commercialization due to its bright color with good fastness values.

INTRODUCTION

With the increasing concern over environmental issues, emphasis is now given to use natural dye sources because they are procured from renewable resources, having no health hazards; in fact sometimes act as health cure and having no or mild chemical reactions involved in their preparation. Though almost all-vegetable matter can give coloring matter, but eventually there are fewer sources, which yield colorants that can be extracted and proved to be commercially viable. However a lot of creativity is required to use these dyes judiciously to avoid wastage of valuable natural resources.

Bixa orellana Linn. (Annatto) Bixaceae

Seeds, Leaves, Bark, Roots, Shoots are the parts used. The plant is distributed in India, Peru, Brazil, South and Central America, Caribbean and in some parts of Mexico. Annatto is a fruiting shrub or small tree that grows 5-10 m in height. Its flowers are beautiful pink and pods are prickly reddish-orange heart-shaped, growing out at the end of the branches. A pod contains approximately 50 seeds inside it. The seeds are covered with a reddish aril that is the source of an orange-yellow dye. Annatto is a high yielding tree which can produce up to 270 kg of seeds.

Annatto is known worldwide for its use in food color. Annatto seed paste produced in South America is reported to export to North America and Europe where it was used as food color. A natural dye extracted from the red pulp of the fruit is known for coloring cotton, silk, wool, ship skin, feathers, wax, ivory, etc. Sometimes it is used in the paint, varnish, lacquer, cosmetic, pharmaceutical ointments, hair oils, shoe polishes and soap industries (Dastur, 1964).

Annatto seed contains 40% to 45% cellulose, 3.5% to 5.5% sucrose, 0.3% to 0.9% essential oil, 3% fixed oil, 4.5% to 5.5% pigments, and 13% protein, alpha and beta-carotenoids and other constituents. The annatto seed contains bixin and norbixin as main source of pigments, which are classified as carotenoids and are extracted from the seed. In addition to bixin and norbixin, annatto contains bixaghanene, bixein, bixol, crocetin, ellagic acid, ishwarane, isobixin, phenylalanine, salicylic acid, threonine, tomentosic acid and tryptophan.

Some tribal communities knew medicinal value of the entire plant from centuries. The plant is useful for skin care and skin anti-aging (for its antioxidant and ultraviolet ray [UV]-protective effect, to tone, balance, and strengthen liver function and for hepatitis and liver inflammation/pain (hepatoprotective), insect, repellant, diuretic, hypocholestremic, hypotensive, expectorant and wound healer.

This research paper reports experimental results of extraction and dyeing conditions for *Bixa orellana,* Linn. (Annatto) seeds for cotton fabric.

MATERIALS AND METHODS

Grey cotton fabric was used for the study, which was scoured and bleached. Raw materials used to extract colorants were Annatto seeds. Acetone, methanol and petroleum ether were used as organic solvents for dye extraction. Basic alum and tannic acid were used as mordants.

The annatto seed powder was extracted with different solvents like water, methanol, acetone and petroleum ether at material-to-liquor ratio 1:30 at 50°C for 30 minutes. Optical density of the solvent, which was found to be maximum were selected for further experimentation.

Scoured and bleached cotton fabric was mordanted with basic alum and tannic acid. The cotton fabric was dyed with the extracted dye liquor at a material-to-liquor ratio of 1:30 at 60°C for 45 minutes. Scoured-bleached and mordanted fabric samples were dyed with the extract by adjusting neutral, acidic and alkaline pH conditions. After dyeing, the samples were rinsed with water and then treated with neutral soap solution at 70°C and finally washed thoroughly with water and dried. The dyed samples were assessed for K/S values and L^* a^* b^* color coordinates on spectrophotometer.

Fastness to washing, dry and wet rubbing and sunlight were tested using standard test methods. The light fastness of the dyed samples was tested by AATCC standard test method and ratings were given. For washing fastness, samples were tested according to the ISO-2 test. Samples were evaluated for change in color and staining on the adjacent fabrics with rating. For rubbing fastness ISO standard test method was followed. Both wet and dry rubbing were tested using crockometer and evaluated for the rating.

RESULTS AND DISCUSSION

The extracts were obtained separately with water, acetone, methanol and petroleum either. Fig. 1 shows the value of optical density of annatto seed powder extract with different solvents. The optical density of the extract was found maximum with acetone solvent.

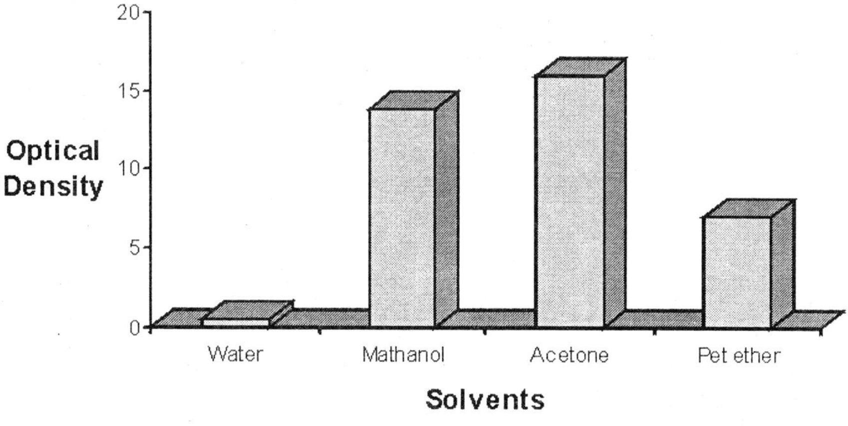

Fig. 1 Optical density of the Annatto seeds extract with different solvents

The extract obtained was used for dyeing of unmordanted cotton fabric as well as for the fabric mordanted with basic alum and tannic acid. All the samples dyed with Annatto seed extract were orange in color. Fig. 2 shows effect of pH condition on K/S values of unmordanted and dyed samples with annatto. Fig. 3 shows effect of pH condition on K/S values of mordanted and dyed samples with annatto. K/S values of dyed samples were found more or less equal for unmordanted-dyed and mordanted-dyed samples. Samples dyed in alkaline pH showed maximum K/S values in both the cases.

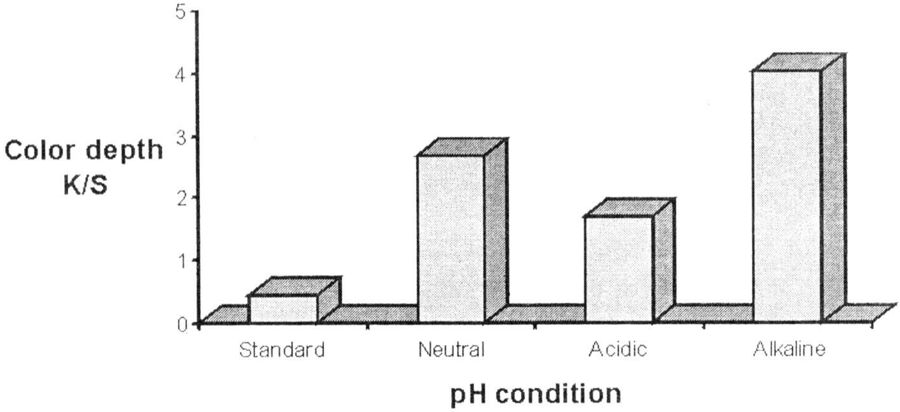

Fig. 2 Effect of pH on the K/S values of unmordanted and dyed samples with annatto

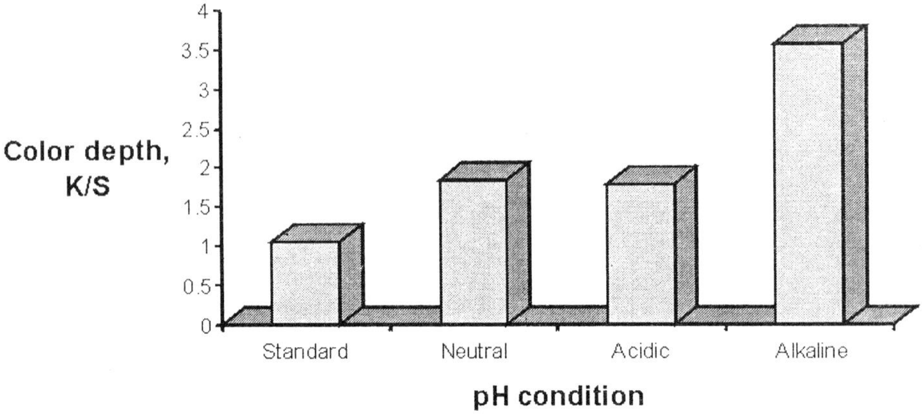

Fig. 3 Effect of pH on the K/S values of mordanted and dyes samples with Annatto

Table 1 shows color characteristics of samples dyed in neutral, acidic and alkaline pH condition. Visually unmordanted-dyed and mordanted-dyed samples did not show any tonal variation. The values for each color coordinates vary between the tolerances limits of ± 5, which is acceptable, except very slightly higher value for the samples dyed in alkaline medium.

Table 1. Color characteristics of mordanted and unmordanted samples

Dyed Samples	L*	a*	b*
UM-Neutral	64.16	29.20	27.81
UM-Acidic	67.46	26.51	21.92
UM-Alkaline	61.10	31.78	31.96
M-Neutral	66.22	25.97	23.75
M-Acidic	66.14	25.58	23.22
M-Alkaline	62.17	31.96	31.58
M-Alkaline	72.94	0.6	32.31

UM-Unmordanted, M-Mordanted

Table 2 shows fastness properties of the dyed samples. It is evident from the result that it is not affected by mordanting treatment with basic alum and tannic acid. Fastness to sunlight and washing was found excellent for both the cases however moderate to good fastness to rubbing was reported.

Table 2. Fastness ratings of the dyed fabrics

Dyed Samples	Rubbing Fastness				Washing fastness			Light fastness
	Warp		Weft		CC	SAF		
	Dry	Wet	Dry	Wet		Cotton	Wool	
UM-Neutral	4-5	4-5	4-5	4-5	4-5	4-5	4-5	4
UM-Acidic	4	4	4	4	4-5	4-5	4-5	4-5
UM-Alkaline	4	4	4-5	5	4-5	4-5	4-5	4
M-Neutral	3-4	4	3-4	3-4	5	4-5	4-5	4-5
M-Acidic	4	4	4	4	5	4-5	4-5	45
M-Alkaline	4-5	4	4	5	4-5	4-5	4-5	4-5

UM-Unmordanted, M-Mordanted, CC-Change in color, SAF-Staining on adjacent fabric

CONCLUSIONS

From the results it can be concluded that acetone is a suitable solvent for dye extraction from annatto seeds. Annatto dyes can be used for dyeing cotton fabric with out any pretreatment with mordant. The annatto dye itself has good quality of dye fixation on cotton fabric. This quality of annatto dye makes it less expensive and less time consuming in application, provides higher opportunity for commercialization due to its bright color with good fastness values.

Acknowledgements to University Grants Commission, College of Home Science Nirmala Niketan and Anjali Srivasthava.

REFERENCES

Dastur J F, *Useful Plants of India and Pakistan*, Tarporva Sons and Co. Private Ltd. Mumbai, 1964-36.

http://yucca.standardout.com/pics/Mitjo_1058203730_398.jpg@imgrefurl or

http://plantsdatabase.com/showimage/20497.html&h.

26

DYEING OF COTTON, SILK AND WOOL YARNS BY BABUL FLOWERS (*PROSOPIS JULIFLORA*)

V. Luniya and S. Arora

Department of Home Science, J.N.V. University, Jodhpur

ABSTRACT

Natural dyes gain importance in the present era of revivalism extraction of natural dyes from abundantly occurring material. In this study, extraction of natural dye from largely available and waste product namely Babul flowers have been done. This results in production of wealth from waste. Different mordant and mordant concentrations were adopted to get different shades. Myrobalan treatment is carried out to increase the dye affinity of the yarns. Dyeing was done using open bath (exhaust) method. Pre-mordanting technique has been used. Effect of the mordants used for mordanting has been studied on the dye uptake and the fastness properties of the dyed samples.

INTRODUCTION

Dyeing has played an important part in adding beauty to the world. The use of natural dyes or vegetable dyes is having an age-old history since ancient periods. India had a distinction of being the colour box of the ancient world. India was once, one of the richest countries in the dyes.

Today, people all over the world have come to accept the fact that natural dyes are much more traditional and hygienic. In the present global scenario, if the textile industry wishes to bring prosperity through diversification and modernization they have to manufacture eco friendly product. With this in mind researches are being initiated by the various national and international institutes for extracting colors from natural sources to dye textile materials. Now a days natural dye is not only limited to dyeing of fibers but natural dyes is used in preparation of Herbal - gulals. Some of the commonly available natural dyes have been explored to get variety of colours on leather. The availability of natural dyes historically has dependent on three sources - specialized plant and animal sources by products especially in food colours, chemicals synthesis. Today a fourth source is tissue or cell culture possibly with DNA transfer biotechnology.

As the world is progressing to natural dyes by using new techniques, raw materials and by various investigations and implications it could be rightly said that the coming dying world will be natural dyes world and the future of textile world will be limited to natural dyes only.

Keeping this in mind it was considered to be of immense importance to explore the natural resources of the region. Though natural dyes are a renewable source - trees cannot be grown overnight. Therefore, in the present study while selecting the parts of the plant as natural dye source, care was taken not to harm or destruct the plant growth. So, for the study Babool flowers were chosen, which is common throughout the Rajasthan region. Babul tree is a small tree with arching branches. Grows very quickly, common along the roadside.

Objectives of the Study

1. To develop different shades on cotton, wool and silk yarn by dying with selected dye i.e. Babool flowers using different mordants.
2. To study the changes in effects of colour on change of mordants and mordant concentration.
3. To evaluate colour fastness properties of dyed yarns.

MATERIAL AND METHODS

For the present study cotton yarn as a cellulosic fibre and silk and wool yarn as a protein fibre were selected for experimentation.

Mordants

Natural Mordant: Myrobalan

Chemical Mordants: Alum, Tartaric Acid, Ferrous Sulphate, Stannous chloride

Preparation of yarns for dyeing

In order to make the yarns absorbent, to obtain level dyeing penetration of dyestuffs, to remove the natural as well as added impurities scouring, bleaching of cotton, wool yarns and degumming of silk yarns was done. Standard procedures were adopted for the process.

Preparation of Dye Extraction

For the preparation of dye solution, required amount of Babool flowers were dried in shade for 4-5 days. The stems from the spikes were removed then 250 gms of dye source was soaked in 1 litre distilled water overnight and then the liquor was allowed to boil for 60-90 minutes for maximum dye extraction. The coloured solution was separated by filtration by a fine muslin cloth to get clear liquid care was taken not to pass fine particles in the filtrate. The solution was then further heated to achieve a concentration. The solution was then transferred to open bath for exhaust dyeing. The pH of the dye solution was scanned and it was found to be acidic i.e. pH4 and was kept as original and prepared the dye bath for dyeing.

Mordanting

Before mordanting the yarns with chemicals they were first treated with myrobalan (Harda fruit). Myrobalan adds the affinity between the fiber and the dye solution. It also increases the fastness properties. Harda was taken 20% for cellulosic and 10% for animal fibers.

Pre-Treatment with chemical salts

Pre mordanting method was used for mordanting the yarns. Two concentrations 1% and 3% of each modrant were used with dye extract. The weighed amount of mordants was first dissolved in water and proper stirring was ensured. Presoaked yarns were immersed in the containers and the temperature of water bath was raised upto 60°-70°C and the treatment was given for 30 minutes each. The samples were than drawn out washed, squeezed to remove the excessive solution and dried in shade.

Dyeing Process

The dye bath was prepared with 100% concentration of the dye solution. Pre mordanted yarns were entered in the prepared dye bath having M: L ratio as 1:40 and temperature of 45°C. After entering of the samples the temperature was raised to boiling point and than dyeing was carried for 45-60 minutes. After dyeing the samples were cooled down in the dye bath itself and finally rinsed with soft water and dried in shade.

Assessment of Colour Fastness

The dyed yarns samples were analysed for their performance properties in terms of light, wash, perspiration and rubbing fastness.

- Colour fastness to washing - ISO -105-C01
- Colour Fastness to perspiration - ISO-105
- Colour Fastness to rubbing - ISO-105
- Colour Fastness to sunlight - AATCC16

Evaluation of the tested samples with geometric grey scale as per ISO 105 standard method.

The testing samples were assessed for colour change and colour staining.

RESULTS AND DISCUSSION

A variety of shades can be obtained from Babool flowers by varying the mordant and their concentrations. Tartaric acid produces light shades as compared to ferrous sulphate, alum and stannous chloride. Lower concentration of mordants produced a lighter shade as compared to high concentration.

But in some cases despite of varying concentration of mordants similar shade were obtained. The shades showed that samples dyed with Babool flowers varied in shades from light yellow to reddish brown. Silk yarns give bright shades because of its original shine as compared to cotton and wool yarns.

The findings of the present study predicts that the Babool dye extract give soft, lustrous and bright colours on cotton, wool and silk yarns (Tables 1-8).

The study also show that 1% concentration of Alum, tartaric acid was optimum for mordanting and dyeing of cotton yarns (Tables 1-5). However in the case of stannous chloride and ferrous sulphate optimum concentration required for mordants for bright shades on cotton was 3 percent (Tables 4, 8).

Table 1. Ratings for wash fastness, light fastness, perspiration and rubbing fastness of yarns dyed with Babool flowers extract with Alum 1%

S. No.	Yarns	Wash Fastness				Rubbing Fastness		Perspiration Fastness								Light Fastness
		Colour	Staining Adjacent Fabric					Alkali				Acidic				
								Colour	Adjacent Fabric Staining			Colour	Adjacent Fabric Staining			
			Cotton	Wool	Silk	Dry	Wet		Cotton	Wool	Silk		Cotton	Wool	Silk	
1	Cotton	5	4-5	5	-	3	3	4-5	4-5	4	-	3-4	5	4-5	-	4-5
2	Wool	5	4-5	4-5	-	3-4	3	4-5	4-5	4-5	-	4-5	5	4-5	-	Tone changes to brown
3	Silk	3-4	4-5	-	4-5	4	4-5	4-5	4-5	-	4-5	5	5	-	4-5	Tone changes to brown

1. **Wash fastness** – colour fastness with alum on cotton and wool is excellent but with silk its good and negligible staining in seen in all the yarns.
2. **Rubbing fastness** – Both dry wet rubbing is very good with silk but to cotton and wool its fair.
3. **Perspiration fastness** – Colour fastness with alkali and acidic medium is very good to excellent in all cases except cotton in acidic medium and staining is seen to be negligible except in cotton where staining can be seen in wool as adjacent fabric.
4. **Light fastness** – Light fastness of cotton is very good as very slight fading was seen but in wool and silk the tone changed to brown – change in colour was more in silk yarn.

Table 2. Ratings for wash fastness, light fastness, perspiration and rubbing fastness of yarns dyed with Babool flowers extract with Alum 3%

S. No.	Yarns	Wash Fastness				Rubbing Fastness		Perspiration Fastness								Light Fastness
		Colour	Staining Adjacent Fabric					Alkali				Acidic				
								Colour	Adjacent Fabric Staining			Colour	Adjacent Fabric Staining			
			Cotton	Wool	Silk	Dry	Wet		Cotton	Wool	Silk		Cotton	Wool	Silk	
1	Cotton	4	4-5	4-5	-	4-5	3	5	4-5	4-5	-	4-5	4-5	4	-	5
2	Wool	4	4-5	4-5	-	4-5	3	4	4-5	4	-	5	5	4-5	-	5
3	Silk	4	4-5	-	5	5	4	5	4-5	-	4-5	5	5	-	4-5	

1. **Wash fastness** – Colour fastness is good but staining is very negligible i.e. very good to excellent.
2. **Rubbing fastness** – Dry fastness is very good to excellent but wet is just good.
3. **Perspiration fastness** – Both alkali and acidic medium has very good to excellent fastness.
4. **Light fastness** – Light fastness is excellent in cotton and wool as no change in colour was noticed but in silk tone changed from yellow to brown.

Table 3. Ratings for wash fastness, light fastness, perspiration and rubbing fastness of yarns dyed with Babool flowers extract with Stannous Chloride 1%

S. No.	Yarns	Wash Fastness				Rubbing Fastness		Perspiration Fastness								Light Fastness
		Colour	Staining Adjacent Fabric					Alkali				Acidic				
								Colour	Adjacent Fabric Staining			Colour	Adjacent Fabric Staining			
			Cotton	Wool	Silk	Dry	Wet		Cotton	Wool	Silk		Cotton	Wool	Silk	
1	Cotton	5	4-5	4-5	-	4	2-3	4-5	4-5	4-5	-	4-5	5	4-5	-	5
2	Wool	4	5	4-5	-	3	2-3	5	4-5	4-5	-	5	5	4-5	-	4-5
3	Silk	4	5	-	5	4-5	4	5	4-5	-	5	4-5	5	-	5	Tone changed

1. **Wash fastness** – Colour fastness is excellent in cotton and in wool and silk it is good but no staining is seen in silk and in cotton and wool also its very slight.
2. **Rubbing fastness** – Wet rubbing is not good but dry is satisfactory in cotton and wool but very good in silk.
3. **Perspiration fastness** – Both alkali and acidic medium has excellent colour fastness and staining is also negligible.
4. **Light fastness** – Light fastness is excellent in cotton and very good in wool but change in tone was noticed in silk.

For wool 1% concentration of Alum, Tartaric was optimum for mordanting of wool yarns. However in the case of ferrous sulphate and stannous chloride optimum concentration required for better dark and appealing shades on wool was 3%.

Table 4. Ratings for wash fastness, light fastness, perspiration and rubbing fastness of yarns dyed with Babool flowers extract with Stannous Chloride 3%

S. No.	Yarns	Wash Fastness				Rubbing Fastness		Perspiration Fastness								Light Fastness
		Colour	Staining Adjacent Fabric					Alkali				Acidic				
								Colour	Adjacent Fabric Staining			Colour	Adjacent Fabric Staining			
			Cotton	Wool	Silk	Dry	Wet		Cotton	Wool	Silk		Cotton	Wool	Silk	
1	Cotton	5	5	5	-	5	4-5	4-5	4-5	4-5	-	4-5	5	4-5	-	5
2	Wool	4	5	5	-	3-4	3-4	5	4-5	4-5	-	4-5	5	4-5	-	4-5
3	Silk	4	5	-	4-5	4-5	4	5	4-5	-	4-5	4-5	5	-	5	Tone changed

1. **Wash fastness** – Colour fastness as observed is excellent in cotton and good for wool and silk but staining is negligible in all the three yarns i.e. excellent.
2. **Rubbing fastness** – Dry fastness is satisfactory in wool but very good to excellent in cotton and silk wet rubbing is good in cotton but fair to good in wool.
3. **Perspiration fastness** – Both acidic and alkali medium showed excellent colour fastness and negligible staining.
4. **Light fastness** – Light fastness as observed is excellent in cotton and very slight fading could be seen in wool whereas tone changed in silk.

Table 5. Ratings for wash fastness, light fastness, perspiration and rubbing fastness of yarns dyed with Babool flowers extract with Tartaric Acid 1%

S. No.	Yarns	Wash Fastness				Rubbing Fastness		Perspiration Fastness								Light Fastness
		Colour	Staining Adjacent Fabric					Alkali				Acidic				
								Colour	Adjacent Fabric Staining			Colour	Adjacent Fabric Staining			
			Cotton	Wool	Silk	Dry	Wet		Cotton	Wool	Silk		Cotton	Wool	Silk	
1	Cotton	5	5	5	-	4	3	5	4-5	4	-	4	5	4	-	5
2	Wool	4	5	5	-	3-4	3-4	4	4	3-4	-	4	4-5	3-4	-	5
3	Silk	3-4	5	-	4	4-5	4	4-5	4	-	3-4	4-5	4-5	-	4-5	5

1. **Wash fastness** – Colour fastness to washing is excellent in cotton and to silk is fair whereas it showed negligible staining in cotton and wool whereas silk and slight staining on silk fabric.
2. **Rubbing fastness** – Rubbing is very good only on silk in dry state where as wet rubbing is fair to satisfactory in all three years.
3. **Perspiration fastness** – Colour fastness is excellent where cotton is concerned whereas in rest it is good and noticeable staining could be seen on wool & silk.
4. **Light fastness** – All the three yarns didn't showed even a slight fading so could be said that the fastness is excellent.

As silk is a natural lustrous fiber so all shades obtained by 1% or 3% mordanting are bright and shiny but stannous and ferrous 3% give better shades than the ferrous and stannous 1%. However all the other mordants give bright shades with little change in hue and tone of colour with the two mordants.

Table 6. Ratings for wash fastness, light fastness, perspiration and rubbing fastness of yarns dyed with Babool flowers extract with Tartaric Acid 3%.

S. No.	Yarns	Wash Fastness				Rubbing		Perspiration Fastness								Light
		Col-our	Staining			Fastness		Alkali				Acidic				Fastness
			Adjacent Fabric					Col-our	Adjacent Fabric Staining			Col-our	Adjacent Fabric Staining			
			Cotton	Wool	Silk	Dry	Wet		Cotton	Wool	Silk		Cotton	Wool	Silk	
1	Cotton	5	4-5	5	-	4-5	4	5	4	4	-	4	5	4	-	5
2	Wool	4	4-5	4-5	-	4	3-4	5	4	3-4	-	4	4-5	3-4	-	5
3	Silk	5	4-5	-	5	4-5	3-4	4-5	4	-	4	4-5	4-5	-	4-5	5

1. **Wash fastness** – Colour fastness is excellent in cotton and silk as compared to wool but all the yarns have negligible staining.
2. **Rubbing fastness** – Wool and silk showed fair stains in wet rubbing whereas cotton and silk has very good dry rubbing as compared to wool.
3. **Perspiration fastness** – Alkali medium have excellent colour fastness as compared to acidic medium and slight staining could be seen in wool as adjacent fabric.
4. **Light fastness** – All the three yarns showed excellent light fastness.

Table 7. Ratings for wash fastness, light fastness, perspiration and rubbing fastness of yarns dyed with Babool flowers extract with Ferrous Sulphate 1%.

S. No.	Yarns	Wash Fastness				Rubbing		Perspiration Fastness								Light
		Col-our	Staining			Fastness		Alkali				Acidic				Fastness
			Adjacent Fabric					Col-our	Adjacent Fabric Staining			Col-our	Adjacent Fabric Staining			
			Cotton	Wool	Silk	Dry	Wet		Cotton	Wool	Silk		Cotton	Wool	Silk	
1	Cotton	3-4	4-5	4-5	-	4	3-4	4	4-5	4-5	-	4	5	4-5	-	5
2	Wool	4	4-5	4-5	-	3	2-3	5	4-5	4-5	-	4-5	5	4-5	-	5
3	Silk	3-4	4	-	4-5	4	3	4	4-5	-	4-5	4-5	5	-	5	5

1. **Wash fastness** – As observed colour fastness is good only in wool whereas in cotton and silk it is fair and staining is negligible in all the three yarns.
2. **Rubbing fastness** – Wet rubbing is poor in all three whereas dry rubbing is fair.
3. **Perspiration fastness** – In Alkali medium colour fastness is excellent with wool while good in cotton and silk and very slight staining could be seen in all three yarns. In acidic medium colour fastness is good in wool and silk as compared to cotton ad no staining is in silk whereas very slight can be seen in cotton and wool.
4. **Light fastness** – Light fastness is excellent in all the three yarns as no colour fading was noticed.

Table 8. Ratings for wash fastness, light fastness, perspiration and rubbing fastness of yarns dyed with Babool flowers extract with Ferrous Sulphate 3%.

S. No.	Yarns	Wash Fastness				Rubbing Fastness		Perspiration Fastness								Light Fastness
		Colour	Staining Adjacent Fabric					Alkali				Acidic				
								Colour	Adjacent Fabric Staining			Colour	Adjacent Fabric Staining			
			Cotton	Wool	Silk	Dry	Wet		Cotton	Wool	Silk		Cotton	Wool	Silk	
1	Cotton	3-4	4-5	4-5	-	4	3	4	4-5	4-5	-	4-5	5	5	-	5
2	Wool	4	4-5	4-5	-	2-3	2	4-5	4-5	4-5	-	4-5	5	4-5	-	5
3	Silk	4	4-5	-	4-5	3-4	5	4-5	4	-	4-5	5	5	-	4-5	5

1. **Wash fastness** – Colour fastness is fair in cotton whereas in wool and silk its good and staining is very slight in all three yarns.
2. **Rubbing fastness** – Dry rubbing is fair to good in cotton and silk whereas in wool its poor and wet rubbing is excellent in silk, poor in wool and fair in cotton.
3. **Perspiration fastness** – Both in alkali & acidic medium colour fastness is very good to excellent and negligible staining was noticed in all the three yarns.
4. **Light fastness** – Light fastness is excellent in all the three yarns.

The assessment of colour fastness was done using ISO and AATCC standards. The test results showed very good to excellent colour fastness and negligible staining by washing in all yarns with all mordants.

Dry rubbing is better in all the three yarns as compared to wet rubbing.

Both acidic and alkali medium perspiration showed very good to excellent colour fastness and negligible staining was seen except in some cases with wool slight staining was seen.

Light fastness was excellent as no fading of colour was noticed but silk showed change in tone in some cases.

Findings of the study led to conclusion that Babul flower can be effectively used to dye cotton, wool and silk yarns with excellent results in context to wash, light, rubbing and perspiration fastness.

27

DYEING OF VEGETABLE DYE ON COTTON AND SILK FABRICS WITH BANANA PEEL (*MUSA PARADISIACA* VAR. *SAPIENTUM*) USING FERMENTATION PROCESS

Inderpal Rai and Santosh Soni

Department of Home Science, J. N. V. University, Jodhpur

ABSTRACT

Natural dyes are being looked at as an "eco-solution" to the ill-effects of synthetic dyes. The Indian textile exporters have resorted to the use of natural dyes so that the products appeal to the green minded consumer. Viewing this trend, it is essential to ensure the environmental adequacy of natural dyes. The paper deals with the application of dye extracted from banana Peel *(Musa paradisiaca* var. *sapientum)* on mordanted cotton and silk substrate. Pre-mordanting technique has been used for the work. Effect of the mordanting technique adopted and various mordants used for mordanting has been studied on the dye uptake and the fastness properties of the dyed samples.

INTRODUCTION

India has a very rich tradition in the use of natural dyes. The tradition of using natural dyes in textile dyeing and printing is very old but there is no historical record available. Natural dyes have been a part of human life since times immemorial. Egyptian mummies, documents of Mughal period bear evidence to the use of these dyes.

Natural dyes are amongst the first useful products that come to mind when wild plants are thought of. The colours they produce are soft, subtle, lustrous, aromatic and best of all alive. Natural dyes comprise of those colorants that are obtained from animal, vegetable or mineral matter without chemical processing.

With the introduction of synthetic dyes in the middle of 19th Century, the decline in the use of natural dyes started. The synthetic dyes have replaced the natural dyes. The use of natural dyes in coloration of textile material is just one of the consequences of increased environmental awareness.

Fermentation is the process activated by micro-organisms, which bring about changes of colour and texture in the material provided. The use of enzymes in textile processing and after care is already the best established example of the application of bio-technology to textiles.

Many micro-organisms produce pigments during their growth, which are substantive as indicated by the permanent staining that is often associated with mildew growth on textiles and plastics. Micro-organisms would therefore seem to offer great potential for the direct production of novel textile dyes or dye intermediates by controlled fermentation techniques replacing chemical synthesis, which has inherent waste disposal problems example toxic heavy metal compounds.

Fermentation is the oldest process and was in use since 2000 B.C. and present traditional dyers of Japan (isolated areas in Nigeria) and some contemporary dyers still use fermentation process.

MATERIALS AND METHODS

2.1 Materials

2.1.1. *Fabric*

Mill scoured and bleached, plain weave cotton and silk fabrics were used for the work.

Preliminary Information of the Fabrics

S. No.	Test	Details
1.	Fibre Content	100% pure cotton and silk fabrics
2.	Weave	Plain
3.	Fabric Count (per inch)	Cotton- 65/61 and silk 90/84

2.1.2 *Natural Colourants*

- Botanical name- *Musa paradisiaca* var. *sapientum*
- Local name- Kela, Kadali, Rampha
- Part of the plant use- Banana peel
- Cultivation area- Tropical Asian Countries, leading countries are Brazil, India, Equador, Indonesia, Thailand and Cuba.

- History- Banana is consumed throughout the world, even though it is cultivated only in tropical countries. In India it is grown in Tamilnadu, Kerala, Maharashtra, Bihar and Andhra Pradesh. Plant has underground fleshy rhizome and an erect stem composed of sheathing bases of petioles with a crown of large leaves at the top. Plant may reach up to 6 meter in height. Banana is seedless fruit and its edible part is mesocarp.

Banana is highly nutritious as it contains carbohydrates, proteins, minerals and vitamins. It is not only cheap but also delicious. Green bananas are cooked as vegetables. Unripe fruits are useful in diabetes. Powdered banana is given to children when they are suffering from carbohydrate specially lactose intolerance.

Dried green powdered bananas are used in preparation of soap. Banana Peel is also used as cattle feed.

2.1.3 *Mordants*

(i) Alum and Myrobalan (Harda) were used as natural mordants.

(ii) $CuSO_4$, $FeSO_4$ and tannic acid were used as Chemical mordants.

2.2 Experimental Methods

2.2.1 *Extraction of the Dye*

The dye is prepared by taking about 250 gm. dried banana peel powder mixed with (125 gm) jaggery and (25 gm) bajra flour taken in earthen or plastic pot, then the mixture was made up with five litre of water and the mouth of pot is covered by tying a fine muslin cloth. The mixture is then allowed to ferment for 15 days in summer and 30-45 days in winter. Then the contents were filtered to remove the unwanted residues and the dye was ready for use.

2.2.2 *Mordanting*

Pre-mordanting technique is used for the study For pre-mordanting, the baths were prepared with required amount of mordant (3% and 5%). The temperature of the bath was raised to 40°C and then the fabric is impregnated into the liquor. Gradually temperature is then raised to 60°C and mordanting continued for 30 min.

The fabric is then rinsed in cold water and dried in shade.

2.2.3 *Dyeing*

The pre-mordanted samples were dyed with 100% concentration of the dye solution on the weight of the samples, maintaining the temperature at 90-100°C for 45 minutes. The pH of dye is maintained at 4-5.

Blank experiments were also performed by dyeing on an un-mordanted cotton and silk fabrics. These samples were considered as control samples for comparison of results with mordanted and dyed samples.

*After treatments

After dyeing, all the dyed samples were rinsed, soaped at 60°C for 15 min. using non-ionic detergent (2 gpl), washed thoroughly and dried at ambient temperature.

2.2.4 *Assessment of Fastness Properties*

Assessment is done by using following ISO standards-

1. Colour fastness to washing - ISO-105- C01: 1989 (E)
2. Colour fastness to rubbing - ISO -105- E04: 1994 (E)
3. Colour fastness to perspiration - ISO -105- E04: 1994 (E)
4. Colour fastness to light - AATCC- Test Method- 16: 1998.

Rating

Staining	Colour
5- Negligible staining	5- Excellent
4- Slight staining	4- Good
3- Noticeable staining	3- Fair
2- Considerable staining	2- Poor
1- Much changed	1- Very Poor

RESULTS AND DISCUSSION

Fastness grades of the Banana peel dye on cotton fabrics-

1. **Washing Fastness -** Cotton samples dyed with harda and tannic acid mordant showed good washing fastness. Neutral samples showed good results as compared to acidic and basic. After washing, negligible staining is there on adjacent fabrics.
2. **Rubbing Fastness -** In both dry and wet state harda pre-mordanted samples showed sight to negligible staining on adjacent fabrics. In case of tannic acid pre-mordanted samples there is negligible staining in dry state where as slight staining in case of wet rubbing.
3. **Perspiration Fastness -** The pre-mordanted samples of harda showed good colour fastness in acidic perspiration where as fair to good in case of alkaline perspiration. Acidic dyed samples showed good results where as in case of basic dye there is change in tone.

In case of tannic acid pre-mordanted samples the result is *vice-versa*.

In case of staining of adjacent fabric both pre-mordanted samples showed negligible staining in case of acidic perspiration and slight to negligible staining in case of alkaline perspiration.

Table 1. Fermented Banana Peel Dye Grading of Cotton Fabric

S.No.	MEDHOD OF MORDANTING	MORDANT	CONS OF MORD-ANT	WASHING FASTNESS					RUBBING FASTNESS		PERSPIRATION FASTNESS									
				CHANGE IN COLOUR			COLOUR STAINING ON ADJACENT FABRIC		DRY	WET	ACIDIC					ALKALINE				
							COTON	WOOL	COLOUR STAINING	COLOUR STAINING	CHANGE IN COLOUR			COLOUR STAINING		CHANGE IN COLOUR			COLOUR STAINING	
														COTTON	WOOL				COTTON	WOOL
				A	B	N					A	B	N			A	B	N		
(1)	PRE-MORDANTING	Harda	5%	¾	2:3	3	4/5	4/5	4/5	4	3	2	2/3	4/5	4/5	2/3	2/3	3	4/5	4/5
(2)		Tannic Acid	5%	3	2/3	3	4/5	4/5	4/5	4	3	2	3	4/5	4/5	3	3	2/3	4/5	4/5

A - Acidic, B - Basic, N- Neutral

Table 2. Fermented Banana Peel Dye Grading of Silk Fabric

S.No.	MEDHOD OF MORDANTING	MORDANT	CONS OF MORD-ANT	WASHING FASTNESS					RUBBING FASTNESS		PERSPIRATION FASTNESS									
				CHANGE IN COLOUR			COLOUR STAINING ON ADJACENT FABRIC		DRY	WET	ACIDIC					ALKALINE				
							COTON	WOOL	COLOUR STAINING	COLOUR STAINING	CHANGE IN COLOUR			COLOUR STAINING		CHANGE IN COLOUR			COLOUR STAINING	
														COTTON	WOOL				COTTON	WOOL
				A	B	N					A	B	N			A	B	N		
(1)	PRE-MORDANTING	Harda	5%	¾	2:3	3	4/5	4/5	4	4	2/3	2/3	2	4/5	4/5	2/3	2	3	4	4/5
(2)		Tannic Acid	5%	3	3	2/3	4/5	4/5	4/5	4	¾	2/3	½	4/5	3	2/3	3	3	4/5	4/5

A - Acidic, B - Basic, N- Neutral

Fastness grades of the Banana peel dye on Silk fabrics—

1. **Washing Fastness** - Samples of silk dyed with Banana peel dye showed good to excellent washing fastness in both harda and tannic acid pre-mordanted samples. After washing, slight staining is there on adjacent fabric. Acidic samples showed good results.
2. **Rubbing Fastness** - Both Pre-mordanted samples showed slight staining in case of wet rubbing and slight to negligible staining in case of dry rubbing.
3. **Perspiration Fastness** - In case of harda pre-mordanted, the samples showed good colour fastness to acidic perspiration with slight staining on adjacent fabrics.

In case of tannic the samples showed fair to good colour fastness to acidic perspiration as compared to alkaline perspiration with negligible staining on adjacent fabrics. In acidic perspiration the neutral dyed samples showed change in tone.

CONCLUSIONS

Fermented dye gave excellent results where as non-fermented dye gave negligible colour to the fabric. Pre-mordanted samples showed good results as compared to post and simultaneous technique. Intensity of colour is maximum in case of tannic acid than myrobalan and minimum in case of $FeSO_4$, $CuSo_4$ and alum. 5% pre-mordanted samples of harda and tannic acid showed good results as compared to 3% samples because they have more intensity of colour. Thus intensity of colour depends up on the amount of mordant used. The shades obtained are brown to grey. The dye showed good colour fastness grading between 3-4.

28

APPLICATION AND EVALUATION OF MICROENCAPSULATED PERFUMES ON THREE TEXTILE FABRICS

Deepa Rathi and Misty Rao.

Department of Textiles & Fashion Technology, College of Home Science, Nirmala Niketan, Mumbai

ABSTRACT

Perfumery is the art of producing fragrance through the combination of odoriferous substances. Throughout history, perfumes have played an important role in human lives, and have been associated with notions of happiness, beauty and satisfaction. The development of the chemistry has allowed the invention of totally new odoriferous molecules as well as the synthesis of natural ones. Microcapsules preserve a substance in a finely divided state and release the substance from the enclosed capsule as required. An interesting application being the encapsulation of fragrances, essential oils, flavors, vitamins, nutritional seaweed extracts antibacterial agents, mosquito-repellents and many other useful ingredients. Taking the advantages of perfumes a study was conducted to evaluate the application of microencapsulated perfumes on different fabrics-cotton, polyester and cotton-polyester. From the experimental procedures good fastness properties associated with perfumes was obtained.

INTRODUCTION

The art of applying fragrance to our possessions for the beautification of body and our personality is something very common to the human species from time primitive. Fragrance attracts and holds attention (Manral and Agarwal, 1998).

1.1 Perfumes

Perfumery is the art of producing fragrance through the combination of odoriferous substances. Early in the history of humanity it appears that men burned various kinds of woods for warmth and light during the nights and cold winter seasons. Thus, they accidentally discovered the pleasant smell of the smoke that drifted through the air over the flames from some of the naturally perfumed woods. The French name 'parfum' comes from the Latin preposition 'per' meaning through and the substantive 'fumus' meaning smoke.

1.1.1 History of perfumery

Perfumes had already moved into commerce by the time of the earliest known records from Egypt. The *Bible* refers to spicy balm, myrrh, galbanum, and other aromatic substances. The *Koran* speaks of perfumery materials and specifically mentions musk's and hyacinth. In Greek mythology, Aphrodite was the first woman described as using aromatics, and man's knowledge of them is attributed to an indiscretion of one of her nymphs (Kirk, 1981).

The actual mixing or blending of perfumes is discussed by Theophrastus, 370 BC, whose writings show that the art form was already well advanced. The contributions of the chemical industry began near the end of the nineteenth century. Catalyzed by the synthesis of odour suitable aldehydes led to the introduction of a rich-smelling floral aldehydic fragrances. Today, perfumers have at their disposal hundreds of synthetic ingredients, as well as the traditional natural materials (Kirk, 1981).

1.1.2 The chemistry of smell

The fragrance doesn't smell the same through the way. When one applies it first, a strong fragrance envelopes, to disappear and is replaced by a lighter lingering fragrance. Fragrances are made up of 'notes'. It's what comprises the whiff and the smell and the aroma. They could be different ingredients such as cinnamon or vanilla or rose or lime (Fernandes, 2001).

(i) Head notes

The first whiff that engulfs one when a bottle is opened is called the Head note or Top note. The strong fragrance (note) is used to mask the scent of the alcohol, which is always present and necessary to carry the fragrance. Head notes generally last for a few minutes before completely evaporating and giving way to the Heart notes or Middle notes (Fernandes, 2001).

(ii) Heart notes

These last for several hours and are considered the real essence of a fragrance. This is why, when trying a fragrance for the first time one should wait a few minutes after applying to decide whether one actually likes it or not (Fernandes, 2001).

(iii) Base notes

This is the final phase and lasts for another few hours. While testing a fragrance, initially one might smell the notes of mint and citrus amongst others. The fragrance develops onto lavender and cinnamon (heart notes) and goes on eventually to the smell of sandalwood, musk and amber (base note).

1.2 Microencapsulation

Microencapsulation is a process in which tiny particles or droplets are surrounded by a coating to give small capsules with many useful properties. In its simplest form, a microcapsule is a small sphere with a uniform wall around it. The material inside the microcapsule is referred to as the core, internal phase, or fill, whereas the wall is sometimes called a shell, coating, or membrane. The microcapsule even may have multiple walls. Most microcapsules have diameters between a few micrometers and a few millimeters (Kirk, 1979).

The contents of the capsule can be made available by mechanical rupture of the capsule wall, by causing its disintegration by electrical or chemical means, or by a leaching action carried in an appropriate liquid environment. Microcapsules range in diameter from 1 to 1000 μm, capsules greater than 1000 μm can be called macrocapsules and those smaller than 1μ are termed as **nanocapsules** (Aggarwal *et al.,* 1998).

Microcapsules preserve a substance in a finely divided state, and release the substance from the enclosed capsule as required. This property has enabled microcapsules to serve many useful functions and find application in different fields to technology. The techniques and practice of microencapsulation cover several scientific disciplines, including colloidal and physical chemistry, polymer chemistry and physics, material science, suspension and drying technology and in controlled-released application, pharmaceutical and pesticide technology. Many of these interact with each other in practical applications and in the resulting commercial implications (Aggarwal *et al.,* 1998).

1.3 Fragrant Fibers

Fragrant fibers contain special fragrant additives such as high-boiling fragrant oils or essence, like lemon oil, bergamot oil, lavender oil, cedar wood oil, and eucalyptus oil, pineapple and strawberry essences. Fragrant substances are mixed with thermoplastic polymers and are incorporated in fiber cavities or in the core or also incorporated in the fiber using the principle of micro encapsulation. The capsules break very gradually during war, so the effect is long lasting and withstands much washing. Micro fibers can be specifically engineered so that beneficial chemicals and vitamins can be suspended along hollow fibers in tiny capsules; to be released gradually into the wearer's body. These substances are invisible to the eye and yet can have a dramatic effect on the wearer's health and well-being. Many of these so called healthy fibers were originally developed for use in space and are now being used in fashion. Perfumed fabrics are made into hosiery and lingerie. Vitamin C and nutritional seaweed extracts can also be incorporated in this way, to be slowly absorbed into the skin (Braddock and Mahony, 1998).

1.4 New Product Developments

With the advancement in technology more and more companies worldwide are experimenting to develop and have developed perfumed fabrics.

ESPRIT DE FLEURS produced by Kanebo in 1987 is made of fibres to which resin made microcapsules containing perfume essences are bound. The microcapsules when pressed are broken and the perfume is released. It lasts for almost two years and can withstand several washings. Applying the science of aromacology various kinds of perfumes such as jasmine, rose, lavender, and lily can be provided and the products such as pull covers, cardigans, neck-ties, scarves, T-shirts, handkerchief, etc. can be made.

British textile innovator Welback Fabrics features intimate clothing embedded with microencapsulated aromas like lavender, lemon, vanilla and rose. The new technology keeps the scents in the fabric for more 40 washes and prevents the smell from contaminating other clothing. Moreover, microencapsulated fabric releases its scent steadily through regular wear and tear. For those who prefer a fruity scent, Welback has dyed the cloth in colors that are inline with their aroma, such as yellow for lemon, pink for rose and ecru for vanilla. The handle remains fine and supple (www.welbeck-fabric.com).

Brookstone Chemicals, UK, application system comprises aqueous dispersions of microencapsulates applied by pad, exhaust or hydro extractive techniques. Products as varied as skin moisturizers, fragrances, vitamins or insect repellents can be applied using the system. Depending on application levels, the capsules will last up to 30 washes. Durability to washing and handle may be improved by incorporating suitable formaldehyde-free binders and softeners.

A French company called Neyret has come out with perfumed lingerie, which would release fragrance upon human contact. The lingerie consists of thousands of microcapsules full of perfume, which when rubbed would release perfume. The company claims that the scent life of a piece of lingerie is five washes.

LTP Group Korea has come up with perfumed textile fabrics with natural herb perfume. The company claims that the perfume lasts long even after many times of washing (www.ec21.net/ltpkorea).

Purpose of Study

On perspiration for some of us an unacceptable odour is obtained or given out. Sandalwood has the property of absorbing moisture, which results in cooling effect. Rose has the ability to relieve premenstrual tension, ease depression, mental fatigue and to alleviate stress. It also tones up the entire system. Lavender regulates the nervous system, eases aches, pains, palpitation, and faintness and is highly antiseptic. Oil of lemon is stimulant and rubefacient when applied externally. It is also useful in the verbena type of perfumes and toilet waters, and in similar soap perfumes (Kirk, 1981).

Therefore, by taking the advantage of some of these properties of various perfumes an attempt has been made in the present study to develop perfumed fabric.

The specific objectives of the study are to

- Study the application of seven different microencapsulated fragrances on cellulosic, polyester and cotton-polyester blend.
- Study the wash fastness of the treated perfumed fabric (five cycles).
- Study the extent of acidic and alkaline perspiration on the treated perfumed fabric (three cycles).
- Study the rub (dry) fastness of the treated perfumed fabric (five cycles).
- Study the tear strength of the treated fabric.

Methodology

The methodology has been divided into experimental materials and experimental procedures.

EXPERIMENTAL MATERIALS

Fabric Used

1. 100% Cotton (muslin)
2. 100% Polyester
3. Cotton-Polyester Blend (50-50)

Perfumes applied

Seven Microencapsulated Perfumes — Lavender, Rose, Musk, Lemon, Sandalwood, Tropical Paradise and Bouquet

Soap Used : Non-ionic Detergent

EXPERIMENTAL PROCEDURES

Procedure for Scouring

For scouring 5% of soap solution was used, keeping M: L ratio of 1:50. The fabric was wetted and entered in the soap solution. Scouring was carried out for 30-40 minutes at 100°C.

Procedure for application of Perfumes

For the application of perfumes M: L ratio used was 1:5. The required amount of perfume was added to the bath. The fabric was entered and kept in the bath at room temperature for 30 minutes. Cotton, cotton-polyester blend, and polyester fabrics were treated using 1% of all seven perfumes.

Procedure for Curing

The treated samples were dried in the oven for 2 minutes at 60°C. (Standardized from the preliminary work).

Procedure for Wash Fastness

(ISO 105/C-1982 Textiles-Tests for Colour Fastness C01)

Procedure for Rub (Dry) Fastness

The treated samples were cut to the size of 20 cm x 6 cm and a small sample of the untreated fabric was cut to the size of 5 cm x 5 cm. Rubbing fastness tests were carried out on crock meter in which the untreated fabric samples were rubbed on the treated fabric samples. Rubbing was done in a straight line along a track of 10 cm on the treated samples 10 times each to and fro along the length for 10 seconds. Each of the treated fabrics was subjected to 5 cycles for dry rub fastness.

In between each cycle the extent of perfume remaining on the fabric was evaluated by a set of same five volunteers.

Procedure for Perspiration Fastness

The tests were carried out in a perspirometer. Glass plates each measuring about 7.5 cm x 6.5 cm were used, between which the samples to be tested were sandwiched. The temperature of the oven was maintained at 37 ± 2°C. Acid test liquor was prepared by dissolving 2.65 gm of sodium chloride and 0.75 gm of urea per litre and pH of the solution was adjusted to 5.6 with the addition of acetic acid. Alkaline test liquor was prepared by dissolving 3 gm of sodium bicarbonate. M:L ratio was taken as 1:50. The samples to be tested were dipped in the solution for 30 minutes at room temperature. Samples were placed between 2 glass plates under force of 4.5 kg. They were placed in the oven for 4 hours at 37 ± 2°C. The samples were dried in air at room temperature not exceeding 60°C.

In between each perspiration cycle the extent of perfume remaining on the samples were evaluated by a set of same five volunteers.

Procedure for Tensile Strength

KMI tensile strength tester was used to determine the breaking load and the elongation percentage. The warp and the weft samples were cut in the size of 30 cm x 6 cm. The constant weight of 100 kg was maintained throughout the test. The breaking load and the elongation for each sample were noted.

Sample Size

Total sample size was 21 samples.

RESULTS AND DISCUSSIONS

Three different fabrics – cotton, polyester, and cotton-polyester blend were treated with seven microencapsulated perfumes. The perfumes used were lavender, rose, musk, lemon, sandalwood, tropical paradise and bouquet. The treated fabric samples were evaluated for the following parameters;

1. Wash fastness
2. Dry rub fastness
3. Perspiration fastness, and
4. Tensile strength

1. Wash fastness

Table 1. Evaluation of treated samples for wash fastness

Perfumes	1st Wash	2nd Wash	3rd Wash	4th Wash	5th Wash
Fabrics Cotton					
Lavender	S	M	F	VF	VF
Rose	S	S	M	F	VF
Musk	S	F	VF	-	-
Lemon	S	S	M	M	F
Sandalwood	S	M	M	M	F
T. Paradise	S	S	M	M	F
Bouquet	S	S	M	M	F
P/C Blend					
Lavender	S	S	M	M	F
Rose	S	M	M	F	VF
Musk	S	S	M	F	VF
Lemon	S	S	M	M	F
Sandalwood	S	M	F	VF	VF
T. Paradise	S	M	M	M	M
Bouquet	S	S	S	S	M
Polyester					
Lavender	S	M	F	F	VF
Rose	M	VF	VF	-	-
Musk	S	M	F	-	-
Lemon	S	S	M	VF	-
Sandalwood	M	M	F	-	-
T. Paradise	S	S	M	F	VF
Bouquet	S	S	M	M	F

T. Paradise = Tropical Paradise; P/C Blend = Cotton-polyester Blend; S= Strong; M= Moderate F= Faint; VF=Very Faint.

Cotton, cotton-polyester blend and polyester gave strong fragrance after the first wash cycle, with the exception of polyester samples treated with rose and sandalwood, which gave mild fragrance. On evaluation of the treated fabric after the second wash cycle it was seen that besides polyester all the other fabrics had strong to moderate fragrance. However, by the fifth cycle fabric samples had faint to very faint fragrance with the exception of cotton-polyester samples. The samples treated with topical paradise and bouquet had moderate fragrance.

Thus is may be said that cotton-polyester gave the best results for wash fastness, also observed that some fabric samples lot their fragrance by the fifth cycle, especially polyester samples. In terms of perfumes bouquet, tropical paradise and lemon gave the best results for wash fastness test followed by lavender.

2. Dry rub fastness

Table 2. Evaluation of Treated samples for dry rub fastness

Perfumes	1st Wash	2nd Wash	3rd Wash	4th Wash	5th Wash
Fabrics Cotton					
Lavender	S	S	M	F	F
Rose	S	S	F	F	F
Musk	S	S	F	F	VF
Lemon	S	S	S	F	VF
Sandalwood	S	S	S	F	F
T. Paradise	S	S	F	F	F
Bouquet	S	S	S	M	M
P/C Blend					
Lavender	S	M	F	F	VF
Rose	S	S	M	F	VF
Musk	S	S	M	F	F
Lemon	S	S	S	M	F
Sandalwood	S	S	M	F	F
T. Paradise	S	M	M	F	F
Bouquet	S	S	S	M	F
Polyester					
Lavender	S	S	M	F	F
Rose	S	S	M	F	F
Musk	S	S	F	F	F
Lemon	S	S	F	F	F
Sandalwood	S	S	M	F	VF
T. Paradise	S	S	M	F	VF
Bouquet	S	S	M	F	VF

T. Paradise = Tropical Paradise; P/C Blend = Cotton-polyester Blend; S= Strong; M= Moderate; F= Faint; VF=Very Faint.

Cotton, polyester and cotton-polyester samples gave strong fragrance for the first and the second rub cycles for all the perfumes with the exception of cotton-polyester blend, which gave moderate fragrance after the second cycle for lavender and tropical paradise. Evaluating the extent of fragrance left water the fifth cycle it was observed that the results varied from faint to very faint for the various perfumes and different fabrics.

Cotton-polyester samples gave best results for dry rub fastness followed by cotton and last by polyester. In terms of perfume, best results were obtained for bouquet, followed by lemon, tropical paradise, lavender and sandalwood. Other perfumes such as rose and musk gave satisfactory results.

3. Perspiration fastness

Table 3. Evaluation of treated samples for perspiration fastness

Perfumes	1st Test		2nd Test		3rd Test	
	Acid	Alkaline	Acid	Alkaline	Acid	Alkaline
Fabrics Cotton						
Lavender	S	S	S	S	M	S
Rose	S	S	S	S	M	S
Musk	S	S	M	M	F	M
Lemon	S	S	S	S	S	S
Sandalwood	S	S	M	M	M	S
T. Paradise	S	S	S	S	M	S
Bouquet	S	S	S	S	S	S
P/C Blend						
Lavender	S	S	S	S	S	S
Rose	S	S	S	S	M	M
Musk	S	S	M	M	M	M
Lemon	S	S	S	S	S	S
Sandalwood	S	S	M	M	M	M
T. Paradise	S	S	S	S	S	S
Bouquet	S	S	S	S	S	S
Polyester						
Lavender	S	S	S	S	F	F
Rose	S	S	S	S	F	F
Musk	S	S	M	M	F	VF
Lemon	S	S	S	S	S	S
Sandalwood	S	S	M	M	F	F
T. Paradise	S	S	S	S	S	S
Bouquet	S	S	S	S	S	S

T. Paradise = Tropical Paradise; P/C Blend = Cotton-polyester Blend; S= Strong; M= Moderate; F= Faint; VF=Very Faint.

Representing the result of the first perspiration test it was observed that strong fragrance remained for all treated samples in case of acidic and alkaline perspiration. However in the second cycle for acidic and alkaline perspiration musk and sandalwood gave mild fragrance in terms of fabric, polyester samples had faint to very faint fragrance for lavender, rose, musk and sandalwood perfumes. Cotton and cotton-polyester blend gave strong to moderate fragrance for all the other perfumes.

Cotton and cotton-polyester blend gave the best results for both acidic and alkaline perspiration test, since the fragrance remained on the fabric even after the third test. Polyester did not give very good results by the third perspiration test. Higher ratings were obtained for alkaline perspiration than for the acidic perspiration for all the treated fabrics by various perfumes. In terms of perfumes it was seen that lemon, bouquet, and tropical paradise gave the best results followed by lavender and rose.

4. Tensile strength test

Table 4. Evaluation of treated samples for Tensile Strength

Perfumes	Breaking strength in kg.		Elongation at break in cm		Percent of elongation at break	
	Warp	Weft	Warp	Weft	Warp	Weft
Fabrics Cotton						
Untreated	17	4.5	1.4	3.5	7	18.9
Lavender	16	4.5	1.3	3.5	6.5	18.9
Rose	16.5	5	1.3	3.5	6.5	18.9
Musk	16	5	1.4	4.5	7	24.3
Lemon	16.5	5	1.5	4.5	7.5	24.3
Sandalwood	16	4.5	1.4	3.5	7	18.9
T. Paradise	16.5	4.5	1.4	3.5	7	18.9
Bouquet	16.5	5	1.5	4.5	7.5	24.3
P/C Blend						
Untreated	79.5	56	8.7	4.9	43.5	6.4
Lavender	80	56.5	8.7	4.9	43.5	26.4
Rose	81	55	9.4	4.9	47	26.4
Musk	81	56.5	9.3	4.9	46.5	26.4
Lemon	80	56	9.1	4.7	45.5	25.4
Sandalwood	79	55	9	4.8	45	25.9
T. Paradise	80	55	9	4.9	45	26.4
Bouquet	80	56.5	9.2	4.9	46	26.4
Polyester						
Untreated	97	42.5	7.5	8.3	37.5	44.8
Lavender	96.5	43	7.3	8.4	36.5	45.4
Rose	97	42	7.5	8.2	37.5	44.3
Musk	97	43	7.4	8.3	37	44.8
Lemon	97	42	7.5	8.3	37.5	44.8
Sandalwood	97	43	7.6	8.4	38	45.4
T. Paradise	96.5	42.5	7.3	8.3	36	44.8
Bouquet	96.5	43	7.3	8.4	36.5	45.4

T. Paradise = Tropical Paradise; P/C Blend = Cotton-polyester Blend; S= Strong; M= Moderate; F= Faint; VF=Very Faint.

The breaking load of the treated cotton, polyester and cotton-polyester blend samples did not differ to a great extent from the breaking load of the untreated fabric. Accordingly for elongation percentage of the treated and untreated sample, no difference was observed in both warp and weft direction.

Thus it may be said that the application of fabric samples with microencapsulated perfumes for this study did not affect the breaking load of the fabrics.

Conclusions of the experimental tests

1. In terms of fastness properties, among the perfumes, Bouquet was the best perfume followed by Tropical Paradise, Lemon and Lavender. Musk and Sandalwood perfumes gave satisfactory results.
2. Among the fabrics, cotton-polyester blend gave the best results for all the fastness tests.
3. Best results of wash, rub (dry), and perspiration fastness was found on Cotton Fabric.
4. 100% Polyester fabric gave the lowest value for all the three fastness properties.
5. Application of perfumes of all fabric samples did not affect the tensile strength as it was observed that the breaking load of the untreated samples and the treated samples was nearly the same.

Thus it may be suggested that microencapsulated perfumes can be applied on blend for such end uses which require restricted laundering for maintenance.

REFERENCES

Aggarwal, A. K., A. Dayal and N. Kumar (1998); *Colorage;* August; 15

Braddock, S and M. Mahony (1998); *Techno textiles – Revolutionary Fabrics for Fashion and Design*; Thames and Hudson; p. 16.

Fernandes, A. (2001), *Femina* February; pp. 8, 10 and 14.

Kirk, O. (1979); *Encyclopedia of Chemical Technology*, John Wiley Interscience, New York; 15; pp 470-471, 473.

Kirk, O. (1981) *Encyclopedia of Chemical Technology*, Published by John Wiley Interscience Publication, New York; 16; pp. 947-951, 959-965.

Kochhar, B. (1994); *Health and Beauty through Aromatherapy*, USB Publishers Distributors Ltd., New Delhi; First Edition; pp 4-7, 11, 19, 41.

Manral, K. and Agarwal, R. (1998); *Femina*; February 1-15; pp. 8-33.

Secondini, O. (1998); *Handbook of Perfumes and Flavors*; East West Press Pvt Ltd, New Delhi; pp. 2, 48-64.

29

MEDICINAL HERBS FOR DYE NORMAL AND GRAY HAIR

Shehnaz Y. Qureshi

Beauty, hair and holistic health therapist, Vadodara

Now a days graying of hair is very common among young persons. Wrong dietary habits, tension and chemical cosmetics have taken over the minds of our new generation. Medicinal herbs are a God-given remedy safe enough to be used by children, young men, and women and even by elderly.

For the best result of dye, hair should be coloured only twice a month. Paste required is dependent on the length and thickness of hair. Paste should be kept on hair for at least two hours. Hair colour in root change is seen only with a number of applications. Mix the paste in iron bowl with 2.5 cups of boiling tea water or *Aloe vera* juice 12 hours before the application. It may be applied by hands or by hairbrush.

Natural dyes (Medicinal herbs) useful for natural brown black hair effect and orange gold effect to gray hair.

For Brown Black Hair

(For Gray hair – Gives colour and shine and Black hair gives only shine)

Materials	Part used	Quantity
Mehandi (*Lawsonia alba*)	Leaf Powder	21 Table Spoon
Amla (*Emblica officinalis*)	Fruit Powder	4 Table Spoon
Bangra (*Eclipta alba*)	Leaves Powder	4 Table Spoon
Indigo (*Indigofera tinctoria*)	Leaves Powder	4 Table Spoon
Jasud (*Hibiscus rosa sinensis*)	Flower Powder	7 Table Spoon

Wheat-germ Oil	Grass Oil	2 Table Spoon
Palmarosa Oil	Grass Oil	1 Table Spoon
Aloe vera	Leaves	Sufficient for paste

For Orange – Gold Hair

(For Gray hair – Gives colour and shine and Black hair gives only shine)

Mehandi (*Lawsonia alba*)	Leaf Powder	25 Table Spoon
Amla (*Emblica officinalis*)	Fruit Powder	4 Table Spoon
Jasud (*Hibiscus rosa sinensis*)	Flower Powder	7 Table Spoon
Aloe vera	Juice	Sufficient for Paste

Mehandi (*Lawsonia alba*)

Easily available in villages, farms as border bushes, well known all over the world, queen of herbs in beauty therapy, darkens only grey hair. Using of mahendi prevents dandruff and greying of hair and reduces harmful effects caused by chemicals.

Amla (*Emblica officinalis*)

This is a medium sized tree, 6 to 10 m height, full bloom with small yellow flowers; fruits are round, smooth, light-greenish yellow colour, seen in the month of October to February. Amla is KESHYA, helps in hair growth, in conditioning and as hair dye, protects acid mental of scalp and gives cooling effects.

Jasud (*Hibiscus rosa-sinensis*)

Medium sized tree, 5 to 9 feet height, bloom with red, pink, white flowers of 17 varieties, red colour is useful in hair care. Jasud is good for hair colour conditioner, and help hair growth and turns it smooth and dark. It also helps in faster hair growth, reduces dandruff and gives pleasant fragrance.

Wheat germ oil

A germinating part of wheat grain (embryo) is the wheat germ. This germinating part contains proteins, fats and vitamins in large amounts. It contains natural vitamin E, which is very beneficial for hair. Wheat germ oil is a good nutrient and helps to keep skin clean, bright, fresh and healthy. It gives nutrition to hair and is useful in the treatment of grey, dry and split hair, stops grey hair in young age and turns it dark.

Palmarosa (Motia Rosa Oil)

This oil is distilled from Rosa grass. It has mild fragrance and helps hair growth, stimulates glands near the hair and improves blood circulation, helps to prevents greying of hair. Palmarosa oil is obtained from the stems and flower heads of Palmarosa grass. It is scientifically cultivated in some parts of India. India holds a prime position in the export of this oil. Gujarat's climate suits its growth.

Aloe vera

The plants are grown full year. *Aloe vera* is extraordinary herb of cactus type. Leaves are smooth, juicy, 15 to 30 cm long, 1-2 cm thick, coming out from root, remains fresh for many days. The margins of leaves are turned inwards, have small thorns on the edge, give cooling effects, helps in healing of wounds. Old dead cells are destroyed and new cells are stimulated by specific ingredients in *Aloe vera* juice. It is good for hair growth and reduces dandruff, completely relieves itching in the scalp and gives a shiny appearance to hair, gives protection from bad effects of chemicals. Beautifully wrapped *Aloe vera* shampoo or cream is kept in big cosmetic shops with imported titles.

C

BIOPESTICIDES

30

BIOCONTROL AGENTS AND THEIR INTERACTION WITH AGRICULTURAL CROPS

Dolly Kumar and Bhumika Naidu

Department of Zoology, Division of Entomology, The Maharaja Sayajirao University of Bāroda, Vadodara, Gujarat 390002

ABSTRACT

Studies were conducted to find out the bio-control agents and their interaction with agricultural crops in various Agricultural fields of Vadodara district. Biological control of insect pests damaging different parts of the economically important crops is one of the effective strategies used in Integrated pest management which can minimize the use of highly toxic insecticides. It involves several biological agencies working in nature such as parasitoids, parasites, predators and pathogens. The main principle behind this strategy is to suppress the reproductive potential of the organisms through other organisms which control crop pest population, keeping the insect pest below economic threshold level by prey – predator interaction. *E.g. Mantis religiosa* feeding on thrips of wheat and paddy *Anax* feeding on *Achoea janata* of castor plant. Thus the purpose of this study is to make farmers aware of microbial and zoological biopesticide agents. This will enable them to identify the parasitic and predatory insects which feed upon insect pests destructive to crop. By using these Zoological biopesticide agents the frequency of spray of chemicals such as Organophosphorus (Lentrek, Guardian) and Synthetic Pyrethroids (Cyperguard, Spark) will be reduced which will avoid accumulation of pesticides in soil and environment.

INTRODUCTION

Biological control is an important tactic of IPM strategy and should be utilized wherever it is feasible (Dhuliya and Yadav, 1994). Its purpose is to establish stable low level equilibrium of pest density (Smith and Reynolds, 1960) this can be achieved by several biocontrol agents working in nature such as parasitoids like *Trichogramma chilonis*, parasites like Ichneumon wasp, predators like Praying mantis and Lady bird beetle, as well as pathogens such as bacteria (like *Bacillus thuringiensis*) and viruses (like Non Polyhedrosis Virus) which suppress the pest population making it less damaging than it would be in the absence of bio control agents. Biological control does not cause immediate reduction in target pest population. It only achieves partial suppression of target pest as residual pest population is necessary to maintain balance in eco system (Picket and Buggs, 1998). For controlling pest, conventional chemical control approach is followed which is not only expensive but leads to residue problem also, to overcome this problem integrated approach with emphasis on bio control is viable option (Garg, *et al.,* 2000).

Biocontrol leaves no pest residue capable of contaminating the crops or environment. Biological control does not disrupt any other pest management practices. This study focuses on the results of the field survey to monitor predator-pest-plant habitat with the view to examine the interaction occurring between them and future potential for the use of biocontrol agents in various agricultural fields of Vadodara district.

MATERIALS AND METHODS

The survey of various agricultural fields was made in and around Vadodara district *i.e.* in Model farm (Gujarat Agriculture fields) (2 km), Chaani (10 kms), Padra (14 km) and Waghodia (26 km) from M.S. University campus. The surveyed fields were of vegetables like cabbage, brinjal, lady's finger and chick pea; crops like wheat, paddy, sugarcane, maize; fruits like banana as well as cotton fields.

Insects were collected using

- Sweep nets,
- Light traps,
- Pheromonal traps,
- Pitfalls and
- Beating method.

Identification was done using taxonomic literature.

Survey area

1. Padra

(a) Paddy, *(Oryza sativa)*: It's common seasonal varieties are:

(i) Aus or summer rice which is cultivated in July and harvested in Aug.-September

(ii) Aman or winter rice which is cultivated in Oct.-Nov., harvested in March - April

(iii) Bore rice, grown between two seasons, which is cultivated in June – July and harvested in September-October.

Caliothrips indicus Bagnall was observed causing damage to leaves and flowers of Paddy. Similar damage was also reported by Saxena (1987). *Hieroglyphus banian* is also found to cause damage by eating fruit, flowers and leaves. It is active from June to November.

(b) Wheat (*Triticum aestivum*): It is cultivated in October to December and harvested from March up to May. It prefers moderately dry climate and thrives under irrigation.

Anapothrips sudanensis Trybom damages tender leaves and flower stalks by sucking cell sap. It is active throughout the year. They have gained increasing importance over the years owing to their tendency to inflict significant damage to agriculture and horticulture crops (Ananthakrishnan, 1971)

(c) Brinjal (*Solanum melongena*): Its cultivation is done in February-march, June-July, and October-November. When fruit is completely ripe harvesting is done.

Nymphs and adults of *Aphis gossipy* are found on leaves and stem. It is present throughout the year and is cosmopolitan in distribution. Caterpillars of *Spodoptera litura* damages the leaves and fruits and making it unfit for human consumption.

(d) Cabbage (*Brassica oleracea* var. *campestris*): cultivation is done from end of October up to first week of November and harvesting is done in March-April. Caterpillars of *Pieris brassicae Linnaeus* and *Plutella maculipenis Linnaeus* were found underside of the leaves cutting several holes thus reducing the yield considerably. In India Diamond back moth was first recorded in 1914 on cruciferous vegetation. *Myzus persicae* is found sucking the sap from lower surface of the leaves. They are present throughout the year and are cosmopolitan they are pest of wide range of crops like, groundnut, green gram, chickpea, lentils, cowpea, pigeon pea, cotton, chilies (Sahoo and Jeena, 1993).

(e) Lady's finger (*Abelmoschus esculentus*): cultivation is done from February to July, harvesting is done after fruit is fully ripened. Caterpillar of *Helicoverpa armigera Hubner*, an American bollworm causes the damage to the plant by destroying the pods. It is a polyphagus pest .In India it has been recorded feeding on 181 cultivated and uncultivated plant species belonging to 45 families (Manjunath *et al.*, 1989).

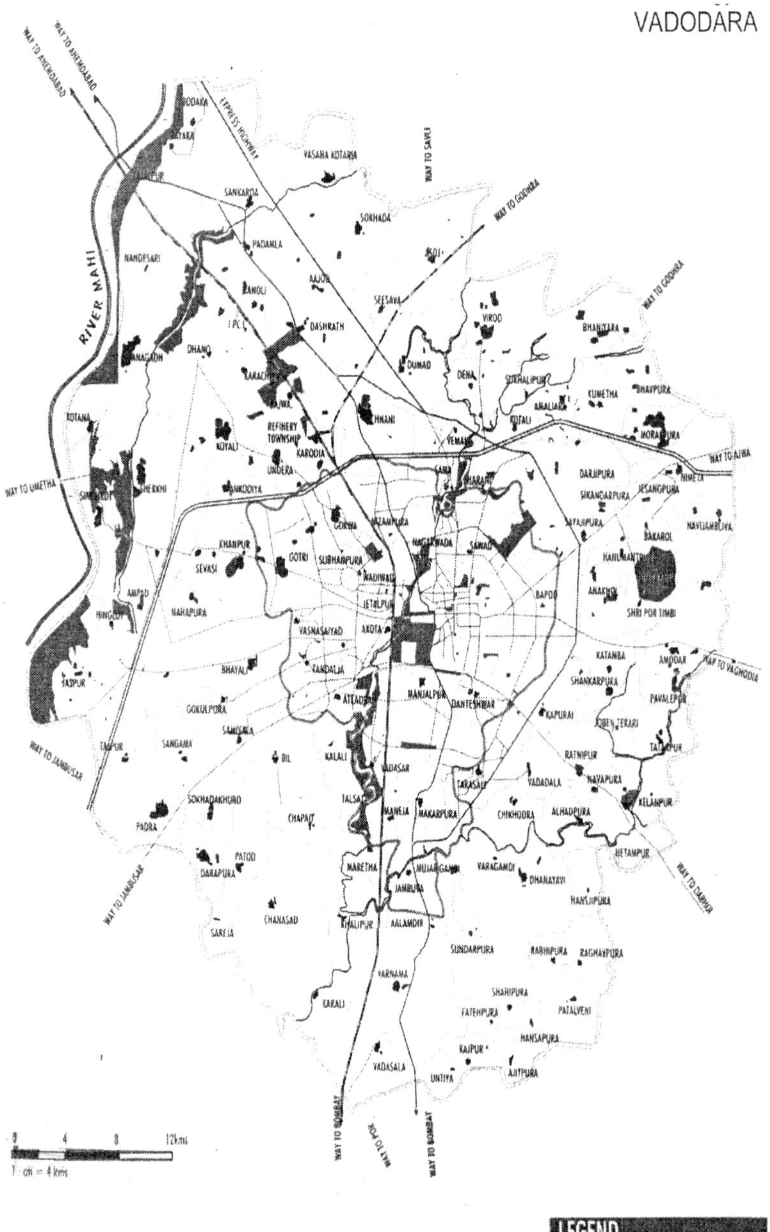

(2) Waghodia

Chickpea (*Cicer arietinum*): cultivated in October-November and harvested in February –March.

Nymphs and adults of *Aphis fabea* Scop. a legume aphid is found feeding on fruit and leaves of chickpea (Siddhanathan *et al.*, 1984). *Tricentrus bicolor*, a leafhopper is also found sucking sap from the plant making it weak.

(3) Chani

(a) Banana (*Musa paradisiaca*): the plantation takes place in the month of June-July and plants are removed after 18 months.

Nymphs and adults of *Aularches miliaris*, a leafhopper is found sucking sap from the shoot and leaves, affecting the growth of plant. Maggot of a fruit fly *Dacus dorsalis* Hendel are found feeding on pulp of fruit. Damage during July-September is heavy.

(b) Sugarcane *(Saccharum officinalis)*: The nymphs and adult of *Pyrilla perpusilla*, a sugarcane leaf hopper, causes damage by sucking sap from leaves which wither and dry up. Due to secretion of honeydew by this pest black fungus develops because of which plant growth is suffered. Quality of the juice also deteriorates. Cane becomes useless for recovery of juice and sugar. It appears in March and active till onset of winter. Serious damage is caused from May up to November.

(c) Cotton *(Gossypium* spp.): life cycle of crop is 8-9 months. Seeds are sowed after first shower and sprout within a week. During harvest cotton is handpicked for over 2 months. The plants are completely removed from the field in February and March. Nymphs and adults of *Dysdercus cingulatus*, Red cotton bug sucks the sap from stem and bolls. Such seeds become useless for sowing and oil extraction. Caterpillars of *Earias vitella* and *Pectinophora gossypiella* damage the tender portion of growing shoot by boring it, resulting into drooping of the plant. They also attack flowers & fruits as a result they are shed or open prematurely and market value are reduced. It is active through out the year, but is at its peak during July and September.

(4) Model Farm

Maize *(Zea mays)*: it is cultivated in April –May, harvested in July-August.

Mylabris pustulata Thunberg, a blister beetle feeds on floral buds and flowers. Single beetle can destroy 20-30 flowers per day. It is serious pest during rainy season.

RESULTS

The crops, the pests and the biocontrol agents observed are presented in Table 1.

Table 1. The pests and their biocontrol agents of some crops of Baroda

Crops	Infested part of plant	Pest spotted	Stage of pest	Biocontrol agents observed.	Stage of bio-agent
Wheat	Flowers	*Anaphothrips sudanensis*	Nymph & adult	*Chrysoperla carnia*	Larva & adult
	Leaves & fruits	*Heiroglyphus banian*	Nymph & adult	*Mantis religiosa*	adult

	Flowers	*Triptocentrus maculatus*	Nymph & adult	*Allocotasia aurata*	adult
Paddy	Leaves & fruits	*Heiroglyphus banian*	Nymph & adult	*Anax species*	Niads & adult
	Roots	*Grylotalpa Africana*	Nymph & adult	*Allocotasia aurata*	adult
	Leaves	*Baliothrips*	Nymph & adult	*Mantis religiosa*	adult
Maize	Leaves	*Spodoptera litura*	larva	*Gongylus gongyloides*	adult
	flowers	*Mylabris pustulata, M. phalereta*	Grubs & adult	*Mantis religiosa*	adult
Cabbage	Leaves	*Pieris brassica, Trichoplusia species*	larva	*Aspongopus janus*	adult
	Leaves	*Aphis gossypii*	Nymph & adult	*Coccinella septumpunctata*	adult
Brinjal	Fruit	*Helicoverpa armigera*	larva	*Chrysopa carnia*	larva
	Leaves & Stem	*Aphis gossypii*	Nymph & adult	*Coccinella septumpunctata*	adult
	Leaves and fruits	*Spodoptera litura*	larva	*Coccinella septumpunctata*	adult
Lady's finger	Leaves & Stem	*Aphis malvacea*	Nymph & adult	*Mantis religiosa*	adult
	Root	*Holotricha insularis, H. serrata*	Grub & adult	*Rhinocoris fuscipes*	adult
Chick pea	Flowers & Leaves	*Myzus persicae, Tricentrus bicolor*	Nymph & adult	*Coccinella septumpunctata*	adult
	Fruit	*Helicoverpa armigera*	Larva	*Chrysoperla carnia*	larva
Cotton	Leaves, seeds, pods	*Dysdercus cingulatus*	Nymph & adult	*Coccinella septumpunctata*	adult
	Fruit, shoots, buds, bolls	*Earias insulana*	larva	*Chrysoperla carnia*	larva
	Fruit, shoots, buds, bolls	*Pectinomorpha gossipiella*	larva	*Mantis religiosa G. gongiloides*	adult
Banana	Leaves	*Aularches miliari*	Nymph & adult	*Mantis religiosa*	adult
	fruits	*Dacus dorsalis*	maggot	*Chrysoperla carnia*	Larva & adult
Sugarcane	Root	*Grylotalpa africana*	Nymph & adult	*Mantis religiosa*	adult
	Leaves	*Pyrilla* species	Nymph & adult	*Gongylus gongiloides, Anthia sexgutata*	adult

Wheat: *Anapothrips sudanensis* Trybom causes damage to leaves and flower stalk. Their population is controlled by *Chrysoperla carnia Stephens* (Fig. 1), a lacewing. *Hieroglyphus banian Fabricius*, a rice grasshopper causes damage to the fruits. It is devoured by *Mantis religiosa*, a praying mantis.

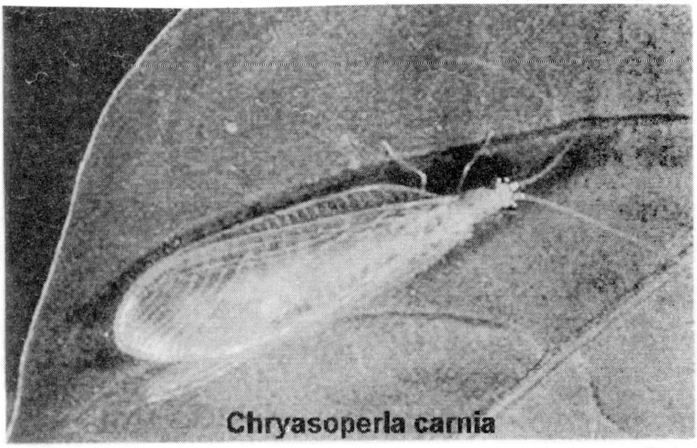

Fig. 1

Paddy: *Hieroglyphus banian Fabricius* causes damage to leaves and flowers of paddy also. It is predated by *Philodicus femoralis* robber fly. *Caliothrips indicus Bagnall*, also causes damage to paddy, it is controlled by *Chrysoperla carnia Stephens*, lacewing.

Maize: *Hieroglyphus banian Fabricius* causes damage to leaves and flowers; it is predated by *Gongylus gongyloides*, (Fig. 2) a praying mantis. *Mylabris* species of blister beetle is found to cause damage to flowers of maize. They are predated by *Allocotasia aurata* (Fig. 3) robber fly.

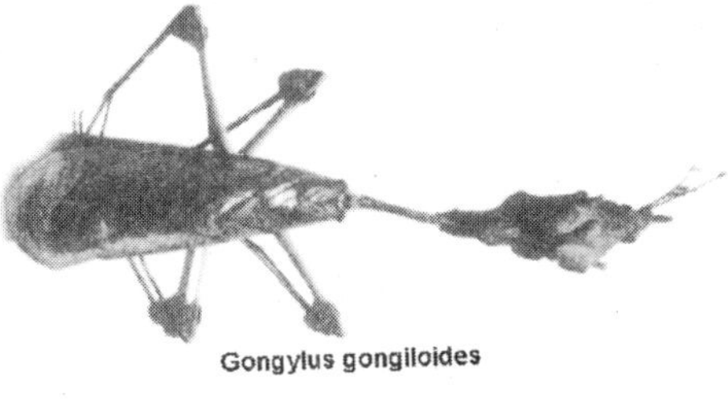

Fig. 2

Fig. 3

Cabbage: Caterpillars of *Pieris brassicae Linnaeus* and *Plutella maculipenis Linnaeus* reduce the yield by damaging the leaves. They are controlled by Reduvid bug, which feeds on 12 larva per day (Annuradha, 1997). *Myzus persicae,* an Aphid sucking the sap and damaging the plant, is controlled by *Coccinella septumpunctata* (Agarwal and Ghosh, 1988)

Brinjal: *Aphis gossypii Glover* damages leaves and stem. *Chrysoperla carnia Stephens* a lacewing feeds on them. *Spodoptera litura Fabricius* is controlled by *Rhinocoris fuscipes*, an assassin bug.

Lady's finger: Caterpillars of *Helicoverpa armigera Hubner*, an American bollworm destroys the pods. *Rhinocoris fuscipes* attacks them and controls their infestation.

Chickpea: *Aphis fabea*, a legume aphid, feeds on fruits and leaves and *Tricentrus bicolor* also sucks sap from the plant. *Coccinella septumpunctata* (Fig. 4) ladybird beetle controls both these pests.

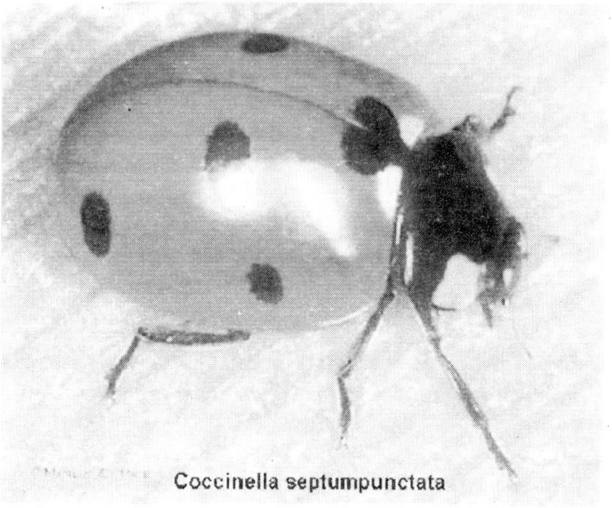

Fig. 4

Cotton: *Dysdercus cingulatus*, a red cotton bug, sucks the sap from the stem and bolls. Its population is controlled by *Coccinella septumpunctata*. *Earias vitella* and *Pectinophora gossypiela* bore the shoot of the plant. They are preyed by *Mantis religiosa* and *Gongylus gongyloides*, two species of praying mantis respectively.

Banana: *Aularches miliaris*, a leafhopper, sucks the sap from leaves and shoots of the plant. It is being predated by *Chrysoperla carnia Stephens*, a green lacewing. Maggot of *Dacus dorsalis Hendel* is found feeding on pulp of fruits. It is predated by *Mantis religiosa*.

Sugarcane: *Pyrilla perpusila*, a leafhopper, causes damage to leaves and it is predated by *Mantis religiosa*.

DISCUSSION

The results of our study shows that relative abundance of *Coccinella septumpunctata* and *Chrysoperla carnia Stephens* is more as compared to other biocontrol agents (Table 1) both are aphidophagous. These observations reveal that these biocontrol agents should be conserved. Increased spray of chemical insecticides destroys their population. Thus their conservation is possible only if insecticide spray is decreased so as to provide a safe ground for these little gladiators to breed and survive. This purpose can be achieved if farmers are made aware of bio control agents and their conservation in the agricultural fields.

Thus biocontrol based IPM with regular monitoring of insect pest and their predators should be adapted with proper motivation and involvement of farmers, which can change the present scenario of Baroda fields, where fertility of land is decreasing due to high use of chemical fertilizers and pesticides.

ACKNOWLEDGEMENTS

Authors are grateful to late Prof. N. Radhakrishnan for his valuable guidance in taxonomic identification of insects.

REFERENCES

Agarwal, B.K and A.K. Ghosh (1988) Tropical *Pest Management* **34**:1.

Ananthkrishnan, T.N. (1971) *J. Sci. and Ind. Res.* **30**:113.

Anuradha. M. (1997) *Possibility on integration of bioagents for management of diamond back moth.* Ph.D. Thesis, University of Agriculture Science, Bangalore.

Dhulia, F.K and D.N. Yadav (1994). *J. Boil. Control.* **8(2)**:94

Garg, D.K.P. Kumar, R.N. Singh and M. Pathak (2002). *Ind. J. Entamo.* **64**(2):117.

Manjunath, T.M., Bhatnagar, V.S., Pawar, C.C. and S. Sithanathan (1989). In *Proceedings of the workshop on bio-control of Heliothes increasing the effective ness of natural enemies.* 11-15, Nov. 1985 (ed. E.G. King), FERRO, USDA, New Delhi, India. p 1997.

Pickett, C.H and R.L. Buggs (1998) *Enhancing biologicalcontrol: Habitat management to promote natural enemies of agricultural pest.* Berkley, California: University of California Press p 422.

Sahoo, B.K and B.C. Jeena (1993) *Ind. J. of Entom.* **55**: 454.

Saxena, R.C. (1987) *Recent advances in Entomology*, p 141.

Siddhanathan, S., Sethi, S.C. and S.P.S Beniwal (1984) *International chickpea Newsletter.* **10** : 19.

Smith R.A and H.T. Reynolds (1996) *Principles, Definition and scope of IPM. Proc.* FAO Symposium, **1**: 11.

31

CONTROL OF TOXICOGENIC FUNGI BY ESSENTIAL OIL FROM PERICARP OF *CITRUS SINENSIS* (MOSAMBI)

Neeta Sharma and Abhishek Tripathi

Botany Department, Lucknow University, Lucknow - 226 007

ABSTRACT

Various stored food products becomes toxic due to contamination by mycotoxin producing fungi. These fungi grow rapidly on a variety of natural substrates and consumption of such contaminated food can pose serious health hazards to human and animals. Essential oil extracted from pericarp of *Citrus sinensis* (L.) Osbeck (Mosambi, Tight skinned orange - Rutaceae) was tested against the test pathogens viz. *Aspergillus flavus* Link. and *A. niger* van Tieghem. It was found that the oil was fungistatic and fungicidal against *A. flavus* at 1000 ppm and 2500 ppm and against *A. niger* at 1000 ppm and 3000 ppm respectively. The oil exhibited 100% inhibition of spore germination at 1500 ppm for both test fungi. Determination of mycelial weight was done and it was found that insignificant growth occurred at 3000 ppm and 2500 ppm for *A. flavus* and *A. niger* respectively. To study the mode of action of essential oil and morphological effects on test fungi light microscopy and scanning electron microscopy (SEM) was done which revealed that the oil induces alteration on cell fungal membranes.

INTRODUCTION

Aspergillus spp., saprophytic molds, are capable of growing upon a wide range of organic substrates, and often cause deterioration of stored food material (Sharma and Verma, 2004). To prevent the deterioration of stored food, the use of synthetic fungicides has a large number of limitations and disadvantages, which include progressively stricter laws, social rejection due to toxicological problems affecting human and environment, and biological limitations due to fungicide resistance. A large number of researchers and health organizations noted that, although synthetic fungicides prevent microbial spoilage, their use in stored products is no longer recommended because of their toxicity and oncogenic risk (Wilson *et al.*, 1991). Effective phytochemicals are expected to be far more advantageous than synthetic pesticides, as they are easily decomposable, not environmental pollutants and possess no residual or phytotoxic properties (Tewari, 1990). Essential oils are naturally occurring terpenic mixtures, whose insecticidal action against specific pests and fungicidal action against some plant pathogens has been recently reviewed (Ismam, 2000). The aim of present work is to focus on the antifungal activity of *Citrus sinensis* (L.) Osbeck (Mosambi, Tight skinned orange- Rutaceae) essential oil, obtained from its pericarp and to investigate its mode of action against the pathogens.

MATERIALS AND METHODS

Extraction of essential oil

For the study, pericarp of *Citrus sinensis* was collected from various juice shops of Lucknow District (Uttar Pradesh), India. The essential oil was extracted from collected material by hydro-distillation for 5 h using Clevenger type apparatus. A clear, light yellow coloured, oily layer was obtained on the top of the aqueous distillate which was separated from the latter and dried with anhydrous sodium sulphate. The obtained essential oil was kept in sealed, air tight glass vials at 4°C until analysis.

Strains of Test Fungi

Strains of *Aspergillus flavus* Link. and *A. niger* van Tieghem from the collection of Mycology and Plant Pathology Division, Botany Department (University of Lucknow) were used. The fungi were maintained on PDA at 28°C.

Studies on the fungi-toxic properties of the essential oil

Antifungal activity assay: Antifungal activity was tested against *A. niger* and *A. flavus* by poisoned food technique.

Poisoned food technique: The fungi-toxicity of the oil was evaluated against the test fungi separately by the poisoned food technique of Grover and Moore (1962). PDA (20 ml) was poured into sterilized Petri dishes and measured amount of oil was added to give desired concentrations of 100, 300, 500, 700, 1000, 1500, 2000, 2500 and 3000 ppm. Tween-80 (0.05%) was mixed in PDA to ensure even distribution of the oil in the medium. Tween-80 mixed with distilled water instead of oil served as the control. After the medium solidified, inoculums of the test fungi were placed in the center of each assay plates, which were incubated at 28 ± 2°C. On the tenth day, the growth of the test fungi were recorded and percent inhibition was computed after comparison with the control.

Fungitoxicity was expressed in terms of percentage of mycelia growth inhibition and calculated as per formula of Pandey *et al.* (1982).

Percentage of mycelial growth inhibition

Where dc = Average diameter of fungal colony in control

dt = Average diameter of fungal colony in treatment

Effect on dry weight : To determine the effect of essential oil on the dry weight of the test fungi different concentrations of oil in PDB medium was prepared in Erlenmeyer flask and inoculated with 5 mm disc of test fungi. In the corresponding control equal amount of distilled water was added. After fifteen days dry weight of mycelium was determined. The mycelia were allowed to dry at 60°C for 6 h and then at 40°C over night. The filter paper containing dry mycelia were weighed and net weight of mycelia was calculated after subtracting the weight of filter paper.

Spore germination assay: Nine concentrations of oil (100, 300, 500, 700, 1000, 1500, 2000, 2500 and 3000 ppm) were tested for spore germination of the test fungi. Fungal spores obtained from 10-day-old cultures of the fungi were taken and placed on glass slides in triplicate. Slides containing the spores were incubated in a moist chamber at 25 ± 2°C for 24 h. Each slide was then fixed in lactophenol-cotton blue and observed under the microscope for spore germination. About 200 spores were counted and the number of spores germinated was scored to calculate the percentage of spore germination (Surender *et al.*, 1987).

Nature of toxicity: The fungitoxicity (fungistatic / fungicidal) of the essential oil was tested by using the technique of Thomson (1989) and Carta and Arras (1987).

Effect of temperature: Experiments were performed to determine if the antifungal factor of the oil was thermo stable or labile. Different glass vials containing 3 ml oil each were stored and subjected to different temperature treatments for three hours in incubators already adjusted to 40, 60, 80, 100°C. Antifungal activity of oil was also tested after autoclaving it at 121°C 15 lbs/in^2 for 15 min. The fungitoxicity of the treated oil from each set was tested against both the test fungi separately at their respective MIC by the usual poisoned food technique.

Microscopy: Light microscopic and scanning electron microscopic studies were done only with *A. niger*.

Light Microscopy: A sample of mycelium was taken from the periphery of the colony that had grown on the PDA media with 0, 1000 and 2000 ppm treatments of oil after four days of incubation. The sample were fixed in lactophenol-cotton blue and examined under the microscope (Nikon, Japan) at 40 X.

Scanning Electron Microscopy: The samples of mycelium were taken as in light microscopy but only for control and 2000 ppm treatment of oil.

Samples were fixed in 2.5% glutaraldehyde for 45 min. at room temperature. They were washed in distilled water thrice for 20 min each. The specimens were then dehydrated in a graded ethanol series (30%, 50%, 70%, 20 min for each alcohol dilution and 45 min in absolute ethanol). Then the samples were coated with gold palladium using an ion sputter-cutter. The preparations were examined using a Cambridge LEO S- 430 microscope at 15 kV.

RESULTS AND DISCUSSION

The results showed that pericarp (waste product) of *C. sinensis* on hydro-distillation yields 1.8% essential oil. Results of inhibitory activity of the essential oil in agar medium are presented in Table 1 and Fig. 1. Radial growth and biomass of both *A. niger* and *A. flavus* were significantly reduced in response to various concentrations of *Citrus* oil ranging from 100 ppm to 3000 ppm. At 700 ppm, growth was reduced to about half that of the control (Table 1), for both the test fungi. It was found that the oil was fungistatic and fungicidal against *A. flavus* at 1000 ppm and 2500 ppm and against *A. niger* at 1000 ppm and 3000 ppm respectively. Determination of mycelial weight was done and it was found that insignificant growth occurred at 3000 ppm and 2500 ppm for *A. flavus* and *A. niger* respectively. The oil exhibited 100 % inhibition of spore germination at 1500 ppm for both test fungi (Table 1).

Table 1. Effect of different concentrations of *Citrus sinensis* pericarp oil on per cent of Radial Growth Inhibition (after 7 days), Dry weight (after 15 days) and inhibition of Spore Germination of fungi *Aspergillus niger* and *A. flavus* at 28±1°C.

Conc of oil (ppm)	% Inhibition		Dry weight (mg)		Spore Germination %	
	A. niger	*A. flavus*	*A. niger.*	*A. flavus*	*A. niger.*	*A. flavus*
Control	-	-	615	694	100.0	100.00
100	6	2	527	600	74.28	73.15
300	14	10	488	527	46.19	49.00
500	40	33	326	368	21.00	19.50
700	47	44	123	229	10.20	9.30
1000	74	77	68	102	4.50	2.00
1500	79	86	51	78	0.00	0.00
2000	87	92	26	44	0.00	0.00
2500	93	100	0.0	18	0.00	0.00
3000	100	100	0.0	0.0	0.00	0.00

The thermostability of the oil was tested and it was found that temperature treatment (40-100°C) and even after autoclaving (121°C 15 lbs/in^2 for 15 min) the oil, its activity was not altered (Table 2).

Table 2. Fungitoxicity of *Citrus sinensis* peel oil treated at different temperature.

Temperature (°C)	Percent inhibition of mycelial growth	
	A. flavus	*A. niger*
40	100	100
60	100	100
80	100	100
100	100	100
121	100	100

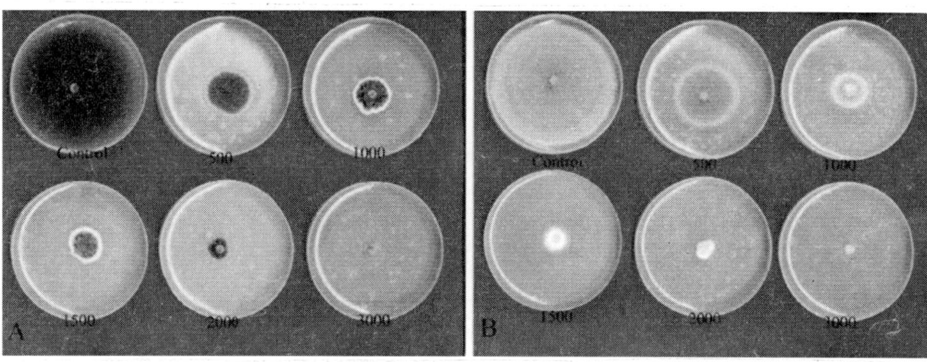

Fig. 1. Effect of *Citrus sinensis* peel oil (Concentration 500, 1000, 1500, 2000 and 3000 ppm) on radial growth, by Poisoned food technique: **A :** *A. niger* and **B :** *A. flavus*.

The observations of *A. niger* examined under light microscope at 40 X, after treatment with different concentrations (1000 ppm and 2000 ppm) of Citrus oil are presented in Fig. 2. Microscopic examination of the control mycelium showed that hyphae had homogenous, clear cytoplasm and profuse conidiation (Fig. 2A). With treatment at 1000 ppm it was found that reduction in cytoplasm started and the hyphal tip became bifurcated, the clear reduction in conidiophore and absence of conidia was observed (Fig. 2B). With 2000 ppm treatment complete absence of cytoplasm and conidiophore formation was observed (Fig. 2C).

The effect of *C. sinensis* essential oil on the morphology of *A. niger* examined by SEM are shown in Fig. 3. The *A. niger* mycelium grown on PDA medium as control presented regular, homogenous hyphae with smooth cell walls. After 4 day of incubation in the treatment (2000 ppm) fungal mycelium showed alterations in the morphology of the hyphae, this appeared severely collapsed and there was marked reduction in conidiation. Squashed hyphae showed that there was complete absence of cytoplasm.

Earlier antifungal study of pericarp of *Citrus sinensis* was done by Patra *et al.* (2003) against nail infection, but the mode of action was not studied. Based on the present study, it could be concluded that essential oil from *Citrus sinensis* possess fungitoxic activities worth exploiting for the biomanagement of stored food commodities.

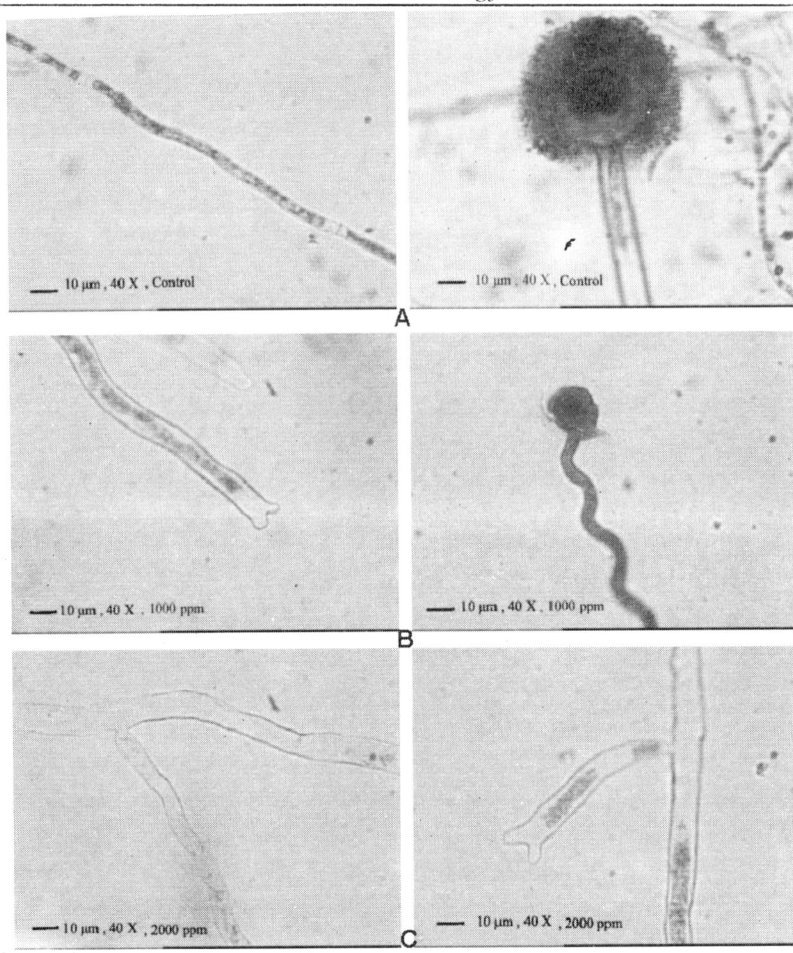

Fig 2. Light microscopy observation of *A. niger* after treatment with 1000 ppm and 2000 ppm of *Citrus sinensis* pericarp oil with compared to control.

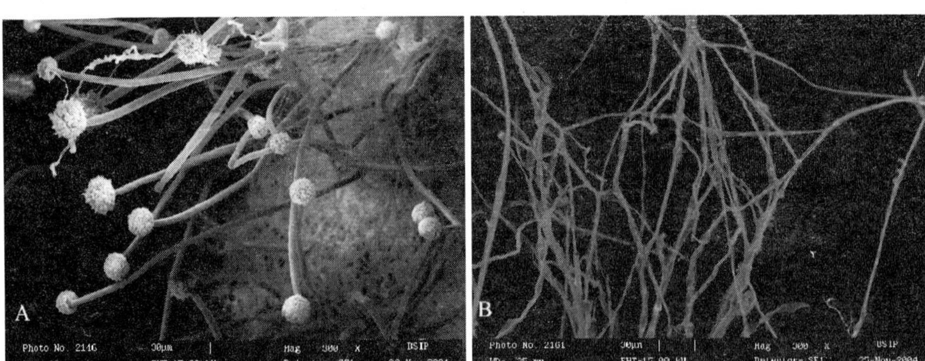

Fig 3. Scanning electron micrograph of *A. niger* grown on PDA with or without *Citrus sinensis* essential oil during 4 day of incubation at 28°C. (A) Control with out essential oil (B) Treatment with 2000 ppm of oil showing absence of conidiophore and conidia and squashed hyphae, cytoplasm escaped from the hyphae as compared with control.

ACKNOWLEDGEMENTS

Tripathi, A. gratefully acknowledges the University Grant Commission (UGC), New Delhi, India for the financial support in the form of Junior Research Fellowship (JRF). Special thanks to staff of Electron Microscopy Unit, Birbal Sahni Institute of Palaeobotany, Lucknow for their help.

REFERENCES

Carta, C., and G. Arras (1987) *La difesa delle plante,* **10** (1): 195

Grover, R.K. and J.D. Moore (1962) *Phytopathology,* **52**:876

Ismam, B.M. (2000) *Crop Prot.* **19**: 603

Pandey, D.K., *et al.* (1982) *Zeit. Pflazenkran. Pflanzensch.* **89**: 344.

Patra, M., Shahi, S.K. and A. Dikshit (2003) *Curr. Sci.* **84** (12): 1512

Sharma, N. and U.V. Verma (2005) In *Plant Diseases Biocontrol Management*. (Ed. N. Sampat) Avishkar Publishers, Jaipur

Surender, P. *et al.* (1987) *Indian J. Exp. Biol.* **25**: 233.

Tewari, S.N. (1990) In *Proc. Symposium Botanical Pesticides in IPM.* (Eds. Chari, M.S. and G. Ramprasad) p. 397. Rajmundry

Thomson, D.P. (1989). *Mycologia,* **81**(1); 151.

Wilson, C.L. *et al.* (1991) *Crop Prot.* **10**: 172

32

ANTIFUNGAL ACTIVITY OF FLORAL EXTRACTS OF *CALOTROPIS PROCERA* AND *MIMUSOPS ELENGI*

Vishal K. Muliya and Arun Arya

Department of Botany, Faculty of Science, The Maharaja Sayajirao University of Baroda, Vadodara 390002

ABSTRACT

Biopesticides are becoming popular recently because of their ecofriendly nature. These do not cause harm to the environment and are comparatively safer than synthetic pesticides as well as biodegradable. Effect of aqueous extracts of flowers of *Calotropis procera* R. Br. and *Mimusops elengi* Linn. was observed on two fungal biodeteriogens i.e. *Myrothecium roridum* Tode ex Fr. (Deutromycotina) and *Chaetomium globosum* Kunze & Schm. (Ascomycotina). Poisoned food technique was used. Three different concentrations i.e. 2, 5 and 10% of *C. procera* and *M. elengi* were used. Flower extract of *C. procera* showed very significant results. *Mimusops* extract at 5 and 10% conc. failed to show any inhibition, on the contrary there was an increase in colony diameter of *Chaetomium.* This increase may be due to presence of certain sugars occurring in floral extract. The use of *Calotropis* flowers as bio-pesticide is recommended as it retarded the growth of *M. roridum* at lower concentration.

INTRODUCTION

In recent years study and uses of chemicals for fungal control from natural sources (particularly plant sources), which are non-toxic and specific in their action, is gaining attention. Search for environmentally sound pesticides received an impetus following the publication of a monumental book *Silent Spring* by Carson (1962). Higher plants are untapped reservoirs of various valuable chemicals awaiting their exploitation for their biological properties for the health of plants and animals (Tripathi, 2005). Higher plants contain a wide spectrum of secondary substances viz. phenols, flavonoids, quinones, tannins, essential oils, alkaloids, saponins and sterols. These plant chemicals may be exploited for their different biological properties (Wain, 1997; Kubo and Nakanishi, 1979; Mahadevan, 1982). A perusal of literature shows that some plant products like protoanemonin (Misra, 1975), allamanoside (Tripathi, 1976) and essential oils of *Cedrus deodara* (Dikshit, 1980) and *Peperomia pellucida* (Singh *et al.*, 1983) have earlier been recorded to be more efficacious than some synthetic fungicides. Oils of *Mentha arvensis, Ocimum canum* and *Zingiber officinale* were found to be more efficacious than the fungicides used *viz.* Benzimidazole, Diphenylamine, Phenylmercuric acetate and Zinc dimethyl dithiocarbamate (Tripathi, 2005). Not much is known about antifungal properties of flowers. However, use of flower extracts of *Jacaranda, Calotropis, Catharanthus* and *Cassia* was reported against *Meloidogyne incognita*, a nematode (Sharma and Trivedi, 1994) and flower extract of china berry (*Melia tosendus* L.) against inhibition of nematode mortality (Sudiqui and Alam, 1985).

Akada or Aak (*Calotropis procera* R. Br.-Asclepiadaceae) is a laticiferous shrub of Indian origin. It reaches to the height of 2-3 m. Pale purple flowers are borne on peduncled umbels (Anonymous, 2002). Its flowers are offered to Lord Shiva and Lord Hanuman by Hindu devotees. The leaves and roots contain a number of cardenolides, the important ones being uscharin (0.45%) and calotoxin (0.15%). The latex contains a proteolytic enzymes Calotropain (similar to papain form papaya) and another bacteriolytic enzyme. Garg and Atal (1963) reported remarkable vermicidal activity of Calotropain. The bark of root mixed with minute doses of arsenic is considered best remedy for leprosy (Sabnis and Daniel, 1990). Arya and Mathew (1993) used leaves of *C. procera* as organic amendment during soil solarization and recorded change in soil mycoflora. Maulshri (*Mimusops elengi* L. – Sapotaceae) is a shade-giving tree with dense foliage, yielding maulsari attar (Chopra *et al.,* 1982). Watt (1889) reported that garland of flowers and seeds of this tree were worn by Indian people. Since flowers are not showy or attractive in this tree their antifungal or medicinal uses may be the possible reasons. Francis jr. (1984) identified 165 plant species used for human adornment, these included flowers of *Calotropis* and *Mimusops*.

MATERIALS AND METHODS

M. roridum and *C. globosum* two fungal biodeteriogens were isolated from a deteriorating objects in Baroda Museum and Picture Gallery, Vadodara (Shah and Arya, 2000). The cultures were identified based on microscopic characters and authenticated from IARI, Delhi. Pure cultures were maintained on Potato Dextrose Agar (PDA) medium (Peeled Potato 200g, Dextrose 20g, Agar 20g, Distilled Water 1000ml) at 25 ± 2°C.

C. globosum is an important cellulose-degrading fungus. It can deteriorate paper, fabric and other cellulose-rich materials. It was causing brown black staining on paper (Aranyanak, 2005) as well as on bleached and unbleached linen (Abdel-Kareem and Szostak-Kotowa, 2005). It was responsible for curd rot of cauliflower and diseases in many plants (Jamaluddin *et al.*, 2004).

M. roridum was causing diseases in many plants like *Morus alba, Prunus avium, Dalbergia sissoo, Aegle marmelos* etc. (Jamaluddin *et al.*, 2004). It is known to produce toxins like ochratoxin A, trichothecens, 3-hydroxyroridin E, roridin A, roridin L, roridin M, isororidin A, epiroridin, 12, 13-deoxyroridin E, 14'-hydroxymytoxin B, 16-hydroxyroridin etc. These toxins are having cytotoxic and mutagenic activity also (Khisal *et al.*, 2002).

Flowers of two plants were collected from botanical garden of Botany department, The Maharaja Sayajirao University of Baroda, Vadodara, 100 g of flowers were washed and crushed in mixer for 20 min and then filtered. Filtrate was collected and centrifuged at 2000 rpm for 20 min. Finally supernatant was taken and the water was added to make it 100 ml so that there was a concentration of 1g/ml.

Bioassay

Poisoned food technique (Nene and Thapliyal, 1979) was used to study antifungal activity of flower extracts. Flower extract was added 2, 5, and 10% v/v to the culture medium (PDA) and the medium was sterilized at 15 lbs pressure p.s.i., for 20 min. Finally culture medium was transferred to Petri plates and then these were inoculated and incubated at 25°C temp for 7 days. The plates were inoculated at the centre with a 6 mm disc of 10 days old culture. The colony diameter was measured to study the fungal growth.

RESULTS AND DISCUSSION

'Sustainable development' is a key concept in discoveries of the global environments. Plant based fungicides offer a new possibility for safer alternative of future disease control programmes. A plant product which has no mammalian toxicity and whose fungitoxic property is broad spectrum and persists for a longer time can be exploited for fungal control *in vivo.*

Numerous flowers contain essential oils. The presence of these oils help in attraction of insects for pollination. Antimicrobial nature of essential oils is reported by numerous workers (Borlchin *et al.*, 1996, Marti *et al.*, 1985, Thakur *et al.*, 1989). Flower extract of *Catharanthus roseus* was found effective against *Alternaria solani* (Narsimhan *et al.*, 1991) Flowers of *Chrysanthemum cinerariaefolium* yield insecticide Pyrethrum.

It is evident from Table 1 that there was inhibition of fungal growth of *M. roridum* with all the concentrations. There was no increase in inhibition by increasing the concentration of extract which indicates that there may be certain compounds having independent effect on fungal growth. At higher (10%) concentration of *C. procera* extract only 6.84% inhibition was observed as compared to 12.32 in 5% concentration however, this may be due to presence of sugars in flower of *C. procera* flower extract of this plant shows negative result in case of *C. globosum*. No literature reports are there on effect of flower extract of *C. procera*. Leaf extract and leaf decoction of *Calotropis* produced good antifungal activity (Khan *et al.*, 1998). Antifungal property of leaves of Aak was noticed against *Alternaria brassicae* on mustard (Patni *et al.,* 2005).

Table 1. showing the effect of cold aqueous floral extract on two different biodeteriogens

	Myrothecium roridum			*Chaetomium globosum*		
Flower Extract	2%	5%	10%	2%	5%	10%
C. procera	12.32**	12.32**	6.84*	-1.63*	-1.63*	-12.29*
M. elengi	2.7**	2.7**	6.84*	-5.73	-6.15	-13.11*

Figure shows the percentage of inhibition in colony diameter in comparison to control
* Shows that there is poor sporulation. ** Shows there is no sporulation

From the Table 1 it can be observed that there was no inhibitory effect of flower extract of *M. elengi* on *C. globosum*. Jaitly (1989) studied effect of saponins from leaf of *Mimusops littoralis* and found them effective against *Ustilago cynodontis, U. scitaminea, U. tritici* and *Cintractia limitata*. The inhibition in germination of smut spores was observed. Methanolic extract from leaves of *M. elengi* showed strong antimicrobial activity against *Sclerospora graminicola* causing downy mildew of pearl millet (Deepak *et al.*, 2005).

REFERENCES

Abdel-Kareem, O. and J. Szostak-Kotowa (2005) In *Biodeterioration of Cultural Property-4* (Eds. Agarwal, O.P. and S. Dhawan), International Council of Biodeterioration of Cultural Property, Lucknow, India, p. 19

Anjaria, J., M. Parabia, G. Bhatt and R. Khamar (2002), *Nature Heals A glossary of selected indigenous medicinal plants of India*, Srishti Innovations, Ahmedabad. p 49

Anonymous, (2002). *Wealth of India*, **3,** CSIR Publication, New Delhi.

Aranyanak, C. (2005) In *Biodeterioration of Cultural Property-4* (Eds. Agrawal, O.P. and S. Dhawan), International Council of Biodeterioration of Cultural Property, Lucknow, India, p.40

Arya, A. and D.S. Mathew (1993) *Indian Phytopathology*, **46,** 39

Balchin, M., S.G. Dean and S. Hort (1996) *Journal of essential oil research,* **8,** 281

Carson, R. (1962), *Silent Spring*, Fawcett Publications Inc. Greenwich Corporation.

Chopra, I.C., K.L. Handa and L.D. Kapur, (1982) *Chopra's Indigenous Drugs of India*, Academic Publishers, Calcutta, p. 814.

Deepak, S.A., G. Oros, S.G. Sathynarayana, N.P. Shetty, H.S. Shetty and S. Sheena (2005) *Arch. Phytopath. Plant Prot.*, **38(1)**, 31

Dikshit, A. (1980), *Ph. D. Thesis*, Gorakhpur University, Gorakhpur

Farrukh, A. and A. Iqbal (2003) *World J. Microbiol. & Biotech.* **19**, 653

Francis, P. Jr. (1984) *Econ. Bot.*, **38(2)**, 194

Garg, L.C. and C. K. Atal (1963) *Indian J. Pharm.*, **25**, 422

Jaitly, N., K. Wadhwani, R. Banerji, G. Mishra and S.K. Nigam (1989) *Curr. Sci.* **58(2)**, 75

Jamaluddin, M.G. Goswami and B.M. Ojha (2004), *Fungi of India 1989-2001*, Scientific Publishers (India), Jodhpur, India, 326

Khan, T.Z., M.A. Gill, M.A. Nasir and S.A.A. Bukhari (1998) *Pak. J. Phytopath.*, **10**, 62.

Khisal, A.A., J. Rabenstein, J. Woodard, D.D. Baker, J.D. Bergthold, J. Lynch, K.L. Lieu and I.A. Braude (2002) *J. Nat. Prod.*, **65**, 742

Kubo, I. and K. Nakamishi (1979) In: *Advances in Pesticide Science* (Part 2) (Eds. Geussbiiler, H.) Pergamon Press, Oxford, 284.

Mahadevan, A. (1982) *Biochemical aspects of plant disease resistance.* (Part 1), Today and Tomorrow Publishers, New Delhi, 400.

Maiti, D., C. R. Cole and C. Sen (1935) *J. Plant Dis. Prot.* **92**, 64.

Misra, S.B. (1975) *Ph. D. Thesis*, Gorakhpur University, Gorakhpur.

Muhammad, Z., S. Y. A. Bokharie, A. T. Muhammad, S. Kamran and A. Abida (2003) *Asian J. Pl. Sci.*, **2 (6)**, 475.

Narsimhan, V. and U. K. P. Pillai (1991), *Indian Phytopath.* supplymentry issue, **45**, XCIX (Abstr.).

Nene, Y. L. & P. N. Thapliyal (1979), *Fungicide in Plant disease control*, Oxford and IBH Pub. Co., New Delhi, p. 507.

Rajyalakshmi, P., K. Venkatalaxmi, K. Venkatalakshamna, Y. Jyothsna, K. Balachandramani and V. Suneetha (2001), *Plant Foods Human Nutr.*, **56**, 225–238.

Ruby, J., P.T. Nathan, J. Balasingh and T.H. Kunz (2000), *J. Chem. Eco.* **26 (12)**, 2825-2841.

Sabnis, S.D. and M. Daniel (1990), *A Phytochemical Approach to Economic Botany*, Kalayani Publishers, New Delhi, 487 pp.

Shah, N.R. and A. Arya (2000), *Studies on Some Fungal Biodeteriogens*, Bharatiya Kala Prakashan, Delhi, 202 pp.

Sharma, R. and P.C. Trivedi (1994) *Proc. Nat. Acad. Sci.* **64 B II**, 231.

Sidiqui, M.A. and M.M. Alam (1985) *Neem Newsletter*, **2**, 43.

Singh, A.K., A. Dikshit and S.N. Dixit (1983) *Beiter. Biol. Pflanz.*, **58**, 357.

Thakur, R.N., P. Singh and M.K. Khosla (1989) *Indian Perfumer*, **33**, 257.

Tripathi, S.C. (1976) *Ph. D. Thesis*, Gorakhpur University, Gorakhpur.

Wain R.L. (1977) *Chemical Aspect of Plant Disease Resistance in Natural Products and the Protection of Plants* (Eds. Marubu Bettika, G.B.) Elsevier Scientific Pub. Co. Amsterdam, p. 483.

D

BIOFERTILISERS

33

ROLE OF MARINE ALGAE AS A BIOFERTILISER

Neelam Pereira, Verlecar, X.N. and V.K. Dhargalkar

National Institute of Oceanography, Goa

ABSTRACT

Marine algae especially seaweeds are rich in Nitrogen and potash and increases humus content of the soil. They also contain natural plant growth regulators (PGR), which control the growth and structural development of plants. Extracts from seaweeds like *Caulerpa racemosa, Enteromorpha instestinalis, Padina tetrastromatica* and *Sargassum tenerrimum* have been tried and tested on growth of crop plants like *Sesamum indicum, Phaseolus vulgaris* and *Vigna catjung*. Seaweeds especially *Sargassum* sp. and *Gracilaria* sp. are used as manure for coconut plantations and paddy crops. Land contaminated with toxic pesticides like DDT could be cleaned up using small amounts of seaweed.

INTRODUCTION

Intensive application of chemical fertilizers in agriculture has caused damage to the ecological state of the agricultural systems (Abetz and Young, 1983). Biological fertilizers do not contaminate the soil and atmosphere thus, helps to produce healthy foods (Blunden, 1973). Organic farming is one of several approaches to sustainable agriculture. Indeed, many of the techniques used in organic farming - such as inter-cropping, mulching and integration of crops and livestock - are practiced under various agricultural systems. Seaweeds and their derivatives are used in agriculture as potential plant growth regulators. The main types of seaweeds are categorized on the basis of their coloration such as, Chlorophyta, Phaeophyta and Rhodophyta. Seaweeds are cryptogrammic plants and are best described as salt or brackish living multicellular forms of algae.

These are found along the inter-tidal zone of the coast. The potential harvest of seaweeds along the Indian coast is about 100,000 metric tones wet weight. About 650 species are known to be present along this coast. Maximum seaweed biomass is reported from Gujarat and Tamil Nadu. Seaweeds are being used as manure in the coastal areas since long. The carbohydrates and other organic matter present in the seaweeds alter the nature of the soil and thus increase the moisture retaining capacity.

Biological regulators are substances that in small concentrations cause a physiological effect in plants inhibiting or stimulating specific function. Some of them are phytohormones (absicines, auxins, cytokines, gibberellins and ethylene) and microbial enhancers. It is observed that abundant use of chemical fertilizers affect soil and plants in due course. Seaweed fertilizers are preferred due to presence of trace elements and metabolites similar to plant growth regulators. These help in controlling the deficiency diseases, and enhance the growth yield (Bukhari and Untawale, 1975; Dhargalkar and Untawale, 1980). They have a suitable content of nitrogen and potassium, but are much lower in phosphorus than traditional animal manures and the typical N: P: K ratios in chemical fertilizers

Algae are relatively high in Nitrogen and potash, but low in phosphorus content. Addition of phosphates could improve the fertility of this manure. The organic matter of seaweeds increases humus content of the soil thereby, ameliorating the soil texture and preservation of its moisture. Seaweed can be used in various forms as a fertilizer. The miscellaneous applications of seaweed as a fertilizer were enumerated and demonstrated at a training workshop held at CMFRI (Central marine fisheries research institute), Mandapam. A survey conducted by CMFRI around the coastal villages of Mandapam in Tamil Nadu unraveled utilization of fresh as well as used-up seaweed. Red algae like *Gracillaria, Gelidiella* and *Gelidium* are extensively used in industries for the production of agar, a phycocolloid. The by-product of this process comprises of seaweed molasses. This mixture is made into cakes and used as manure in the form of cakes.

The drift Seaweeds found along the beach could be used as compost for the soil. The agglutinization of seaweeds help in binding the soil and enhancing the microbial activity needed for fertilizing the soil. A novel way of using the seaweed extract as a fertilizer has been gaining popularity in recent times. This is generally used as a foliar spray. The hormones present in seaweeds accelerate the growth of crops, fruits and plants of economic importance. The seeds soaked in this solution have improved germination rates. The fertilizer is generally prepared by the method of Bhosle *et.al* (1975). Extracts from seaweeds like *Caulerpa racemosa, Enteromorpha instestinalis, Padina tetrastromatica* and *Sargassum tenerrimum* have been tried and tested on growth of crop plants like *Sesamum indicum, Phaseolus vulgaris* and *Vigna catjung.* These have shown beneficial effect on crops.

(Gandhiyappan and Perumal, 2001; Bhosle *et al.*, 1975; Anantharaj and Venkatesalu, 2001). Seaweed extract from *Caulerpa scalpelliformis* reportedly indicated a net increase in xanthophylls, carotene, carbohydrate, protein and lipid content of *Cyamposis tetragonoloba* (Thangam *et al.*, 2003).

Seaweed enhances photosynthesis via increasing chlorophyll levels of plants. Chlorophyll is responsible for green color in plants. By increasing the level of chlorophyll the plant is able to efficiently harness the suns energy. Along with this, seaweed contains a complex range of biological stimulants, nutrients, and carbohydrates. More than 60 different types of nutrients in seaweed have been found.

Seaweed extracts contain natural plant growth regulators (PGR), which control the growth and structural development of plants. The major plant growth regulators are auxins, cytokinins, indoles and hormones. Indole compounds help in development of plant roots and buds. Cytokinins are hormones that promote growth via rapidly speeding up the process of cell division making seaweed extract of value in treating tissue cultures. It helps in stimulation of photosynthesis. The cytokinins in seaweed extract are a major factor when applied to apple and peach trees in promoting the growth of fruiting spurs and reduce premature dropping of fruit. Auxins occur in the roots and stems during cell division. They move to areas of cell elongation where they allow the walls of cells to stretch. Auxins actually give fruits and vegetables a naturally longer shelf life. It helps in delaying of senescence.

Seaweeds especially *Sargassum* sp. and *Gracilaria* sp. are used as manure for coconut plantations and paddy crops in India i.e. along the coasts of Tamil Nadu, Kerala, Maharashtra, Karnataka, etc. In Malvan (Maharashtra) *Hypnea musciformis* and sometimes *Sargassum* are used as manure for coconut plantation (Deshmukh *et al.*, 2001). In South Africa, extensive research on the utilisation of seaweed extract on the crop plants is being done. Experiments on the effect of *Ecklonia* extract on tomato plants revealed that it not only helps in increasing the plant height and yield, but also reduces the manifestation on the tomato plants (Mathiesan, 1969).

Land contaminated with toxic pesticide DDT could be cleaned up using small amounts of seaweed, according to report published in the Journal of Chemical technology and biotechnology. Researchers have found that sprinkling just the right amount of powdered seaweed on contaminated soil can accelerate the breakdown of the deadly insecticide DDT. The organisms have enzymes that convert DDT into other substances that deteriorate quickly. They discovered the right dose of seaweed caused 80 per cent of the chemical to vanish within six weeks (Kantachote *et al.*, 2004).

A few developed countries (e.g. Germany, France) compel or subsidize farmers to use organic techniques as a solution to water pollution problems. In Austria and Switzerland, organic agriculture has come to represent as much as 10% of the food system, while USA, France, Japan and Singapore are experiencing growth rates that exceed 20% annually.

Organic farming is one of the several approaches to sustainable agriculture. Indeed, many of the techniques used in organic farming - such as inter-cropping, mulching, and integration of crops and livestock - are practiced under various agricultural systems. What makes organic agriculture unique is that, under various laws and certification programmes, almost all synthetic inputs are prohibited and "soil building" crop rotations are mandatory. Properly managed, organic farming reduces or eliminates water pollution and helps conserve water and soil on the farm (Fresco, 2003).

REFERENCES

Abetz, P. and C.L Young (1983). *Botanica Marina.* **26 :** 487.

Anantharaj, M. and V. Venkatesalu (2001). *Seaweed Research and Utilization.* **23 (1-2):** 33-39.

Bhosle, N.B., Untawale, A.G. and V.K. Dhargalkar (1975). *Ind. J. Mar. Sci.* **4:** 208.

Blunden, G. (1973). *Proc. Seventh International Seaweed Symposium.* School of Pharmacy, England. pp. 23.

Bukhari S. and A.G. Untawale (1975). *Seaweed Research and Utilization.* **3**: 71.

Deshmukh, G.V., Dhargalkar, V.K. and A.G. Untawale (2001). In: *The Indian Ocean: a perspective* Vol-2 (Desa, E. and Sen, R. Eds.).

Dhargalkar V.K. and A.G. Untawale (1980). *Proc. Natl. Workshop on Algal System.* (Sheshadri, P.C. Thomas and N. Jeeji Bai, Ed.) p. 63.

Fresco, L.O. (2003). *Fertilizer and the future. Agriculture 21.* FAO Publications.

Gandhiyappan, K. and P. Perumal (2001). *Seaweed Res. Utilin.* **23 (1&2)**, 23.

Kantachote, D. *et al.* (2004). *J. Chem. Technol. Biotechnol.* **79**, 632.

Mathiesan A.C. (1969) *Oceanology International.* 37.

Thangam, R.T., S.M.V. Rani and P. Mariam 2003. *Seaweed Research and Utilization.* **25(1-2)**, 99.

34

EFFICACY OF BIOFERTILIZERS WITH NITROGEN LEVELS ON PRODUCTIVITY OF WHEAT (*TRITICUM AESTIVUM* L.)

H.P. Ponkia, K.H. Ribadia and D.R. Gondalia

Wheat Research Station, Junagadh Agricultural University, Junagadh (Gujarat)

ABSTRACT

A field experiment was conducted during the winter (*rabi*) seasons of 1999-2000 to 2000-2001 at Junagadh, Gujarat to study the effect of nitrogen levels (60, 90, 120 and 150 kg ha^{-1}) and bio fertilizers (control, *Azotobacter chroococcum* and *Azospirillum brasilense*) on yield and yield attributes of wheat (*Triticum aestivum* L.) under normal sown conditions. Application of 150 kg nitrogen ha^{-1} recorded significantly the highest grain and straw yields along with improved yield attributes viz., ear heads m^{-2}, grains per ear head and test weight, but remained at par with the recommended level of 120 kg nitrogen ha^{-1} in all these regards. Seed inoculation by *Azotobacter chroococcum* was found marginally superior followed by *Azospirillum brasilense* in respect of yield and yield components, however did not attend the level of significance over uninoculated control.

INTRODUCTION

Wheat is a major winter crop grown in India. Nitrogen management is one of the major factors to attain higher productivity of wheat. The current emphasis is on the integrated use of different sources of plant nutrients such as organic manures and biofertilizers in combination with chemical fertilizers. Favorable effects of *Azotobacter* and *Azospirillum* bacteria in supplementing plant nutrients to several crops as well as in saving of expensive inorganic fertilizers have been observed. Therefore, the present investigation was conducted to find out the effectiveness of biofertilizers and chemical fertilizers on yield and yield attributing characters of wheat.

MATERIALS AND METHODS

A field experiment was conducted during the winter seasons of 1999-2000 and 2000-2001 at the Wheat Research Station, Junagadh Agricultural University, and Junagadh to evaluate the efficiency of biofertilizers in increasing the efficacy of nitrogen applied to wheat. The soil was medium black clayey in texture and slightly alkaline (pH 8.0) in reaction having medium status of available N and P and high status of available K. Two biofertilizers viz., *Azotobacter chroococcum* and *Azospirillum brasilense* along with uninoculated control were allotted in main plots and four levels of nitrogen (60, 90, 120 and 150 kg N ha^{-1}) were allocated in sub plots arranged in split plot design with three replications. The basal dose of 60 kg P ha^{-1} and half dose of N (60 kg ha^{-1}) in form of diammonium phosphate and urea was applied at the time of sowing and remaining half dose of N as per treatments was top dressed through urea at 25 days after sowing. The wheat variety GW 273 was used as test crop. Seeds were inoculated with biofertilizers and sown on 14th and 20th November in respective seasons at 23 cm row spacing using seed rate of 120 kg ha^{-1}. The yield attributing characters like, tillers m^{-2}, ear head m^{-2}, grains per ear head, test weight (g) and grain and straw yields (q ha^{-1}) were recorded. The crop was harvested on 23rd February and 7th March in respective years.

RESULTS AND DISCUSSION

Nitrogen levels

A significant variation was found during both the years among N levers towards yield attributes of wheat (Table 1), except tillers m^{-2} in both the years and grains per ear head in 2000-2001. Application of 150 kg/ha^{-1} significantly enhanced the yield attributes viz., ear heads m^{-2}, grains per ear head and test weight, but remained statistically at per with recommended level of N (120 kg/ha^{-1}). The improved yield attributes with these levels ultimately resulted in significantly higher grain and straw yields over 60 and 90 kg/ha^{-1} (Table 2) during individual years and in pooled results too. The results corroborate with earlier findings of Reddy *et al.* (1981), Sushila and Giri (2000) and Ray *et al.* (1989).

Biofertilizers

The seed inoculation with *Azotobacter chroococcum* or *Azospirillum brasilense* resulted in slightly improved yield attributes over uninoculated control, however it did not reach at the level of significance during both the years (Table 1). Akin to these, grain and straw yields also showed similar trend in independent and combined results (Table 2). Mehrotra and Lehri

(1971) and Ram *et al.* (1985) also reported non-response of wheat to biofertilizers. It could be inferred from the results that under normal sown conditions of South Saurashtra Agro-Climatic Zone, application of 120 kg N ha^{-1} was found to be optimum in respect of yield and its yield attributes of wheat variety GW 273.

Table 1. Effect of nitrogen and biofertilizers on yield and its attributes of wheat.

Treatments	Tillers/ m^2		Earheads/ m^2		Grains/ Ear head		1000-grain wt (g.)	
	1999-2000	2000-2001	1999-2000	2000-2001	1999-2000	2000-2001	1999-2000	2000-2001
				Nitrogen kg ha^{-1}				
60	137.00	127.00	253.89	331.44	21.25	26.87	42.33	41.65
90	149.33	124.78	250.22	343.44	28.03	28.17	42.75	42.38
120	150.11	126.89	254.89	359.78	30.23	28.62	43.25	44.49
150	152.33	124.89	260.56	362.00	30.70	30.09	43.45	44.70
C.D. at 5%	NS	NS	4.89	7.93	1.72	NS	0.50	0.92
				Biofertilizers				
Control	148.92	126.08	251.75	347.83	27.07	28.17	41.88	43.18
Azotobacter	141.08	125.25	257.50	351.67	28.26	28.65	44.17	43.39
Azospirillum	151.58	126.33	255.42	348.00	27.32	28.49	42.79	43.34
C.D. at 5%	NS	NS	NS	NS	NS	NS	0.89	NS

Table 2. Grain and straw yield of wheat as affected by nitrogen and biofertilizers.

Treatments	Straw yield (q/ha^{-1})			Grain yield (q/ha^{-1})		
	1999-2000	2000-2001	Pooled	1999-2000	2000-2001	Pooled
			Nitrogen kg/ha^{-1}			
60	59.06	78.68	68.87	23.37	36.01	29.69
90	63.79	87.79	75.79	29.66	40.75	35.21
120	64.97	92.48	78.73	32.93	43.69	38.31
150	67.59	97.66	82.63	33.81	44.73	39.27
CD at 5%	2.85	7.32	5.59	2.29	3.75	3.88
			Biofertilizers			
Control	62.93	85.25	74.09	29.39	39.29	34.34
Azotobactor	64.37	92.65	78.51	30.80	42.62	35.82
Azospirillum	64.26	89.55	76.91	29.64	42.00	36.71
CD at 5%	NS	NS	NS	NS	NS	NS

REFERENCES

Mehrotra, C. L. and L.K Lehri (1971) *J. Indian Soc. (Soil Sci.)* **19** (3): 243-248.

Ram, G., B.S. Chandrakar and R.K. Katre (1985) *J. Indian Soc. (Soil Sci.)* **33** (1): 424-426.

Ray, R., R. K. Ghosh and S.S. Mondal (1989) *Environment and Ecology* **7** (3): 630-633.

Reddy, M.R., G.B. Reddy and A.V. Chari (1981) *Indian J. Agron.* **26** (3): 317- 321.

Sushila, R. and G. Giri (2000) *Indian J. Agron.* **45** (3): 590-595.

35

YIELD POTENTIAL OF BGA BIOFERTILIZERS FOR PADDY CROP

S.N. Nandan, K.B. Thakre and S.P. Khairnar*

Dept. of Botany, S.S.V.P.S.L.K. Dr P.R. Ghogrey, Science College, Dhule - 424 005
**Dept. Botany, Lokneta Vyankatrao Hiray College, Panchavati, Nashik - 422 003*

ABSTRACT

Blue Green Algae (BGA) are found to be most economic biological nitrogen supplying bio-fertilizers for the wetland paddy in recent years. The water logging conditions of paddy fields and paddy crop canopy favors luxuriant growth of BGA and there by increases the fertility of soil by adding nitrogen and other growth substances. In present study impact of BGA bio-fertilizers on grain yield of paddy was studied by arranging field trials in fields of 24 farmers. For the purpose the field trial plots were divided into experimental plots and control plots. For experimental plots BGA and chemical fertilizers (Urea) was used while for control plots only chemical fertilizers was used. Both the plots were exposed to uniform environmental conditions. The results of present study showed that in first year, there was no significant difference in yield of BGA treated and untreated plots but in second year there was statistically significant increase in paddy yield of BGA treated plots.

INTRODUCTION

We have experienced in past the 'Green Revolution' and 'White Revolution'. The success of green revolution depended upon the availability of Fertilizers, high yielding variety of seeds, improved agronomic practices and timely availability of water. The demand for nitrogenous fertilizer has gone up while supply has fallen short. So in such situation it is need of hour to find alternative source for nitrogen fertilizer supply. Use of biological source of nitrogen (Bio-fertilizer) is the only solution.

Among the various bio-fertilizers, blue-green algae (BGA) bio-fertilizers are most economic bio-fertilizers for traditional crops like wet paddy. Rice is staple food of nearly world's half population. The crop also occupies highest area under cultivation compared with other food crops. The per hectare yield fluctuation directly affects living standards and family budget of farming community. The irregular rain pattern, high cost of chemical fertilizers, pest and diseases etc. acts as main constrains in paddy farming. Use of chemical fertilizers for wet paddy is found to be uneconomic when compared with cost benefit ratio. So development of alternative techniques, which are low cost, situational, economic and eco-friendly, is need of hour. Use of bio-fertilizers like BGA is found to be most suitable agro-biotechnology for wet paddy. The water logging conditions throughout the crop season and paddy crop canopy favors the luxuriant growth and establishment of blue green algae in paddy fields.

The results of earlier workers with BGA bio-fertilizers for paddy showed positive results. Arora *et al.* (1986) and Dubey (1991) showed that algae inoculation increases paddy yield by 10-25 per cent. It was also pointed out that application of BGA bio-fertilizers increase rice yield even by reducing the recommended dose of chemical nitrogen fertilizers (Venkataraman, 1982). Thus BGA has highest potential, as fertilizers for traditional crop like paddy. It needs to diffuse the technology information towards end users for wider adoption. Present study was undertaken with main objective of communication of information of BGA bio-fertilizer for sample farmers by organizing field demonstrations in open fields of farmers. It was also objective to find the effect of BGA inoculation on grain yield of treated plots.

MATERIALS AND METHODS

For present study 24 farmers from 12 villages of two tehsils – Peth and Trimbak of Nashik District (M.S.) were selected as sample population. For each sample farmer 4 kg of BGA bio-fertilizers were provided to use of half acre of transplanted paddy. The half-acre BGA treated plot was named as 'experimental plot'. For experimental plots chemical fertilizers in the form of urea was also allowed to use. The dose of urea for experimental plot crop and control plot crop was kept uniform. Field demonstration on BGA bio-fertilizer application schedule were organized at representative fields of farmers. Keeping all the environment, soil and climatic condition uniform for both, the effect of BGA inoculation on per hectare yield of paddy was measured and statistically compared with yield of untreated plots.

RESULTS AND DISCUSSION

The data given in the Table 1 reveals that per hectare yield of paddy for two successive years of treated (experimental) and untreated (control) plot of 24 samples farmers. The table reveals that in year 2000, the average yield of untreated plots was 11.12 quintals per hectare with 0.96 quintals increase over untreated plots. In year 2001 the average yield of untreated plots was 12.92 quintals per hectare and for BGA treated plots it as 16.08 quintals per hectare with 3.16 quintal increase.

Table 1. Yield data of paddy by use of BGA and chemical fertilizers together

Sr. No.	Name of Farmer	Yield in quintals per hectare			
		Year 2000		Year 2001	
		Untreated	Treated	Untreated	Treated
1	Chimna Gangaram Gavali	10	10.05	11.33	13.63
2	Parashram Bhika Kamdi	10.05	11.25	12.03	14.34
3	Chandar Mahadu Natue	10.05	12.05	12.21	14.68
4	Smt. Mangala Mahadu Nathe	10.68	11.65	11.78	13.42
5	Muralidhar Ananda Nathe	11.05	13.25	12.58	15.43
6	Deoram Govinda Boke	10	10.06	12.13	14.82
7	Gangaram Narayan Tople	12.07	15	18.17	21.22
8	Jagdish Krishna Sahare	14	14.26	12.72	14.87
9	Chandrakant Tulshiram Mahale	13	15	13.42	17.36
10	Jayram Kashiram Mahale	15	15.75	13.02	17.22
11	Madhykar Kalu Mahale	14	14.04	12.37	16.28
12	Kalu Dharma Mahale	20	20.78	20.42	24.31
13	Madhusudan Muralidhar Bhoraskar	10	10.38	11.48	15
14	Prakash Soma Udar	8.05	9.75	12.24	16.14
15	Bhagirat Raoji Bhoe	8.65	9.03	12.62	14.76
16	Vaman Skharam Bhoe	8.65	10.05	12.03	15.44
17	Parashram Shivram Wakhare	9.08	10.15	15.07	19.05
18	Kantilal Muralidhar Lokhande	13.03	14.45	15.35	19.02
19	Kashinath Pandurang Bhoe	8.08	9.03	11.14	14
20	Asok Laxman Bhoe	9.01	10	10.08	14
21	Namdeo Ragho Malekar	12	12.05	13.48	18.03
22	Gopal Tukaram Melekar	7.07	10.04	10.32	12.72
23	Muralidhar Gopal Shende	8.08	9.01	11.85	16.62
24	Savliram Sonu Shende	8.05	9.25	10.18	12.08

The data given in the Table 2 show t-test results. It can be observed from the data that in the first year of experiments there was no significant difference in the yield of treated and untreated plots. The t-test value was low (1.2) than expected value (2.02) at 5% level of significance. While in second year of experiments the t-test calculated values was high (4.13) than expected (2.02) at 5% level showing statistically significant increase in the yield due to inoculation of BGA bio-fertilizers.

Table 2. Statistical t-test results by use of BGA and chemical fertilizers together

Year 2000				Year 2001			
Untreated		Treated		Untreated		Treated	
M_1	11.12	M_2	12.08	M_1	12.92	M_2	16.08
ΣX_1^2	176.2	ΣX_2^2	183.15	ΣX_1^2	129.87	ΣX_2^2	179.21
$б_1$	2.70	$б_2$	2.76	$б_1$	2.32	$б_2$	2.73
Calculated t = 1.2				Calculated t = 4.13			
Expected t = 2.02 (at 5% level)				Expected t = 2.02 (at 5% level)			
Thus $t_{cal} < t_{expect}$				Thus $t_{cal} > t_{expect}$			

Singh *et al.* (1992) conducted about 500 field demonstrations in Kharif season in 11 districts of U.P. and study the effect of various doses of Urea + BGA bio-fertilizer on paddy yield. They observed that there was 0.27 to 6.25% increase in yield due to application of BGA. Subramanyan *et al.* (1964), Venkataraman (1982), and Pachpande (1990) reported that BGA inoculation significantly increases paddy yield at field conditions and in pot culture. In present study, in first year about 0.52 per cent and in second year 3.33 percent increase in per hectare yield was observed. Thus results of present study also confirm the findings of earlier researchers.

CONCLUSIONS

The above results to conclude that:

1. BGA bio-fertilizer inoculation is responsible for significant increase in per hectare yield of paddy.
2. The non significant difference in yield and overall low yield in first year of experiments might be due to overall near drought like situation in the study area during crop season.
3. The results also led to conclude that continuous two years of BGA inoculation is necessary for significant increase of the yield.
4. BGA bio-fertilizers have greater yield potential for paddy.

ACKNOWLEDGEMENT

The author (SPK) expresses his sincere thanks to UGC (WRO) for providing financial assistance and their Principal for providing facilities.

REFERENCES

Arora, J., Garcha, H.S., Pandher, M.S. & R.P. Gupta (1986). *Res. Dev. Reptr.,* **3(2)**:72.

Dubey, S.K. (1991). *Farmer and Parliament.* **XXXVI (7)** : 15.

Pachpande R.R. (1990). *Proc. Nat. Symp. on Cyanobacterial Nitrogen Fixation,* IARI, New Delhi. 297.

Singh, A.N., Sinha, S.K., Pandey, V.P., Dubey, A.K. and D.C. Verma (1992). *New Agriculturist* **3(2):** 189.

Subramanyan, R., Relwani, L.L. and G.B. Manna (1964) *Curr. Sci.* **33**:485.

Venkataraman, G.S. (1982). *Int. Cong. of Soil Sci., New Delhi,* **12**:69-82.

36

OCCURRENCE OF AM FUNGI- A PHOSPHATE FERTILIZER, IN SOILS OF VADODARA

Tanushree Chatterjee and Arun Arya

Department of Botany, Faculty of Science, The Maharaja Sayajirao University of Baroda, Vadodara - 390 009

ABSTRACT

The biofertiliser having specific microbial inoculants like bacteria, cyano-bacteria and fungi act as perpetually renewably inputs helping in better crop nutrient management and maintenance of soil health. Eco-friendly use of biofertilizer is gaining importance over synthetic or chemical fertilizer. AM fungi not only provide phosphorus to the plants but helps in maintaining the porosity of soil. The increase in photosynthesis and change in stomatal behaviour and concentration of chlorophyll due to inoculation of the AM fungi in different plant species is well documented. Extensive survey was carried out in and adjoining areas of Vadodara. Soil sample of 43 places were collected and spore population/100 g soil was calculated. Soil of Sayajipark (Ajwa road) and Mahi river bank harbored 511 and 342 spores/ 100g soil respectively. Two spp. of *Acaulospora* and five species of *Glomus* have been identified from different samples.

INTRODUCTION

Long term sustainability in agriculture and social forestry is possible only through the use of low cost farm grown inputs which work in harmony with the nature. In contrast to costly chemicals and organic fertilizers, economically effective and ecofriendly biofertilizers certainly hold the key of success. The biofertilizers having specific microbial inoculants like bacteria, cyanobacteria, or different micro fungi act as perpetually renewable inputs for maintaining better soil health.

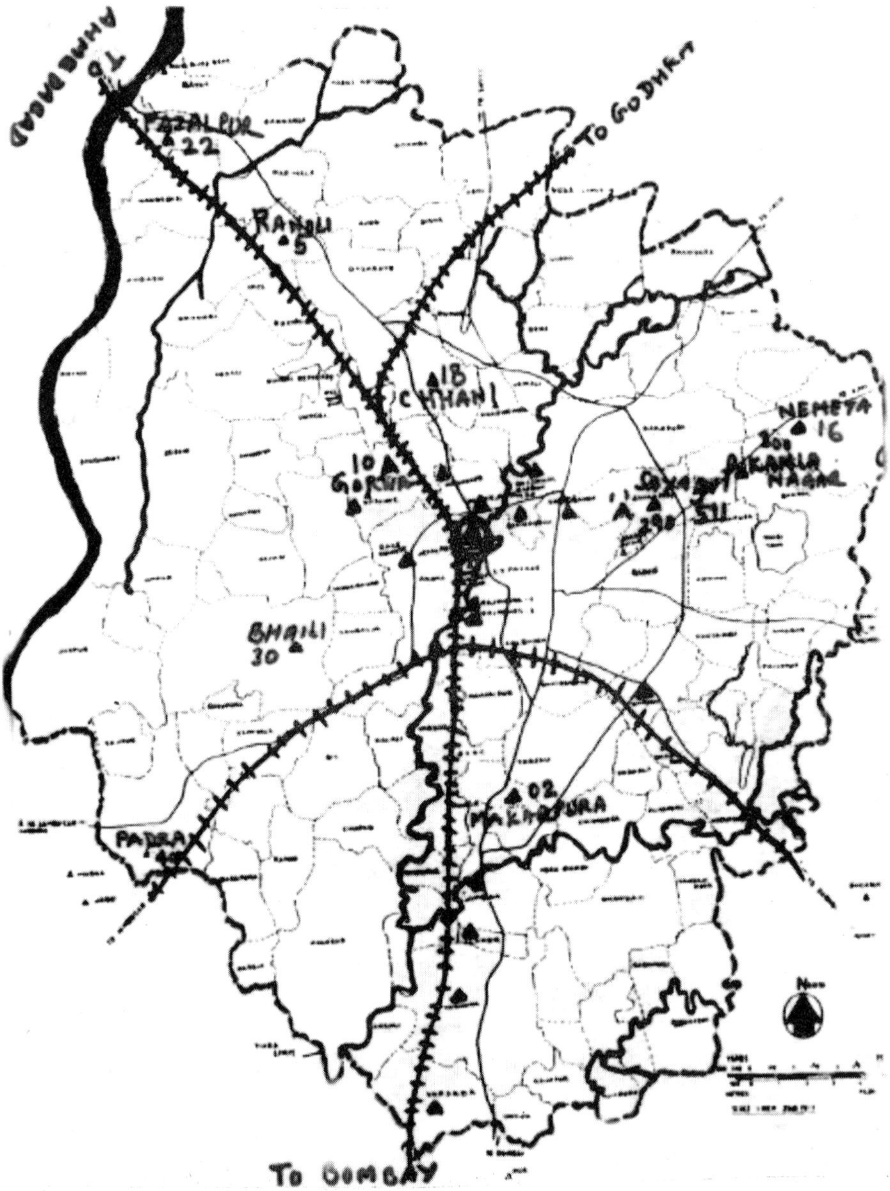

Plate 1

Arbuscular Mycorrhizal Fungi (AMF) have great potential for sustainable agriculture because they improve

- Plant growth
- Effect the levels of plant hormones and exudation form roots
- Reduce soil borne diseases etc.

AMF are non host specific endophytic fungi. Efforts have been made to develop mycorrhizal plant mutants. Mercy *et al.* (1990) found that AMF colonization is a heritable trait. This has opened up new possibilities of tailoring plant fungus combination for maximum benefit. Use of AM fungi for the improvement of crop productivity requires reselection of an efficient and appropriate fungus, the right type of soil and identification of crop (Menge, 1983). An efficient mycorrhizal association is required for every crop for P uptake in nearly all soils. The two of the criteria suggested for appropriateness are infectivity and efficacy (Abbott and Robson 1982, Haas *et al.* 1985) Chatterjee *et al.* (1995) and Arya (2006) studied response of different AM fungi on growth of *Ageratum* and neem respectively. An increase in growth of these two plants was observed. As a consequence of extreme stress under the degraded ecosystem, the uses of *Glomus caledonicum* was found to improve resistance of the host plant through accumulation of proline and higher rates of photorespiration of the AM and P-amended stress plants (Johri, 2005). The AMF not only help plants in better absorption of P but also help indirectly in increased uptake of other macro and micro nutrients (Safir *et al.*, 1972) Hayman (1982) and Panwar (1991) reported increase in concentration of chlorophyll and increased rate of photosynthesis in plants. Baqual *et al.*, (2005) observed increase in chlorophyll content of mulberry due to effect of arbuscular mycorrhizal fungi and other microbial inoculants like *Azotobacter, Bacillus megaterium* and *Aspergillus awamori.*

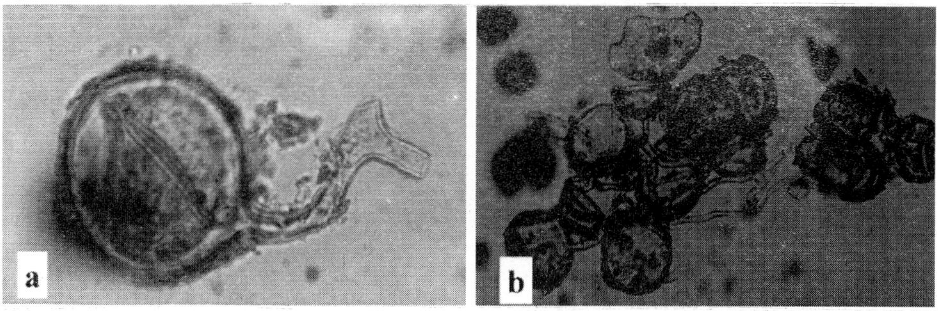

Plate 2

According to Singh and Adholeya (2001) degradation of arable land is a major threat world wide to the sustainability of agriculture production. During the last 40 years, nearly one third of the world's curable land has been lost by erosion (Pimental *et al.*, 1995). It is certainly of great importance for India where nearly half of the districts have been classified as being highly deficient in plant available phosphorus (Pingali *et al.*, 1997)

Arbuscular Mycorrhizal (AM) fungi colonize plant roots and the surrounding bulk soil. They transport mineral nutrients form the soil to the plant and carbon compound from the plant to the soil, and have pervasive effects on plant form and function, and on the composition of the soil microbiota under the growing plants, the hyphae of AM fungi were shown to bind micro aggregates of an alfisol into water stable macro aggregates (Tisdall and Oades, 1979). The AM hyphae appeared to entangle micro aggregates firmly adhered. Miller and Jastrow (1992) proposed that AM hyphae form and stabilize aggregates of soil through three distinct processes

1. The AM hyphae physically entangle primary particles of soil.
2. Roots and AM hyphae create conditions that enable micro aggregate to form in soil
3. Roots and AM hyphae enmesh and bind micro aggregates and smaller macro aggregates into larger macro aggregates.

MATERIALS AND METHODS

The mycorrhizal status of plant species were studied by making an extensive survey of such plant growing areas in Vadodara district (Fig. 1). The sites were visited periodically and sampling of plant roots with adhering soil was done. The samples were collected in polythene bags and brought to the laboratory. Mycorrhizal spores from 100 g soil were isolated by wet sieving and decanting method (Gerdemann and Nicolson, 1963). The residues were collected after passing sample through 60, 90, 150, 210, and 500 μm sieves. The residue collected on 210 μm and 150 μm sieves were examined for sporocarps and large spores. The residues from 90 μm and 60 μm sieves were observed for small and detached spores. Most of the spores were trapped in the sieves. After the spore count, selected spores representing each type were isolated. AM fungi were identified using the key of Schenck and Perez (1990) and Mehrotra and Baijal (1994).

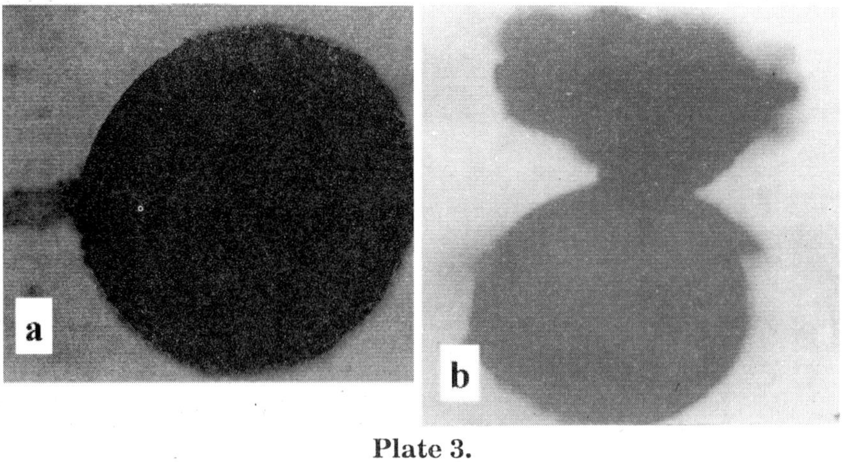

Plate 3.

RESULTS AND DISCUSSION

Arbuscular Mycorrhizal Fungi (AMF) are non-septate zygomycetous fungi. The six genera namely *Glomus, Gigaspora, Acaulospora, Entrophospora, Archacspora, Paraglomns* and *Scutellospora* belong to the order Glomales. These obligate biotrophs are known to show increased plant growth of a large number of agricultural and forest corps.

Although the presence of AM fungi with roots of different plants is the common feature, but the performance of corps and plant-yield may be increased by inoculating exogenous mycorrhizae. It is evident form Table 1 and Plate 1 that soils of Sayajipark (Ajwa-road) and Mahi river bank harbored 511 and 342 spores/ 100 g of soil respectively. Which are alluvial soil that is river transported. On the basis of spore size and morphological characters, two species of *Acaulospora, A. laevis* and *A. nicolsoni* and five species of *Glomus, G. mosseae, G. fasciculatum, G, albidum, G. aggregatum* and *G. macrocarpum* were identified.

Table 1. Occurrence of AMF in different soil samples of Vadodara

Sl. No.	Place	No. of spore per 100 g of soil
1	Akash Ganga	20
2	Alamgir	16
3	Amravati Society	22
4	Anand Baug	16
5	Arboratum (Botany Department)	85
6	Ajwa	106
7	Bhayli	30
8	Botanical Garden (Science Faculty)	08
9	Boys Hostel (MSU)	10
10	Fazalpur	22
11	Gorwa Road	10
12	Gotri Road	31
13	Harni Road	12
14	Jampura	18
15	Kamlanagar 1	300
16	Kamlanagar 2	260
17	Kamlanagar 3	42
18	Karelibaug	06
19	Laxmipura	20
20	Lalbaug	15
21	Mahi river bank	342
22	Makarpura	02
23	Manjalpura	08
24	Nandesari	10
25	Nawagarden	20
26	Nawayard	12
27	Nizampura	15
28	Old Padra Road	15

29	Polytechnic campus	10
30	Pratapgunj	11
30	Pratap Nagar	11
32	Race Course	07
33	Rajmahal	10
34	Ranoli	05
35	Sayaji Park	511
36	Science Faculty campus	07
37	Subhanpura	18
38	Tara Baug	10
39	Varnama	12
40	Varsada	60
41	Vikram Baug	22
42	VIP Road (Sardar estate)	10
43	Yoginagar	17

Minimum difference required for significance (C.D.)/ ± 0.432 at 5% level

Iqbal and Gautam (1995) studied status of arbuscular mycorrhizal fungi in different localities of Aligarh district in U.P. they found dominating nature of *G. mosseae* and *G. constrictum.* They concluded that *Glomus* is more in Aligarh soil, which is distributed irrespective of soil conditions. crop grown, and environmental factors. Species of three different AM genera i.e. *Glomus, Gigaspora* and *Sclerocystis* (*Glomus*) were reported form desert soils of Rajasthan (Mohan and Mishra 1998).

Species of *Glomus* such as *G. fasciculatum, G. monosporum* and *G. macrocarpum* were dominant in Mangalore region of Karnataka (Bhat and Daveriappa, 2005). This might be due to the high sporulation capacity and high viability of *Glomus* spp., while others were scanty due to the adverse edaphic conditions, longer reproductive times and short viability. Similar observations were made by Sieverding (1991). All species did not respond in the same way. Another important factor in the AMF distribution is related to the edaphic and climatic conditions against which every species struggles for existence and the best suited species multiplies quickly and gets established in soil. Sludge dumps and other sites such as coal mines or waste lands are devoid of the AMF due to soil disturbance, which results in failure of reclamation practices. Soil amendments with fertile soil served as a source of inoculum for rapid development of fungi, which in turn, efficiently nourished the plants with essential elements. Biological activities in soil also increased the absorbing capacity of roots.

The details of six AM fungi isolated from Vadodara are as follows

Acaulospora: The word Acaulospora is derived form 3 Greek words, A = without; Kaulos = Stem and Spora = spore. Chlamydospores formed singly in soil or in sporocarps. Sporocarps may attain several cm in length. Spore borne laterally on the subtending hypha with large terminal relatively thin walled sporogenous saccule, which may be twice the size of the spore.

(1) *Acaulospora laevis* Gerdemann and Trappe

The fungus had globose spores, which ranged form 120-130 μm. The spores were yellow to brown in colour.

(2) *A. nicolsoni* Walker, Reed and Sanders

Spores were globose, size 112-120 μm, compound wall thickness 3 μm. They were hyaline in colour, outer wall was thick and brown coloured.

Glomus: *Glomus* in Latin means ball of thread. In this genera chalymdospores borne terminally, intercalarily or laterally on undifferentiated non-gametangial hyphae, formed individually in soil or inside the cortex of the root. Chlamydospores are generally formed on a single hypha, but may occasionally have more than one hyphal attachment. Most common species, *G. fasciculatum,* is known to posses three wall layers.

(3) *Glomus aggregatum* (Schenck and Smith)emend. Koske

Spores were globose, size of spores 126 to 132 μm, compound wall thickness 3 μm, subtending hypha straight or curved, yellow to brown coloured spores. Aggregates of intraradical spores frequently formed.

(4) *G. albidium* Walker and Rhodes

Spores were globose to subglobose. Size of spores 168 to 176 μm, compound wall thickness 2 μm, single cylindrical hypha, Hyphal pore closed with septum, white to light yellow coloured spores, outer wall thick, coarsely rough.

(5) *G. fasciculatum* (Thaxter) Gerdemann and Trappe emend. Walker and Koske

Spores were globose to flattened, size of spores 120 to 140 μm, compound wall thickness 3 μm with one cylindrical hypha, spore smooth, hyaline, outer wall of spore thin.

(6) *G. macrocarpum* (Tul. and Tul.) Berch and Fortin

Spores were globose to subglobose, size of spores 100-120 μm, compound wall thickness 3 μm, single cylindrical hyphal pore occluded by wall thickening, spores yellow coloured, outer wall thick brown, outer surface smooth to dull rough.

(7) *G. mosseae* (Nicol. and Gerd.) Gerdemann and Trappe

Spores were globose to subglobose, size of spores 112 to 125 μm, compound wall thickness 2 μm, single cylindrical hypha, pore occluded by plug. Smooth, reddish brown coloured, outer wall thin.

Change in soil pH, temperature (the samples were collected in different months) and inorganic nutrients availability may be the possible reasons of change in number and variety of AM spores in different soils of Vadodara.

ACKNOWLEDGEMENTS

Authors are thankful to the CSIR sanction No. 38 (0881) 94/EMR II, New Delhi for financial assistance to carry out this work and the Head, Department of Botany, The Maharaja Sayajirao University of Baroda, Vadodara for providing laboratory facilities.

REFERENCES

Abbott L.K. and A.D. Robson (1982) *Aust. J. Agric. Res.* **33:** 1049.

Arya A. (2006) In *Mycorrhiza* (Eds. Prakash, A. and Mehrotra, V.S.) Scientific Publishers (India) Jodhpur. p. 259.

Baqual M.F., P.K. Das and R.S. Kagtiyar (2005) *Mycorrhiza News* **17(3)**: 12.

Bhat P.R. and K.M. Kaveriappa (2005) *Mycorrhiza News* **17(2):** 10.

Chatterjee T., S. Chandra and A. Arya (1995) *Res. J. Pl. Environ.* **11**: 49.

Gerdemann J.W. and T.H. Nicolson (1963) *Trans. Brit. Mycol. Soc.*, **46**: 235.

Haas J.H., A. Bar-Tal, B. Bar-Yosof and Krikum (1985) *Ann. Appl. Biol.* **108:** 171.

Iqbal, J. and A. Gautam (1996) *Mycorrhiza News*, **8(3)**: 12.

Johri B.N. (2005) *Mycorrhiza News* **17(3)**: 17.

Mehrotra V. S. and U. Baijal (1994) In *Biotechnology in India*. (Eds. Dwivedi E.K. and G. Pandey) Allahabad: Bioved research society. pp 227.

Menge J. A. (1983) *Can. J. Bot.* **61**: 1015.

Merey, M.A., G. Shivashankar and D.J. Bagyaraj (1990) *Plant and Soil* **121**:292.

Miller, R.M. and J.D. Jastrow (1992) In *Mycorrhizae in Sustainable Agriculture* (Eds. Bethlenfalvay C.J. and R.G. Linderman), Corp Science Society and Soil Science Society of America, Madison, WI. p. 29.

Mohan V. and R.N. Mishra (1998) *Mycorrhiza News*, **10(1)**: 16.

Panwar J.D.S. (1991) *Indian J. of Pl. Physiol.* **34**: 357.

Pimentel D.C. *et al.* (1995) *Science* **267:** 1117-1123.

Pingali P.L., M. Hossain and R.V. Gerpacio (1997) *Asian rice bowls: the returning crisis?* CAB International 341 pp.

Safir G.R., J.S. Boyer and J.W. Gerdemann (1972) *Plant Physiology,* **49**: 700-703.

Schenck N.C. and Y. Perez (1990) *Mannual for the identification of mycorrhizal fungi*. University of Florida. Gaineswille, Florida, 241 p.

Sieverding E. (1991) *Vesicular-arbuscular Mycorrhiza Management in Tropical agrosystems,* Germany, Eschborn: Technical Corporation.

Singh R. and A. Adholaya (2001) *Mycorrhiza News* **13(1)**: 23-24.

Tisdal J.M. and J.M. Oades (1979) *Aust. J. of Soil Res.* **17**: 429-441.

E

BIOFUEL

37

BIOGAS – A POTENT BIOFUEL: EFFECT OF DIFFERENT SOLID-TO-WATER RATIONS ON ITS PRODUCTION

Darshna Patel and P. Shilpkar

Biogas Research and Extension Centre, Gujarat Vidyapith
At & Post-Sadra, Dist. – Gandhinagar – 382 320 (Gujarat)

ABSTRACT

A laboratory experiment was conducted to study the effect of 1:1 and 1:1.28 solid-to-water ratios on quality and quantity of biogas production. Results show that 1:1 solid-to-water ratio gives higher amount of biogas with good fuel value at mesophilic environmental temperature and at neutral pH of the digestion material.

INTRODUCTION

In the Rural India the energy need of domestic sector are location specific and depends upon living standards and cooking conditions because among the household activities cooking consumes maximum energy. The chief energy sources are fire woods, agro-wastes and dry dung cakes. Malhotre and Pal (1998) concluded that when 25 kg. fresh cattle dung is used for direct burning then it produces 1046 kcal useful energy with 10% device efficiency and no manure is produced, whereas when it is used for Biogas production in 1 cu.m. plant then it produces 2592 kcal useful energy with 55% device efficiency and also gives 10 kg. air dried manure. So it is economic to use dung cakes for Biogas production instead of direct burning. The Biogas is odorless, colorless and burns with a clear blue flame without giving out smoke. It can give a hygienic, clean and safe atmosphere in and around dwelling and populated areas.

Biogas production requires dung slurry as a feeding material whose microbial decomposition produces methane, so it is essential to optimize solid-to-water ratio in feeding slurry to obtain higher methane content. The present experiment was carried out to study the effect of two different solid-to-water ratios *viz.* 1:1 (8% TS) and 1:1.28 (7% TS) and of environmental temperature on quantity and quality of Biogas production.

MATERIALS AND METHODS

Four 5 L capacity glass digester bottles were marked at 4.8L and joined with 2 L capacity of gas holder bottle filled up to the mark with acidified colored water. Each gas holder bottle further joined with another 2 L capacity liquid displacement bottle filled with the same acidified colored water to replace the displaced liquid in gas holder bottle. Displacement of Liquid in gas holder bottle reflects the quantity of biogas production in the digester. Since the Hydraulic Retention Time (HRT) for Gujarat region is 40 days so to fill the digesters up to 4.8 L marks, 120 ml material was fed daily into them for 40 days the two digesters were fed with 1:1 solid-to-water ratio (60 g cattle dung + 60 ml tap water) and remaining two digesters were fed with 1:1.28 (52.5 g cattle dung + 67.5 ml tap water) solid-to-water ratio feeding slurry. Mixed consortia were also added in all the four digesters @ 10% of there working volume from the running Biogas plant of our institute. All the digesters were run up to 89 days and the quantity of daily gas production was measured at 12 hrs. intervals after 40 days. Produced gas was also analyzed for its quality in terms of its methane content through orset apparatus. The pH of the digested slurry was checked by pH meter.

RESULTS AND DISCUSSION

Table 1 shows the effect of different solid-to-water ratios on daily Biogas production. Data shows that 1:1 solid-to-water ratio gives 11.61% higher total biogas production over 1:1.28 solid-to-water ratio during 49 days period of study. Data further show that the daily gas production in 1:1 solid-to-water ratio was also high than compared to 1:1.28 solid-to-water ratio. The production of biogas depends upon the microbial conversion of total solids present in the feeding material. Higher flowable organic loading supplies sufficient amount of nutrients to the microorganisms whose optimum growth may results into higher biogas production. The results are in tune with Anonymous (1990) who suggest high volumetric gas production rate, gas productivity and digestion efficiency at cattle-waste to water ratio raging from 1:1 to 3:1.

Data presented in Table 2 reflects the effect of day (9:00 to 21:00 hrs.) and night (21:00 to 9:00 hrs.) environmental temperature and solid-to-water ratios on corresponding amount of daily biogas production during day and night time. Since the biogas production is a microbial activity and microbial action depends on their optimum growth temperature so variation in environmental temperature results in varied amount of biogas production.

Stable gas production depends on stable environmental temperature. Summers and Bousfield (1980) suggested more efficient digestion is obtained when the digester temperature does not fluctuate more than a degree or two. The maximum and minimum amount of biogas production in 1:1 and 1:1.28 solid-to-water ratios were recorded at 41 and 89 days age of digester, respectively. Data further show that the comparatively higher environmental temperature during day time results in 46.93% higher amount of biogas production over night time in treatment receiving 1:1 solid-to-water ratio and 57.98% higher in treatment receiving 1:1.28% solid-to-water ratio.

Table 1. Effect of different solid-to-water ratios on daily biogas production (ml d^{-1})

Age of digester (days)	Gas Production		% increase in gas production in 1:1 solid-to-water ratio over 1:1.28 solid-to-water ratio
	1 : 1	1 : 1.28	
1	2	3	4
41	1899.7	1844.5	299
42	1757.0	1717.1	2.32
43	1770.3	1725.3	2.60
44	1753.6	1639.6	6.95
45	1823.5	1641.3	11.10
46	1773.1	1466.8	20.88
47	1746.7	1443.7	20.98
48	1506.4	1299.6	15.91
49	1706.1	1429.5	19.34
50	1524.4	1390.2	9.65
51	1766.3	1626.1	8.62
52	1558.5	1515.7	2.82
53	1584.8	1279.6	23.85
54	1621.4	1359.6	19.25
55	1595.9	1306.2	22.17
56	1693.0	1419.5	19.26
57	1531.9	1137.1	34.71
58	1781.4	1550.4	14.89
59	1222.2	1175.5	3.97
60	1219.7	1112.9	9.59
61	1166.4	1059.6	10.07
62	1193.0	1093.0	9.14
63	1179.6	1079.6	9.26
64	1210.3	1110.5	8.98
65	1263.6	1130.5	11.77
66	1157.1	1090.5	6.10
67	1064.0	937.6	13.48
68	1010.8	937.6	7.80
69	1064.0	997.4	6.67
70	1117.2	1030.7	8.39

71	1338.2	1240.2	7.90
72	1276.8	1134.8	12.51
73	1396.5	1290.1	8.24
74	1529.5	1496.3	2.21
75	1409.8	1369.9	2.91
76	1288.1	1259.6	2.26
77	1356.6	1332.6	1.80
78	1429.2	1311.3	8.99
79	1290.1	1163.4	10.89
80	1255.0	1196.9	4.85
81	1250.2	1183.7	5.61
82	1196.9	1170.4	2.26
83	1189.7	1147.1	3.71
84	1164.0	1070.8	8.70
85	1163.8	1084.6	7.30
86	1171.7	1105.0	6.03
87	1088.5	1020.0	6.71
88	1013.7	927.9	9.24
89	944.3	784.8	20.32
Total	68014.5	60936.6	11.61
Average	1388.05	1243.60	

Table 2. Effect of Environmental Temperature and Solid-to-water ratios on day (9:00-21:00 hrs.) and night (21:00 to 9:00 hrs.) Biogas production (ml d^{-1})

Age of digester (days)	Temperature (°C)		Gas Production						% increase in gas production in 1:1 solid-to-water ratio over 1:1.28 solid-to-water ratio	
			1:1 solid-to-water ratio			1:1.28 solid-to-water ratio				
1	2	3	4	5	6	7	8	9	10	11
	Day	Night	Day	Night	Total	Day	Night	Total	Day	Night
41	26	29	1020.76	879.0	1899.76	990.7	853.8	1844.5	3.03	3.03
42	23	26	878.5	878.5	1757	851.2	865.9	1717.1	3.20	1.43
43	23.5	26	880.0	890.3	1770.3	859.4	865.9	1725.3	2.40	2.80
44	25	27	857.2	896.4	1753.6	830.7	808.9	1639.6	3.20	10.8
45	25	26	876.2	947.3	1823.5	826.4	814.9	1641.3	6.03	16.2
46	23.5	27	827.9	945.2	1773.1	808.3	658.5	1466.8	2.42	43.56
47	26	28	813.1	933.6	1746.7	790.6	653.1	1443.7	2.84	42.94
48	26.5	29	899.8	606.6	1506.4	806.4	493.3	1299.7	11.58	22.99
49	25.5	29	1086.5	619.6	1706.1	913.1	516.4	1429.5	18.99	19.98
50	26	29	933.2	591.2	1524.4	866.4	523.8	1390.2	7.71	12.86
51	25	29	999.8	766.5	1766.3	986.3	639.8	1626.1	1.36	19.80

52	29	27	946.5	612.0	1558.5	939.7	576.0	1515.7	0.72	6.25
53	27	27	858.2	726.6	1584.8	753.1	526.5	1279.6	13.95	38.00
54	25	28	879.9	741.5	1621.4	813.1	546.5	1359.6	8.21	35.68
55	25	27	916.0	679.9	1595.9	893.0	413.2	1306.2	2.57	64.54
56	25	27	866.5	826.5	1693.0	746.4	673.1	1419.5	16.09	22.79
57	25	28	882.4	649.5	1531.9	771.6	365.5	1137.1	14.3	77.7
58	24.5	28.5	879.9	901.5	1781.4	799.8	750.6	1550.4	10.0	20.10
59	23	27	732.3	489.9	1222.2	702.3	473.2	1175.5	4.27	3.52
60	23	27	766.5	453.2	1219.7	726.4	386.5	1112.9	5.52	17.25
61	23	27	739.9	426.5	1166.4	693.1	366.5	1059.6	6.75	16.37
62	23	27	726.5	466.5	1193	699.8	393.2	1093	3.81	18.64
63	23	27	739.8	439.8	1179.6	706.4	373.2	1079.6	4.72	17.84
64	24.5	27	799.8	410.5	1210.3	759.7	350.8	1110.5	5.27	17.01
65	24	28	704.9	558.7	1263.6	684.9	445.5	1130.4	2.92	25.38
66	23.5	24.5	625.1	532.0	1157.1	578.5	512.0	1090.5	8.05	3.90
67	24	26	638.4	425.6	1064.0	591.8	345.8	937.6	7.87	23.07
68	21.5	25.5	638.4	372.4	1010.8	611.8	325.8	937.6	4.34	14.30
69	22	25.5	678.3	385.7	1064.0	645.0	352.4	997.4	5.16	9.44
70	22.5	25.5	731.5	385.7	1117.2	678.3	352.4	1030.7	7.84	9.44
71	23	26.5	889.4	448.8	1338.2	871.0	369.2	1240.2	2.11	21.56
72	27	30	864.7	412.1	1276.8	799.0	335.8	1134.8	8.22	22.72
73	25	28	912.3	484.2	1396.5	868.7	421.4	1290.1	5.01	14.90
74	27	29	946.8	582.7	1529.5	921.8	574.5	1496.3	2.71	1.42
75	26	29	938.1	471.7	1409.8	912.3	457.6	1369.9	2.82	3.08
76	29	31	865.3	422.8	1288.1	846.2	413.4	1259.6	2.25	2.27
77	24	26	908.1	448.5	1356.6	893.8	438.8	1332.6	1.59	2.21
78	25	29	924.9	504.3	1429.2	827.0	484.3	1311.3	11.83	4.12
79	25.5	27	863.6	426.5	1290.1	757.8	405.6	1163.4	13.96	5.15
80	23	28	839.3	415.7	1255.0	791.3	405.6	1196.9	6.06	2.49
81	24	27	798.0	452.2	1250.2	784.7	399.0	1183.7	1.69	13.33
82	24.5	27	817.9	379.0	1196.9	798.0	372.4	1170.4	2.49	1.77
83	23	26.5	811.3	378.4	1189.7	771.4	375.7	1147.1	5.17	0.71
84	23	26	731.8	432.2	1164.0	698.4	372.4	1070.8	4.78	16.05
85	22.5	26.5	724.9	438.9	1163.8	712.2	372.4	1084.6	1.78	17.8
86	22	26	751.5	420.0	1171.7	744.8	360.2	1105.0	0.89	16.6
87	22	26	724.3	364.2	1088.5	713.2	306.8	1020.0	1.55	18.7
88	21	25	685.3	328.4	1013.7	641.7	286.2	927.9	6.79	14.74
89	20	24.5	650.1	294.2	944.3	538.7	246.1	784.8	20.67	19.54
Total			40471.7	27543.2	68014.5	37316.2	23620.4	60936.6		
Average			825.95	562.10	1388.05	761.55	482.04	1243.60		

The fuel value of biogas does not depends on its quantity produced but it depends on its quality in terms of methane content. Data presented in Table 3 show that the produced biogas, on an average, contains 81.7% methane in 1:1 solid-to-water ratio and 81.2% methane in 1:1.28% solid-to-water ratio treatment. The higher methane content reflects good fuel value of the produced biogas in 1:1 solid-to-water ratio digester.

Table 3. Qualitative analysis of Biogas

Age of digester (days)	Proportion of gases					
	1:1 solid-to-water ratio			1:1.28 solid-to-water ratio		
	CH_4	CO_2	O_2	CH_4	CO_2	O_2
65	85.2	10.6	4.2	84.6	10.6	4.7
66	85.8	10.4	3.8	85.6	10.5	4.0
80	74.2	21.6	4.2	73.2	22.5	4.3
Average	81.7	14.2	4.1	81.2	14.5	4.3

Besides environmental temperature the second important factor affecting microbial activity and hence the quantity and quality of biogas production, is pH in the digester. Methane producing bacteria are highly sensitive to the pH of digestion material and optimum activities require neutral pH. Near neutral pH of digestion material (Table 4) supports higher methanogenic activity which ultimately gives higher methane content in the produced biogas (Table 3).

Table 4. pH of digestion material

Age of digester (days)	pH	
	1:1 solid-to-water ratio	1:1.28 solid-to-water ratio
46	7.10	7.15
50	7.04	7.07
54	7.15	7.10
57	7.2	7.05
59	6.95	6.94
62	9.87	6.87
64	7.00	6.95
67	7.10	7.10
70	6.99	6.95
79	7.09	7.07
Average	7.1	7.0

From the foregone discussion we can conclude that the biogas production in given digester volume depends on solid-to-water ratio, environmental temperature and pH of digestion material. Higher biogas production with good fuel value requires 1:1 solid-to-water ratio, mesophilic temperature and neutral pH.

REFERENCES

Anonymous (1990) *Status Report* (1983-89). Punjab Agricultural University. Ludhiana Centre, AICRP on RES for agriculture and agro based Industries, PAU Ludhiana.

Malhotre, P.D. and R.C. Pal (1998). *House hold fuels in developing countries*. TERI.

Summers, R. and S. Bousfield (1980). *Agric. Wastes*, **2:** 61-78.

38

SOME NEW VEGETABLE OILS AS SOURCES OF BIOFUEL

Divya R. Nair, Purvi, A. Patel and M. Daniel

Department of Botany, faculty of Science,
The Maharaja Sayajirao University of Baroda, Vadodara - 390 002 (Gujarat)

ABSTRACT

Seeds of sixty two plants growing wild as weeds in and around Baroda or collected from forests have been screened for their fatty oils which could be converted to plant oil methyl ester (PME) to be used as a biofuel. At least 10 plants were found to contain more than 20% oil and 13 plants between 10-20%. *Jatropha gossypifolia, Cordia sebastena, Balanites roxburghii, Anisomeles ovata* and *Citrus limon* were found to be the rich sources of seed oils along with *Jatropha curcas*.

INTRODUCTION

The impending depletion of non-renewable sources of energy like petroleum, has forced seeking various renewable sources (of energy) such as petro-crops, liquid resins, vegetable oils, alcohols and biological hydrogen. Biofuel which is the most successful of these are tipped to be the best alternative to petroleum. A few edible oils such as soybean oil, colza oil, corn oil etc have been used thus in the developed countries. But in developing countries like India where edible oil is scarce and costly, non edible oils like those from *Jatropha curcas, Pongamia, Calophyllum* etc. are used as a motor fuel as they are or converted to plant oil methyl ester (PME). Though a number of other trees like *Salvadora, Schleichera* etc. have been tipped as alternative sources, no attempt have been made to use them as biofuel (Daniel, 2006).

India being a large country, with one of the greatest variety of plants, it is rather strange that we have not located a good member of potential oil sources. Almost every one of the 258000 Angiosperms (as well as Gymnosperms) invariably contain oils in their seeds, the percentage varies from 1 to 55%. The present state of the scarcity of seed oils stems from the fact that only a few thousands of these plants have been screened for their oil content. It is therefore felt that a systematic screening of the plants for their seed oils is the need of the hour and the present paper is a step in this direction. In this paper results of estimating oil content of seeds of 62 plants which are growing wild as weeds in and around Baroda and a few forest trees have been presented and the promising sources are selected.

MATERIALS AND METHODS

Most of the seeds which were screened have been collected from plants growing in and around Baroda. The seeds of forest trees have been provided by Gujarat State Forest Development Corporation Ltd. All the seeds have been dried to a constant weight and the oils were extracted in a Soxhlet's extractor using hexane as a solvent. The hexane was distilled off and the remaining oil was transferred quantitatively to a preweighed beaker and the weight of beaker with oil was taken and from this the wt. of oil was determined.

RESULTS AND DISCUSSION

The results of screening seeds of sixty two Angiosperms have been presented in Table 1. All the plants are found to contain oils in seeds and the percentage content of oil varied from 0.67% (*Canna*) to 50.25% (*Jatropha curcas*). The second highest content was observed in *Balanites roxburghii* (43.34%) closely followed by *Cordia sebastena* (40.59%). *Jatropha gossypifolia* (24.68%) *Chrozophora brachiata* (24.4%), *Gynandropsis pentaphylla* (20.88%), *Citrus limon* (33.78%), *Anisomeles ovata* (24.75%) *Caesalpinia crista* (20.5%) and *Sida acuta* (20%) were the other plants containing 20% and more oil in seeds. Thirteen plants were found to contain more than 10% oils. *Melothria maderaspatana* (15.93%), *Salvadora persica* (17.06%), *Butea monosperma* (16.09%) and *Casurina equisetifolia* (15.24%) were notable among them.

Table 1. List of plants and percentage of oil present in them

Sr. No.	Name of Plants	Percentage of Oil
1	*Jatropha gossypifolia* Linn.	24.68
2	*Croton sparsiflorus* Dalz.	10.80
3	*Chrozophora brachiata* Juss.	24.4
4	*Datura inoxia* Linn.	9.14

5	*Gynandropsis pentaphylla* DC.	20.88
6	*Cassia renigera* Benth.	1.18
7	*Grewia villosa* Willd.	1.61
8	*Abutilon indicum* Sweet.	11.5
9	*Cordia sebastena* Linn.	40.59
10	*Ricinus communis* Linn.	14.5
11	*Citrus limon* Burm.f.	33.78
12	*Cassia occidentalis* Linn.	1.59
13	*Cassia tora* Linn.	5.66
14	*Cassia fistula* Linn.	2.02
15	*Balanites roxburghii* Planch.	43.34
16	*Corchorus acutangula* Lam.	5.97
17	*Hibiscus* sp.	12
18	*Ipomoea* sp.	10.16
19	*Hibiscus* sp. 2	12.6
20	*Anisomeles ovata* Br.	24.75
21	*Lagerstroemia flos-reginae* Roxb.	1.30
22	*Caesalpinia crista* Linn.	20.50
23	*Acacia catechu* Willd.	3.27
24	*Eucalyptus hybrida*	1.54
25	*Jatropha curcas* Linn	50.25
26	*Melothria maderaspatana* Cogn.	15.93
27	*Salvadora persica* Linn.	17.06
28	*Acacia chundra* DC.	3.64
29	*Butea monosperma* Roxb.	16.09
30	*Cassia leucophloa* Willd.	1.22
31	*Bauhinia racemesa* Lam.	0.81
32	*Nyctanthes arbour-tristis* Linn.	5.6
33	*Mucuna prurita* Hook.	2.9
34	*Bauhinia purpurea Linn.*	4
35	*Casurina equisetifolia* Forst.	15.24
36	*Dalbergia sissoo* Roxb.	1.39
37	*Lagerstroemia parviflora* Roxb.	1.3
38	*Diospyros melanoxylon* Roxb.	1.22
39	*Adina cordifolia* Benth & Hook	4.42
40	*Dalbergia latifolia* Roxb.	7.44
41	*Haplophragma adenophyllum* Wall	1.16
42	*Sida acuta* Burm.	21

43	*Abutilon indicum* Linn.	2.67
44	*Lawsonia inermis* Linn.	7.9
45	*Nicotiana tabaccum* Linn	13.28
46	*Clitoria ternatea* Linn.	7.71
47	*Canna indica* Linn.	0.67
48	*Teramnus labialis* (L.) Spreng.	2.22
49	*Mimusops elengi* Linn.	3.02
50	*Tephrosia villosa* Pers.	7.20
51	*Anona squamosa* Linn.	11
52	*Merremia dissecta* Hall.	2.73
53	*Leucerna leucocephala* Benth.	4.31
54	*Bridelia retusa* Spr.	0.897
55	*Albizzia lebbeck* Benth.	2.37
56	*Acacia nilotica* Willd.	2.75
57	*Tamarindus indica* Linn.	3.85
58	*Peltophorum ferrugineum* Benth.	3.4
59	*Samanea saman* (Jacq) Merrill	6.44
60	*Acacia Senegal* Willd.	2.64
61	*Pithecellobium dulce* Benth.	11.75
62	*Cassia auriculata* Linn.	4.57

The results obtained here are greatly encouraging. Apart from *Jatropha curcas,* 2 out of 62 i.e. 3% of plants contain 40% and above oils, and these two, *Balanites* and *Cordia* can be profitably used as alternatives of *Jatropha curcas*. 10 plants out 62 (more than 16%) which contain 20% or more oils in seeds can be studied further for elite plants or for agronomic methods to improve oil yield.

The highly promising results of the present study warrants similar studies in all plants (more than 17,000) of our country and at all geographic zones which will bring forth very high promising sources of oils which when properly managed will provide a large amount of vegetable oils (or PMEs) and thus will decrease the shortage of biofuel in this country and reduce the burden on import on costly petroleum. PMEs coupled with other sources of energy such as alcohols (especially ethanol which could be easily produced from the starch which is greatly abundant in our country), biogas, biological hydrogen and of course atomic energy will make India, once, the leader in alternative sources of energy.

REFERENCES

Daniel, M. (2006). *Herbal Technology: Concepts and Approaches*. Satish Serial Publishers, New Delhi.

INDEX